The Stability of Planetary Systems

Proceedings of the Alexander von Humboldt Colloquium on Celestial Mechanics, held at Ramsau, Styria, March 25-31, 1984

Edited by

R. L. DUNCOMBE
Department of Aerospace Engineering, University of Texas, Austin, Texas, U.S.A.

R. DVORAK
Institut für Astronomie, Universität Wien, Wien, Österreich

and

P. J. MESSAGE
Department of Applied Mathematics and Theoretical Physics, Liverpool University, England

Reprinted from

Celestial Mechanics, Vol. 34, Nos. 1-4

D. Reidel Publishing Company

Dordrecht / Boston

ISBN-13: 978-94-010-8854-1 e-ISBN-13: 978-94-009-5331-4
DOI: 10.1007/978-94-009-5331-4

Softcover reprint of the hardcover 1st edition 1984

TABLE OF CONTENTS

The Stability of Planetary Systems

PLANETARY MOTIONS
(Chairman of Session: R. Dvorak)

SATELLITE MOTIONS
(Chairman of Session: J. Chapront)

COSMOGONY
(Chairman of Session: G. Marchal)

PERIODIC ORBITS, RESONANCES, AND ASTEROID MOTIONS
(Chairman of Session: H. Eichhorn)

* Invited Review

Preface

The Alexander von Humboldt Colloquium on Celestial Mechanics (subtitled "The Stability of Planetary Systems") was held in Ramsau, Styria, in the Austrian Alps, from March the 25th to the 31st, 1984. The dedication of the meeting to Alexander von Humboldt presented participants with the challenge that the discussions during the week should reflect the spirit of that great scientist of the last century, that the very many interesting ideas presented and developed during the sessions should be interpreted in the light of a broad view of astronomy and astrophysics. The topics of the meeting ranged from astrometric questions relating to the specification of inertial reference systems, motion of planets (including minor planets) and satellites, with the recurring topic of the search for criteria of stability of the systems, resonances, periodic orbits, and to the origin of the systems. Each session began with one or more invited review papers, followed by offered contributions and discussion. Three evening discussions were held, devoted respectively to inertial systems, to numerical integration techniques, and to cosmogonic problems and ring systems.

On the evening of Wednesday, March 28th, a recital of chamber music was given by Bernhard Piberauer, on the violin, and Meinhard Prinz, on the piano. On the Friday evening, March 30th, participants were entertained at a Styrian Evening, hearing Styrian and Alpine folk songs performed by the Steiner Singers from Ramsau, and also part of "Das Lichte Land" ("The Heavenly Country") by the Stvrian poet Peter Rosegger, read by Professor J. F. G. Grosser.

First of all we have to thank the Oesterreichischen Bundesministerium fuer Wissenschaft und Forschung for financial support. We also have to thank the Oesterreichischen Forschungsgemeinschaft and the Steiermaerkischen Wissenschafts- und Forschungslandesfond. We believe that, due to the co-operative effort of all concerned, everyone present had an enjoyable, as well as a useful and instructive, time at Ramsau. Many thanks are due to the Chairmen of the various sessions, each of whom acted as Editor for the proceedings of his session. Thanks are also due to those colleagues who refereed the reviews and papers presented at the sessions. Last, but by no means least, we are very glad to record our thanks to our hosts at the Alpengasthof Peter Rosegger, Fritz and Barbara Walcher, for their hospitality and kindness to us, making our stay in Ramsau such a pleasant one.

R.L.D.

R.D.

P.J.M.

THE MEANING OF CELESTIAL MECHANICS IN THE KOSMOS OF ALEXANDER VON HUMBOLDT

F. Schmeidler
Universitats-Sternwarte Munchen
D-8000 Munchen
Germany, F.R.

The generous assistance in problems of organization which was given to our colloquium by the Interalp-Institut is justification to say some words about the Academia Cosmologica Nova, since the Interalp-Institut is strongly connected with this academy. One more reason to mention the Academia Cosmologica Nova is its work on Alexander von Humboldt whose name occurs in the title of the colloquium which begins today. The general attitude taken by Alexander von Humboldt in all of his works is characterized by many ideas which, according to the opinion of the Academia Cosmologica Nova, are still important in our time.

The Academia Cosmologica Nova was founded in 1976. Its aim is to promote understanding for scientific questions in the general public. It intends also to create harmonic relations between humanities and natural sciences, which can only by achieved if narrow-minded limitation to special problems is strictly avoided. Interdisciplinary work is the only way which produced a chance to find a general understanding for all branches of science.

It was for these reasons that the Academia Cosmologica Nova was prepared to follow a suggestion which came from the "Sammelstiftung der Preussischen Akademie der Wissenschaften" to work out a modern edition of the "Kosmos" of Alexander von Humboldt: the idea of this suggestion was that the work of Humboldt should be brought to a level corresponding to the scientific standard of our time.

Celestial Mechanics **34** (1984) 7–10. 0008–8714/84.15

Alexander von Humboldt was born in 1776 and lived until 1859; he and his brother Wilhelm, who was two years older, were famous men in German cultural life in the 19th century. Alexander started in 1799 a long journey through the continent of South America about which little was known in that time. Five years later he returned to Europe. Afterwards he lived about twenty years in Paris, where he worked out the scientific results of his long journey. In 1827 he came back to Berlin. Now he undertook the compilation of a comprehensive book which was intended to become a general description of the physical world. The idea to write such a book had come to him long years before. At last, the book became the famous work entitled "Kosmos" which was, according to reports given by experts of this time, the most widely known book next to the Bible in 19th century Germany.

The first volume of the modern edition of the "Kosmos" to be published by the Academia Cosmologica Nova, is ready to be printed; it contains astronomy or, according to the expression used by Humboldt, the uranological part of the "Kosmos". We have cancelled all those passages in Humboldt's work which are today either wrong or irrelevant. We have added footnotes or commentaries, wherever new scientific findings exist which Humboldt could not yet know. Our colleague and member of the Academia Cosmologica Nova, Dr. Dvorak, has done most of this work as main contributor. The two other volumes of the new edition have been treated in the same way and are nearing completion; we hope that all volumes will be published in 1984 or 1985.

It is clear that in a book like the "Kosmos" many questions of celestial mechanics are discussed, because celestial mechanics was one of the most important parts of astronomy in the time of Alexander von Humboldt. Furthermore, Humboldt had friendly relations with many French mathematicians and astronomers who worked on celestial mechanics during the long years of his stay at Paris. So it is not surprising that the following sentences can be found in the introductory chapter of the first volume of the "Kosmos":

> Our neighbors on the other side of the Rhine possess an immortal work, Laplace's exposition of the system of the world. In this book, the results of the most thoughtful mathematical and astronomical researches of several centuries are exposed without the proof. The construction of the universe is considered in this work as the simple solution of a problem of mechanics.

Similar statements may be found in other parts of the "Kosmos" too. Also, in those chapters which describe celestial bodies, the results of celestial mechanics about their orbits and about the laws of motion are reported according to the state of astronomy in the 19th century.

All these considerations were no reason for Humboldt to forget the general conception which he had intended to impose to the whole "Kosmos"; the work was thought to be a description of the entire nature and not an exposition of isolated parts of the natural sciences. This point of view was form him also valid, when questions of celestial mechanics were discussed. That means, that if the name of Alexander von Humboldt is used for any scientific meeting, the idea should be kept in mind that all individual sciences play a useful part in human culture only if their connection with other natural sciences is fully recognized.

It is possible to quote several sentences in Humboldt's "Kosmos" which express this view; I take as example a passage contained in the introductory chapter of the first volume:

> In scientific considerations about the Kosmos special questions are only considered, as far as their relation to the whole world are concerned.

In a somewhat more detailed form, the same idea is expressed in the second volume of the "Kosmos":

> The history of the physical description of the world is the history of knowledge of the whole nature. It is the exposition of all attempts undertaken by mankind to understand all forces

acting on earth and in the universe. Therefore, it describes progress in generalization of scientific theories, and it is part of the history of human thinking.

These general views are also true in smaller fields of research for instance when relations between different parts of a special science are discussed. Therefore, celestial mechanics will only be able to make meaningful contributions, if its connection with the whole of astronomy is not neglected. Investigations of special questions of celestial mechanics are, of course, as necessary as special research in any other parts of astronomy. It should, however, be avoided that astronomers who make special research in some part of astronomy, neglect the rest of astronomy.

Unfortunately, the situation of astronomy in our time is in this respect not as satisfactory as I should like it to be. Especially, the importance of celestial mechanics is not fully recognized by all astronomers who work in astrophysics. We should, however, be honest enough to admit that the reverse is also true. Some colleagues who work about astrometry or celestial mechanics, are not fully aware about the enormous progress which astronomy has made due to modern astrophysics.

I very much hope that the name Alexander von Humboldt will encourage all participants of the discussions which will take place in the next days, not to neglect other parts of astronomy. If all colleagues who work in celestial mechanics, keep this in mind, this field of research will continue to play its traditional part for scientific astronomical research in future.

Inertial Systems - Definitions and Realizations

Heinrich Eichhorn
University of Florida

Abstract:

A frame of reference is defined as inertial if the coordinates of bodies, predicted with respect to it on the basis of some mechanical principles, agree with those actually observed. It is pointed out that within the framework of Newtonian mechanics the concept of an inertial frame of reference is global, but within the framework of general relativity only local. The relativistic local inertial frame of the Solar System moves with respect to the Newtonian global frame by the amount of "geodetic precession". In all rigor, the concept of a globally privileged inertial system has been rendered meaningless by general relativity, even though the concept of an inertial reference frame remains useful in contemporary celestial mechanics. It is further pointed out that none of the practically determined approximations to an inertial frame of reference (e.g., the dynamical reference frame of celestial mechanics, the FK5) must be regarded as a definition of an inertial system itself.

The zero longitude direction of the equatorial coordinate system is critiqued, and because a rigorous conceptual definition of the ecliptic is impossible, it is suggested to specify the intersection of the invariable plane with the equator as the direction of longitude zero. Finally, it is shown that a body in the Solar System which is still unaccounted for could have produced the pattern of discrepancy between the computed and observed rates of the obliquity which has been noted in the observations of the last two centuries.

Definition of an Inertial System

For any observer, a frame of reference is defined by a set of (physical) points which constitutes a well-defined spatial* configuration. We may assume that this refers to a particular observer, and to a somehow synchronized instant. To specify the location of a point which is not an element of the defining set

*We shall, for this purpose, consider the concept of space as a physical entity that is accessible to immediate experimental verification and thus not in need of further definition.

Celestial Mechanics 34 (1984) 11–18. 0008–8714/84.15

within this frame of reference, one needs a construct which allows one to specify completely and uniquely the location of a point with respect to a specific frame of reference. Such a construct is called a coordinate system. The location of a point is therein specified by the assignment of numerical values to certain parameters - the coordinates. The minimum number of coordinates necessary to specify, in a given space, the location of a point is the (physical) dimension of this space. Any point will of course have different coordinates depending on the orientation of the coordinate system in that reference frame with respect to which its location is considered. A coordinate system, in order to be useful for defining locations of points with respect to a particular reference frame, must, at any instant, itself be in a well-defined relationship with the frame of reference under consideration. A given frame may usually be represented by any one from an infinite manifold of coordinate systems.

A point's motion, in the sense of kinematics is the change of its coordinates. The concept of absolute motion would be meaningful only if there were a privileged reference frame representing "absolute" space. Apparently, however, such a system does not exist. Motion can therefore only be considered with respect to a specified frame of reference.

The laws and postulates of that system of mechanics which are invoked for describing the motion of bodies provide (usually differential) equations whose solution (integration) establishes those locations in space-time which are described by the bodies in question in the course of their motion.

In Newtonian mechanics this amounts to the establishment and integration of the so-called equations of motion. Generalizing the definition by Anding (1905), an inertial system in Newtonian mechanics is defined as one in which the motion of bodies follows Newton's postulates. Note that the forces involved are not restricted to Newton's law of gravitation. Within the framework of general relativity equations of motion which are valid only in a privileged set of reference frames do not exist; their place is taken by the field equations whose solutions determine the "world lines" in space-time which the bodies occupy. The concept of gravitational force as completely independent of inertia loses its meaning in general relativity; gravity is rather (by the "principles of equivalence") regarded as a manifestation of inertia in a space which is always curved in the presence of masses. The field equations themselves are invariant to any coordinate transformations and therefore there cannot be, in contrast to the situation in Newtonian mechanics, a privileged set of reference frames with respect to which the field equations are valid. General relativity thus cannot recognize any globally privileged frame of reference even though locally privileged "local inertial systems" exist. Moritz (1981), has pointed out in his excellent review paper that in the (infinitely small) mathematical neighborhood of any observer (located at a point in space-time) there exists exactly one flat "tangent" space-time which approximates, at this point, the curved space-time of the observer. This locally privileged system varies, however, from point to point in space-time. Moreover, the local heliocentric relativistic inertial frame and the Newtonian inertial frame rotate, curiously, with respect to each other by the amount of geodetic precession (cf. Eddington, 1924, p. 99). Even the "harmonic" coordinate systems mentioned by Moritz which are globally privileged are constructed under the assumption that the mass density goes to zero at infinity. While a harmonic system may thus be an excellent frame for the construction of the theory of motion of the bodies in the planetary system, it is still an engineering tool rather than a concept of science (cf., Young, 1983, p. 946) and, even as an ever so accurate, but still only approximate representation of reality, cannot rigorously be used to define a real concept.

While the passing of the globally privileged reference frames from the inventory of scientifically meaningful concepts may be lamented by some, the concept of inertial reference frames retains its scientific legitimacy in the form of the relativistic local heliocentric inertial reference system. In view of the large ratio of the Sun's to the Earth's mass, it will probably be a long time before available measuring precision will enable us to measure on Earth, or anywhere else in the Solar System the relative motion of the local (e.g., geocentric) relativistic inertial frame with aspect to the heliocentric one.

The inertial system — one of the concepts inextricably tied to the Newtonian postulates — is meaningful at least in the same framework within which Newtonian mechanics is meaningful as long as the motions of aggregates of bodies are investigated on the basis of this Newtonian mechanics. Mechanics based on Newton's three postulates of motion thus still remains the prime tool for treating the engineering aspects of celestial mechanics.

Newton's postulates are evidently not valid in all reference frames. Take, as an example Newton's first law (so-called; actually postulate) of motion which states that a point mass, not affected by any force, moves uniformly. Now it is clear that, regardless of how a particle moves with respect to any one coordinate system, one can always specify another coordinate system in such a way that a given single particle has any specified orbit with respect to this coordinate system. If, for instance, a particle is decreed to be always at rest, this can be accomplished by placing the origin of the new coordinate system in the particle itself. Newton's first postulate of motion is thus valid only for a particular class of coordinate systems.

The Establishment of Empirical Approximations

The integration of the equations of motion (in the general sense — in general relativity, for example, their place is taken by the field equations) leads to precepts which define the coordinates of moving entities as functions of time (or more generally, determine their world lines). Astrometry, in general, may thus be defined as the empirical determination of the world lines of celestial objects. The task of astrometry is therefore twofold: first, to realize empirically certain (ideally well-defined) coordinate systems; and second, to establish empirically estimates of (not necessarily complete sets of) the coordinates themselves of certain bodies. Traditional usage calls these the "observed coordinates".

Empirical, as it were, celestial mechanics will be able to perform two further tasks by matching the observed coordinates with those derived from whatever mechanical theory was used for their prediction.

The first of these is the estimation of the constants of integration and other relevant parameters (such as the masses of the target bodies).

The second, of primary interest for the present considerations, is the establishment of that transformation which will transform the coordinate system in which the observations were carried out to that which was used in the theory from which the predicted coordinates were derived. It is not a priori clear that such a transformation exists. If and only if it does, the transformed system, now equivalent to the system to which the predicted coordinates refer, is called the empirical realization of an inertial system and approximates (at least locally) an inertial frame of reference. An approximation for the local inertial frame of general relativity can also be found by this procedure.

One frequently encounters other definitions. These are impractical either because they are based on gedanken experiments to whose performance there is an obstacle in principle, or because they elevate the empirical realization of a special case of the definition given above to be the general definition of an inertial system.

To illustrate the first point, consider the often presented definition (so-called) of an inertial system as one whose axes are defined by the three non-coplanar orbits (presumed to be straight) of mass points in a force-free field. This is not a sound definition because there cannot, in reality, be a field free of forces in the presence of any masses anywhere in the universe; it is even most likely that "space" itself cannot be meaningfully defined in the absence of mass. Regardless of how accurate any particular experiment would realize an inertial system, it is philosophically objectionable to define a rigorous concept by what cannot be anything but an approxmation (cf. Eichhorn, 1983).

The fallacy in the second class of "practical" definitions of an inertial system is more difficult to trace. It has recently been widely suggested to "define" as "the" inertial frame of reference (as if there were only one!) that frame with respect to which the quasars show no proper motions. The presumed enormous distances of the quasars are the basis for this approach. Even in the absence of measurements, it appears fairly safe to assume that the lateral velocity components of the quasars at a distance of 4 gigaparsecs are no more than 4×10^3 AU/yr, which corresponds to about 3% of the speed of light. Even then it would take 100 years for a quasar to change its direction with respect to a genuine inertial frame by 0ʺ001 — a quantity which cannot be measured at the present state of technology.

However, it is also safe to say that in another 100 years the precision of angular separation estimates measured by the techniques which are now the most precise (VLBI) will be regarded as crude. Therefore one may even today predict that the "definition" of an inertial frame through quasars will have to be abandoned in the forseeable future. This does not, of course, imply that observations of quasars are not now enormously useful for the practical estimation of an inertial frame of reference.

Another current "definition", so-called, of "the" inertial system is what has come to be called the "dynamical reference frame of spherical astronomy". Essentially, this is the ideal equatorial (right ascension - declination) reference frame, estimated on the basis of celestial mechanics, in other words, the $\alpha - \delta$ frame adjusted in such a way that the positions predicted for the planets on the basis of specified (celestial mechanical) theories are on the whole (on the average) correct.

Useful though it may be for the practical realization of an inertial frame of reference, the result of this procedure cannot serve as a definition because it is by necessity an estimate of an inertial frame of reference, and therefore cannot be an inertial frame itself. (Any competent statistician will warn against confusing a quanitity with an — even its best available — estimate.) Furthermore, it is well known that there are theoretically still not satisfactorily explained discrepancies between predicted and observed positions of the outer planets (Uranus and Neptune); in decreeing the inertial systems of celestial mechanics to be that which, on the average, best reproduces the positions of the planets, predicted by even the best currently available theory, one is running the (admittedly very small) risk — at least in principle — that what are perceived to be correct predictions are systematically wrong (due to some overlooked effect).

Critique of Present Practice

The determination of the dynamical reference frame, finally, is based on only a fraction of the material which is in principle available for the estimation of an inertial system. Input, such as is derived from investigating the motions — proper motions as well as radial velocities — of stars, coupled with certain models purporting to describe the distribution function of the (implicitly inertial) velocities of the stars in the galaxy, are completely ignored in the quantification of this "dynamical reference system"; they are rather treated separately and were used chiefly for what has become another estimate for an inertial system — basically the ideal system of the FK4, which is unavoidably accelerated with respect to the "dynamical" reference system.

When it becomes available, the "system" of the FK5, ostensibly the dynamical reference frame, will be an excellent and maybe even the then best available approximation to an inertial reference frame. In contrast to entrenched terminology, it should, however not be "defined" as "the inertial system of spherical astronomy", but for the benefit of the non-specialists one should clearly state that the FK5 system in fact only approximates an inertial frame of reference.

One cannot simply decree any — even ever so accurate — estimate for a well-defined physical concept to be the concept itself. The various fundamental systems of spherical astronomy, even disregarding their deviations from self-consistency all differ from and are accelerated with respect to each other, even if ever so little. None of them is more than a more or less accurate approximation to an inertial system and none must therefore be decreed to be an inertial system.

One wonders how there can be several independent "best" approximations to an inertial system. The hapless user is only interested in the best obtainable or maybe even just a good approximation to an inertial system, regardless of how any of them was obtained and he is not acquainted with all the quasi-legalistics that went into their definition, but he is still forced to choose between them. In the absence of definite criteria as to which is the most accurate approximation, one would expect their (weighted, if necessary) average to be more accurate than any of them individually. It is somewhat puzzling that no one has constructed such an average — even more so, that no one has announced the intention to construct such an average and offer it as the most accurate available approximation to an inertial system. A component of the explanation of this strange phenomenon may be the heritage of the need for computational shortcuts, which was one of the realities of practical life for any computing astronomer trained in the pre-electronic-computer era. Though the ready availability of computers has eliminated the necessity for constantly being alert for computational shortcuts, the basic philosophy of computing parsimony inculcated into the researchers' minds is extremely difficult to shed, even for the most ingenious and imaginative investigator. In addition, it appears that the very definition of one of the fundamental concepts of the equatorial system is in principle conceptually impossible: There exists no conceptual definition of the ecliptic.

Critique of the Concept of the Ecliptic

Remember the actual definition, so called, of the ecliptic: The barycentric osculating plane of the Earth-Moon system's barycenter, obtained by numerical integration of the equations of motion of the Solar System's known bodies (assuming certain previously determined values for their masses, and certain initial conditions which are also presumed to be known), is — again numerically — subjected to a smoothing process which removes all identifiable periodic effects in the normal vector to this plane. What is left is the ecliptic.

In practice, this procedure gives useful results and thus satisfies the engineers, but from the scholar's point of view there is everything wrong with it.

The ecliptic -- even if the rest of the definition were philosophically sound — is thus defined in terms of the initial conditions and the masses used; this amounts again to the confusion between an entity itself and one of its estimates.

Another objection is this. Most commonly, this process by which the ecliptic is in practice specified (erroneously regarded as defined) is defended by the statement that "ideally" the ecliptic is defined as that part of the osculating plane of the barycentric orbit of the Earth-Moon system's barycenter which is given by the secular terms only in the quantities describing its orientation, while the periodic terms are left out. This statement implies that there is a canonic way in which to find approximations to the equations which govern the orientation parameters of the osculating plane of the Earth-Moon system's barycentric orbit, and that this approximation consists only of secular and periodic terms which can always be clearly separated. Poincaré(1893, pp 110-111) points out that such a unique separation is not possible. Even more so, if the values of the orientation parameters were bounded, none of the infinitely many possible approximation solutions could contain any secular terms. While contemporary practice thus specifies the ecliptic perfectly satisfactorily, it cannot function as its conceptual definition.

There is, in any case, no privileged analytical approximation theory. The theory actually used at this time for specifying the ecliptic is numerical rather than analytical. Furthermore, any planetary theory is ultimately inaccurate because it retains its character as an accurate approximation only for a finite time interval after which it deviates intolerably from the rigorous solution of the equations of motion.

The third objectionable feature is that the results depend on the time interval used or, more generally, on the weight function applied to the formal solution for the smoothing process.

For all these reasons the ecliptic cannot be conceptually defined. The specification itself depends not only on the particular approximation solution employed but also on the details of the weighting function used for smoothing out the periodic terms (such as they are).

But even if there were a rigorous, i.e., non-approximative physical (mechanical) principle available on which a true, conceptual definition of the ecliptic could be based, the resulting plane obviously still would not have a constant orientation with respect to an inertial frame of reference. This well recognized fact gives rise to a phenomenon described as "planetary precession", which introduces complications into the theory of general precession and, among other things, creates an artificial nonuniformity in sidereal time, (albeit not significant in terms of presently available precision) this is to say a component which is not strictly proportional to the (model) Earth's (inertial) angular velocity.

The Relationship Between Invariable Plane and Ecliptic

A perfectly rigorous concept, independent of all approximation theories, is, however, available: The invariable plane of the planetary system. Note that the presently used and accepted specification of the ecliptic is based (except for the value of the obliquity at a certain epoch) in its entirety on current theory of celestial mechanics, the calculative prediction of the orientation changes of the ecliptic with respect to an inertial system is thus completely and rigorously equivalent to the calculative prediction of its orientation changes with respect to the invariable plane which is by its very definition motionless within an inertial system. Any objections to using the invariable plane as that fundamental plane whose intersection with the equator would be used as the marker for the zero longitude direction in the equatorial system — a role now played by the ecliptic by way of the vernal equinox -- must apply in exactly the same way to the ecliptic.

While the ecliptic cannot be directly observed, the orientation of the ecliptic at a given instant is mostly determined by observations of the Sun and other bodies in the solar system and not very sensitive to small changes in the assumed masses of the planets. At a particular instant this is unfortunately not the case for the calculated orientation of the invariable plane, whose orientation is quite sensitive to the mass estimates of the planets used in the calculations. One might be tempted to derive from this fact a powerful argument for preferring a not conceptually definable but still very accurately determinable ecliptic to the conceptually definable invariable plane whose orientation is much more elusive. However, this objection is seen to be physically irrelevant when one considers that the orientation of any plane at any instant enters only the definition of the direction of zero longitude at this instant. This is physically in no way privileged and thus of no significance. The determination of the physically significant motion of this zero direction is, within the framework of present convention, carried out theoretically only, without, for instance, taking into account the directly observed obliquities of the ecliptic over long time intervals. Note that this means that the motion of the direction of zero longitude must be governed by the very same theory regardless whether it is defined by ecliptic or invariable plane.

At this point, one must ask the question whether the possibility to get a precise hold on the non-privileged and thus physically irrelevant zero longitude direction of a particular coordinate system — even though one sanctioned by two millennia of continuous use — is worth the introduction of as elusive a concept as the ecliptic which was introduced — lest we forget it — when the Newtonian principles of mechanics and with them, the recognition of the existence of an invariable plane, were still one and a half millennium in the future.

The essential insight is that the motion of the ecliptic and the motion (as it were) of the invariable plane are both calculated from identical theories.

Unfortunately, the influence of an unconsidered body in the planetary system on the orientation of the invariable plane grows proportionally to the square root of the semi-major axis of its orbit. A body whose heliocentric orbit has a semi-major axis of 100 AU (and thus a period of 1000 years) will cause a "wobble" in the computed estimates of the invariable plane with respect to the physically true one, whose amplitude depends on mass and orbital inclination of the neglected body. The wobble will be at a standstill when the motion of the neglected body is parallel to the invariable plane; this will take place when it has its largest distance from this plane, and will be at its fastest 250 years afterwards when it is close to it. We have stated above that the discrepancies between observed and computed ecliptic are identical to the discrepancies between actual and computed invariable plane. It is well known that a discrepancy between the theoretically predicted and observed rates of the obliquity of the ecliptic displayed itself in older observations but is no longer noticeable in modern observations. This phenomenon, noticed over the last 200 years or so, is normally (and quite convincingly) laid at the feet of systematic errors, especially the magnitude dependence of the personal equation. Whether a mechanism as described above, i.e. an unrecognized and unaccounted for body in the Solar System which, during the last 200 years moved out of the invariable plane and now moves in a portion of its orbit which is parallel to the invariable plane, might provide for alternate explanation for the discrepancy between the computed and the observed rates of the obliquity is a question which must be answered in the future.

Acknowledgements

The author is indebted to many of his colleagues with whom he has discussed the matters treated in this paper; and who, by their criticisms and comments, have forced him to an even more precise formulation of his statements. Foremost among these are George Contopoulos, James Ipser, Jean Kovalevsky, Robert Leacock, Andrew Murray, Ken Seidelmann, Myles Standish, and Carol Williams.

References

Anding, E. 1905. Enz. d. math. Wiss. vol VI, pt. 2, 1st half, p. 3-15.

Eddington, A. S. 1924. Mathematical Theory of Relativity. Cambridge University Press.

Eichhorn, H. 1983. Philosophia Naturalis 20, p. 147.

Moritz, H. 1981. Proceedings of the IAU Colloquium No. 56, eds. E. M. Gaposchkin and B. Kolaczek (D. Reidel), p. 43.

Poincaré, H. 1893, Les Niethodes Nouvelles de lu Mecanique Celeste, vol. II, reprinted, 1957, by Dover, New York.

Young, A. 1983. Publ. Ast. Soc. Pac. 95, 939.

MINOR PLANET OBSERVATIONS AND THE FUNDAMENTAL REFERENCE SYSTEM

Raynor L. Duncombe
Center for Space Research
and Department of Aerospace Engineering

Paul D. Hemenway
Center for Space Research
and Department of Astronomy

Arthur L. Whipple
Center for Space Reseach
and Department of Aerospace Engineering

The University of Texas at Austin
Austin, TX

ABSTRACT. A 15 year project to establish a dynamical reference system utilizing ground-based and Space Telescope observations of 34 minor planets is being undertaken. The orbits of these minor planets will be knit into a common system through the use of "crossing point" observations. The system of orbits thus established can be used to measure long arcs in the sky (similar to the function of a transit circle) and can be used to detect individual star errors as well as residual periodic effects in the fundamental reference system. The minor planet dynamical reference system will also provide an independent method to establish the zero point and the solid-body rotation of the HIPPARCOS reference system.

INTRODUCTION

Newcomb, in his 1877-1897 tour-de-force through the solar system, used the motions of the inner planets to derive the equinox and equator of his fundamental reference system; a system realized for practical purposes through a related catalog of star places and motions. Transit circle observations of day objects, however, are subject to sources of error, both systematic and random, which may be minimal or even non-existent in night observations. The process of transferring the reference frame derived from day observations of planets to a catalog of star positions based on night observations is complex and lacking in rigor. The use of minor planets for the determination of a dynamical reference against which to compare the systems of fundamental catalogs

Celestial Mechanics 34 (1984) 19–36. 0008–8714/84.15

was proposed by Dyson in 1928. Numerov (1933) suggested a list of ten minor planets to be observed by transit circles and photographic telescopes. An extended list of sixteen minor planets was given by Brouwer (1935,1941). While the transit circle has traditionally been the instrument used to establish the astronomical reference frame, photographic astrometry offers some advantages. Transit circle observations are restricted to brighter minor planets, near opposition; photographic astrometry allows observation of fainter minor planets over a more extended span of their apparent orbits, thus providing a better solution for the fiducial points of the reference frame. Pierce (1971) analyzed the results of the Brouwer program, and Fricke (1982) has discussed the equinox and equator results in comparison with the FK4.

MINOR PLANET PROGRAM

In the past few years, the development of the NASA Space Telescope and the European HIPPARCOS telescope, in addition to the application of new measurement and reduction techniques to ground based observations, holds the promise for a significant improvement in astrometric accuracy. This has led to the present effort to combine Space Telescope and ground-based observations to establish through the use of "crossing point" observations, an interlocking net of orbits of minor planets whose dynamical motion can be used to probe the systematic accuracy of the fundamental reference system. A "crossing point" is defined as a position in the sky where two apparent paths of minor planets intersect. Differential measurements of the positions of minor planets with respect to common faint background stars at crossing points provide a single observational accuracy of 0.02 arcseconds or better (Duncombe and Hemenway, 1982). Thus, this observation type has weight an order of magnitude better than the classical technique of determining the absolute position with respect to the existing Fundamental System. Observations of several minor planets with respect to some reference frame may be used to determine the overall rotation of that reference frame (Branham, 1979), (Branham, 1980), (Kristensen, 1980a), (Orelskaya, 1980). Questions exist about the differences between the tabular motions of the mean sun compared with the computation of Universal Time derived from Sidereal Time and with the general precession in longitude. The time systems have apparent uncertainties of the order of 1" per century (Stumpff and Lieske, 1984). Since observations of minor planets are independent of the earth's rotation axis to a much higher order (of the order of 1 earth radius/1 AU) than transit circle observations, for example, the use of accurate minor planet observations of the 0.002 arcsec range over 15 years or of the order of a few times 0.1 arcsec range over 100 years may add significantly to our knowledge of the vagaries of the timing system, particularly earlier in this century. Further, a program of accurate minor planet observations will provide a direct comparison between any coordinate system defined with respect to extragalactic objects and a dynamically determined coordinate system. In particular, VLBI observations are beginning to dominate the field of positional precision, c.f. (Eichhorn, 1984), and references therein, but the problem of relating any VLBI system to optical objects in general and dynamical objects in particular is fraught with potential errors. Thus, in the first instance, determining any systematic motion of a dynamical zero-point with respect to extragalactic objects might demonstrate unforeseen systematic effects. However, in the final analysis, a dynamical reference frame is expected to rotate with respect to an extragalactic frame due to general relativistic and cosmological effects. Any positive measure of such a rotation would be of great astrophysical interest. While the sensitivity of these observations may not be as great as some other programs, the measurement would be <u>direct</u> over a time scale of 20 years, and <u>any</u> such measurement should be made because of the possibility of detecting an unknown effect (Schwarzschild, 1967).

SELECTION OF MINOR PLANETS

Consideration of the objectives of a full-scale minor planet program leads to several criteria to be applied to the selection of a set of minor planets for the program. Below are listed some of these objectives and their attendant criteria:

- <u>objective</u>: the apparent paths of the minor planets should be well distributed in the sky during the period of observation, in order to cover as much of the celestial sphere as possible.

- criteria: the inclinations should be reasonably high (20° or larger) and the orbits should be evenly distributed in the longitude of the ascending node.

- result: coverage of the half-sphere centered on the <u>ecliptic</u>. 18 smaller and 9 brighter minor planets have been selected accordingly. (See below for diameter and magnitude considerations.)

- objective: The paths should provide adequate crossings to link all the objects together.

- criteria: Include 4 minor planets with inclination close to zero.

- result: 4 minor planets whose paths cross the other minor planets' paths near the ecliptic at the longitudes of the nodes.

- objective: the orbits should contain some minor planets which pass repeatedly through the same star fields, so that long-term motions of the reference frame may be determined with respect to the background stars.

- criterion: include (619) Triberga and the 4 minor planets with low inclination.

- result: 5 minor planets which can be used to tie the system together directly.

- objective: provide crossings which will maximize the determination of the dynamical zero point.

- criteria: two high inclination minor planets with longitudes of the ascending nodes separated by about 180°.

- result: for a well distributed set of minor planets described above, several pairs will be available with the appropriate orbital conditions. The final solution will not rest on one pair, but on the relative weights of all the orbits together. Observations near the ecliptic will be extremely valuable for this purpose, however.

- objective: provide a direct tie to the present Fundamental System.

- criterion: pick a minor planet with good observing history, small diameter, and previous extensive use for Fundamental System studies.

- result: include (51) Nemausa.

Consideration of the observing requirements are outlined below:

- objective: select minor planets which may be observed with the Space Telescope.

- criterion: the angular diameter must be less than the resolving power of the telescope (0.04 arcseconds) because the astrometric guiding system uses an interferometer whose entrance aperture is the 2.4 meter primary mirror.

- result: 22 minor planets were selected with diameters less than 30 kilometers or whose absolute magnitudes indicate that they are very likely smaller than 30 kilometers in diameter. These minor planets range in magnitude from about 14 to 18 apparent magnitude.

- objective: due to the demand for large telescope time, minor planets which may be observed with ground-based telescopes in the 1 to 1.5 meter class have also been selected. This objective conflicts with the previous objective in many cases, but the objects are included to insure adequate observational coverage.

- criterion: select minor planets with diameters of about 100 kilometers, with high albedos, and with the same orbital characteristics as the fainter minor planets.

- result: 10 brighter minor planets relatively evenly distributed as the previous group, have been selected.

The 34 minor planets of various characteristics, selected to encompass the objectives of the project are shown in Table I.

HIPPARCOS OBSERVATIONS

The HIPPARCOS satellite is expected to produce relative positions, proper motions, and parallaxes to an rms accuracy of 0.002 arcseconds. HIPPARCOS field of view will follow a preplanned continuous scan of the celestial sphere due to the satellite's rotation. Therefore, systematic observations of solar system objects to provide a dynamical zero-point and orientation will not be possible.

HIPPARCOS will observe a selection of the brighter minor planets but the observations will be randomly distributed during the mission. Using a simulated error of a single HIPPARCOS observation of 0.01 to 0.02 arcseconds, 20 observations/mp/year, 25 minor planets, and a 2.5 year mission, the ecliptic plane should be determined to 2.5 milliarcsec, and the rotation of the system should be determined to 1 milliarcsec/year about any axis in the plane of the ecliptic, and 4 milliarcsec/year about the ecliptic polar axis. These observations are subject to large phase and shape effects, simply because the minor planets must be large. For a reasonable distribution of observations, however, the phase effects have been shown to cancel to a large extent (Soderheim and Lindergren, 1982). On the other hand, the observations of this project have the advantages of higher weight at crossing points, tying whole orbits together (reducing systematic effects from one part of the sky to another), and smaller minor planets (30km instead of 150 km, typical numbers) so that the phase effects are

Table 1 SELECTED MINOR PLANETS

MP #	NAME	d_{km}	$i(^{\circ})$	a(AU)	e	$\Omega(^{\circ})$	B(1,0)
(373)b	Melusina	90.3	15.4	3.13	.137	4.4	10.08
(1453)	Fennia	11.2	23.7	1.90	.028	6.7	13.84
(475)	Ocllo		18.8	2.59	.380	35.0	12.32
(599)b	Luisa	65	16.7	2.77	.295	44.6	9.38
(965)	Angelica		21.5	3.15	.288	41.0	11.29
(692)b	Hippodamia	46.5	26.1	3.36	.193	64.1	10.13
(1320)	Impala		19.2	2.98	.232	72.1	11.89
(652)	Jubilatrix		15.7	2.56	.125	85.9	12.53
(1276)	Ucclia		23.4	3.16	.112	114.5	11.83
(387)b	Aquitania	113.	18.1	2.74	.238	128.1	8.46
(502)	Sigune	20.9	25.0	2.38	.180	132.8	11.87
(148)b	Gallia	92.0	25.3	2.77	.186	145.1	8.64
(1252)	Celestia	19.5	33.9	2.69	.203	140.8	12.04
(218)b	Bianca	58.9	15.2	2.67	.117	170.7	9.61
(434)	Hungaria	11.6	22.5	1.94	.074	174.9	11.91
(950)	Ahrensa		23.5	2.37	.160	181.9	12.41
(25)b	Phocaea	72.8	21.6	2.40	.254	213.8	9.01
(391)	Ingeborg	17.5	23.1	2.32	.307	212.6	12.24
(849)b	Ara	72.4	19.6	3.17	.179	228.9	9.00
(1108)	Demeter		24.9	2.43	.260	234.2	12.34
(1222)	Tina		19.7	2.79	.249	245.9	13.16
(654)b	Zelinda	73.5	18.1	2.30	.231	278.2	9.47
(1626)	Sadeya		25.3	2.36	.274	279.2	12.87
(2000)	Herschel	17.1	22.8	2.38	.300	291.7	12.33
(1584)	Fuji	25.7	26.7	2.37	.195	305.0	12.02
(61)b	Danae	87.9	18.2	2.98	.166	333.6	8.73
(1474)	Beira	11.2	26.8	2.73	.490	324.9	13.50
(1310)	Villigera		21.0	2.39	.358	357.4	12.69
$i \cong 0^{\circ}$							
(637)	Chrysothemis		0.288	3.15	.145	355.5	11.86
(846)	Lipperta	52.0	.263	3.13	.184	262.4	11.31
(1340)	Yvette		.428	3.17	.144	346.1	12.56
(1383)	Limburgia		0.014	3.08	.191	204.4	12.84
Special Objects							
(51)	Nemausa	156.	9.97	2.37	.066	175.7	8.45
(619)	Triberga	31.8	13.72	2.52	.076	187.3	10.95

b = selected for brightness

reduced ab initio. The program is expected to last for 15 years at high accuracy and will utilize older observations to improve the mean motion. Therefore, our program should have a much higher accuracy in the determination of the systematic motion of the HIPPARCOS reference frame with respect to the dynamical system than will come from the HIPPARCOS minor planet observations themselves. For example, if 30 observations/mp were obtained at crossing points during the program, and the error of a single observation were the 0.01 arcsecond expected from the ground-based observations, then the error expected in the rotation around the ecliptic might be expected to be 0.1 milliarcsec/year. While this estimate is extremely optimistic, it demonstrates the power of having an accurate program for an extended period of time. The lack of overlap in the selection of minor planets between the HIPPARCOS project and this project is due mainly to the magnitude and size constraints imposed by Space Telescope, and considerations such as the correction from center of light to center of mass.

PROGRAM PHASES AND STEPS

The full-scale minor planet program is planned in three phases:

Phase I. An initial phase to reduce the errors in our knowledge of the orbital elements of the selected minor planets, to allow extrapolation at the 0.1 to 0.2 arcsecond level over succeeding years. Such an initial program will cover slightly more than one orbit for each minor planet, relying on previous observations to provide the necessary accuracy in the mean motion.

Phase II. The observational phase, 15 years starting at 1986.0, will allow sufficient coverage to determine the orbits to the 0.02 arcsecond level, and to allow adequate coverage of the celestial sphere. Boiko (1975) has shown that a minor planet program should cover at least 3 orbits of each minor planet, and that observations should extend well beyond opposition to obtain adequate separation of the unknown parameters of the solution.

Phase III. The reduction and analysis phase, in which the corrections to the coordinate system and orbital elements are simultaneously derived from the observations.

Because phase II covers a 15 year period, the reductions of phase III may be made at regular intervals to provide updated analysis of the motion of the reference system and systematic corrections to selected

Table 3

Reference: AN. 158 346 Minor planet: (148) GALLIA

Observatory and Observer: Alger, M. Rombaud

Year of Obs.		T.M. Alger							Reference Star						Authority
1901					Δα		Δδ		α(1901.0)			δ (1901.0)			
March	28	10^h	52^m	49^s	-0^m	$32^s.35$	+ 8'	38".9	13^h	44^m	$55^s.33$	$+18^\circ$	52'	12".7	A.G. Berlin 5022
	28	11	16	21	−0	32.97	+ 8	49.0							
	29	10	31	59	−1	40.13	+ 3	18.0	13	45	22.92	+19	7	16.5	A.G. Berlin 5027
April	5	10	12	39	+2	0.18	+ 1	9.5							
	5	10	34	23	+1	59.41	+ 1	17.0	13	36	41.49	+20	13	17.8	A.G. Berlin 4856
	13	8	43	26	+1	21.26	+ 3	3.7							
	13	9	15	24	+1	20.37	+ 3	12.5	13	31	16.32	+21	11	59.8	A.G. Berlin 4824
	15	9	11	23	+0	39.29	+10	45.0							
	15	9	40	4	+0	38.25	+10	52.8	13	30	24.94	+21	17	23.1	A.G. Berlin 4819
	16	8	37	5	+2	37.66	+10	49.9							
	16	9	16	4	+2	36.45	+10	59.9	13	27	41.13	+21	23	14.7	A.G. Berlin 4812
	18	9	0	1	−0	25.69	− 8	55.7							
	18	9	51	35	−0	27.46	− 8	46.3	13	29	11.74	+21	54	36.6	A.G. Berlin 4817
	27	8	53	33	+0	30.28	+ 8	0.5							
	27	9	25	49	+0	29.18	+ 8	4.6	13	21	38.36	+22	16	15.7	A.G. Berlin 4782
	29	9	30	55	−0	53.10	+13	49.3							
	29	9	52	51	−0	53.57	+13	50.0							
May	1	9	15	45	−2	33.89	−15	42.6	13	21	59.01	+22	50	25.6	A.G. Berlin 4785

portions of the sky and/or individual objects.

In addition to selecting the 34 minor planets, the following Phase I steps have been taken:

1. To produce ephemerides accurate to the 0.2 arcsecond level for several years after the orbital correction phase, about 20 exposures per minor planet distributed around the orbit will be needed. Thus about 4 observations per minor planet per year, are being obtained over the 5 year initial phase. The observations are being made primarily at the McDonald Observatory using the 2.1 meter Otto Struve Telescope, at a plate scale of 7 arcsec/mm. Initially, the technique of measuring secondary reference stars on the National Geographic Society-Palomar Sky Survey will be used, c.f. (Benedict and Shelus, 1978) and reference therein. This technique yields an accuracy of 0.5 arcseconds, limited by the accuracy of the reference stars and the proper motions of the secondary reference stars. However, the observations will be made when the minor planets are near crossing points whenever possible, and eventually microdensitometer scans of the images will be used to obtain the 0.02 arcsecond accuracy which has been demonstrated (Duncombe and Hemenway, 1982) as part of the final program. The status of the observations to refine the orbits of the 34 minor planets is given in Table 2. Due to poor observing weather during the first year of the project, the coverage is not as complete as desired. Over 300 exposures of the program minor planets, have been made with positive identifications. We have determined that observations can be made in bright moon using a IIIaF emulsion and an RG610 filter, without loss of accuracy at the few tenths of an arcsecond level. This development allows us to obtain significantly more telescope time than previously. We have not yet had a chance to test effects at the micron level of the IIIaF emulsion/RG610 filter combination, but have obtained some plates for such tests.

2. The orbital refinement task requires observations distributed over each minor planet orbit, in order to accurately determine five of the six orbital elements. The sixth, the mean motion, will be most accurately determined from old observations used in conjunction with the new data. Over 3000 references to old observations (earlier than 1950) of the 34 minor planets have been identified. Steps have begun to extract the data necessary

Table 2 MINOR PLANET WORK SHEET

as of 2/25/84

No of M.P.	No of Epochs	No of Exposures	No of Years	Comments
25	1	3	1	well studied
51	many	many	3	well studied
61	-	-	-	well studied
148	4	9	1	
218	3	6	1/2	
373	3	11	1 1/2	
387	5	14	1 1/2	
391	5	9	3	
434	3	9	3	
475	5	7	5	
502	4	13	3	
599	4	7	1 1/2	
619	6	15	3	
637	5	9	3	
652	3	8	1 1/2	
654	3	8	1/2	
692	4	8	1 1/2	
846	5	8	5	
849	4	14	1 1/2	
950	5	12	5	
965	4	9	5	
1108	3	5	1/2	
1222	6	21	2 1/2	
1252	9	26	5 1/2	
1276	4	8	2 1/2	
1310	4	6	1 1/2	
1320	4	10	2 1/2	
1340	5	7	3	
1383	4	5	1/2	
1453	4	6	1 1/2	
1474	0	0	0	Far South Now
1584	5	12	3	
1626	3	5	1/2	
2000	3	18	1 1/2	

to obtain reliable positions on the FK5 system for those observations amenable to such reduction, and to eliminate the useless observations. Table 3 is a typical worksheet for one of these references which give possibly useful information including the reference star name and position (the position of the reference star will have to be redetermined, but it provides an unambiguous identification of the star) and the measured offset in right ascension and declination. Other corrections will have to be determined and taken into account, including atmospheric refraction and dispersion at the time of observation, the reduction of the time of observation to a meaningful system, and the reduction of the reference star to a modern coordinate system, as examples.

3. A new Cassegrain Camera has been fabricated. The driving software has been designed and is being implemented. Some of the electronics remains to be integrated. The computer-camera interface was completed in February 1984. On-telescope tests of the Camera are expected at McDonald Observatory in April. The new camera will be controlled by a NOVA computer, and will automatically drive the tracking stage at the appropriate ephemeris rate and angle. Exposure times will be measured to a small fraction of a second (we expect 0.1 second accuracy).

4. Software Development. A program has been developed for the reduction of PDS x, y measured positions to above the earth's atmosphere accounting for atmospheric refraction and dispersion, using an effective wavelength approach based on other work (Chiu, 1978). The transmission of the atmosphere, the reflection of the optical surfaces, the transmission of any filters, the spectral response of the emulsion, and the spectral characteristics of the background stars are taken into account.

   A software system of programs for the simultaneous refinement of the orbits and the determination of coordinate system irregularities has been designed using the Program Design Language. The software being developed includes the following:

   i. A set of routines has been programmed and tested that perform the transformations between Keplerian orbital elements and cartesian equatorial coordinates.

ii. Paired seventh- and eighth-order Runga-Kutta integrators have been programmed and tested for the computation of the minor planet ephemerides.

iii. The programs to perform the numerical computation of the partial derivatives necessary for the correction of orbits to the kilometer level have been written.

iv. An orbit correction procedure following the generalized least squares theory of Jefferys, (1980) for the reduction of the observations has been programmed.

v. An orbital integrator using the DE series from JPL has also been programmed and tested against the integrations of Duncombe, (1969).

5. Minor planet (51) Nemausa has a long history of observations for fundamental system work. Kristensen and Moller have reduced 2240 observations of Nemausa to derive corrections to the FK4 equator point and equinox (Kristensen, 1980a). The corrections had errors of 0.03 arcseconds and 0.005 seconds of time respectively. Relative observations of Nemausa at a retrograde loop crossing point cannot separate effects of equator and equinox. However, they can put stringent constraints on the relation between Nemausa's orbit and the Earth's orbit. In particular, the orientation of Nemausa's orbit (as derived from 2240 observations) with respect to the Earth's orbit (as given by the JPL-USNO ephemerides) will provide a direct link from the stellar coordinate system to the inner solar system derived from radar and orbital-lander-range measurements.

An international observing campaign was undertaken with Kristensen to obtain photographic plates of Nemausa as it passed through the same star field on two different occasions in September, 1980 and March 1981 (Kristensen, 1980b), see Figure 1. Six observatories participated and 117 exposures were obtained. The relative measurement accuracy of the minor planet with respect to the background stars should allow determination of the relation between Nemausa's orbit and the Earth's orbit to 0.003 arcseconds, which corresponds to about 5 km at the distance of Nemausa. Measurement has been delayed awaiting the completion of an upgrading of the Texas PDS (including a new computer system

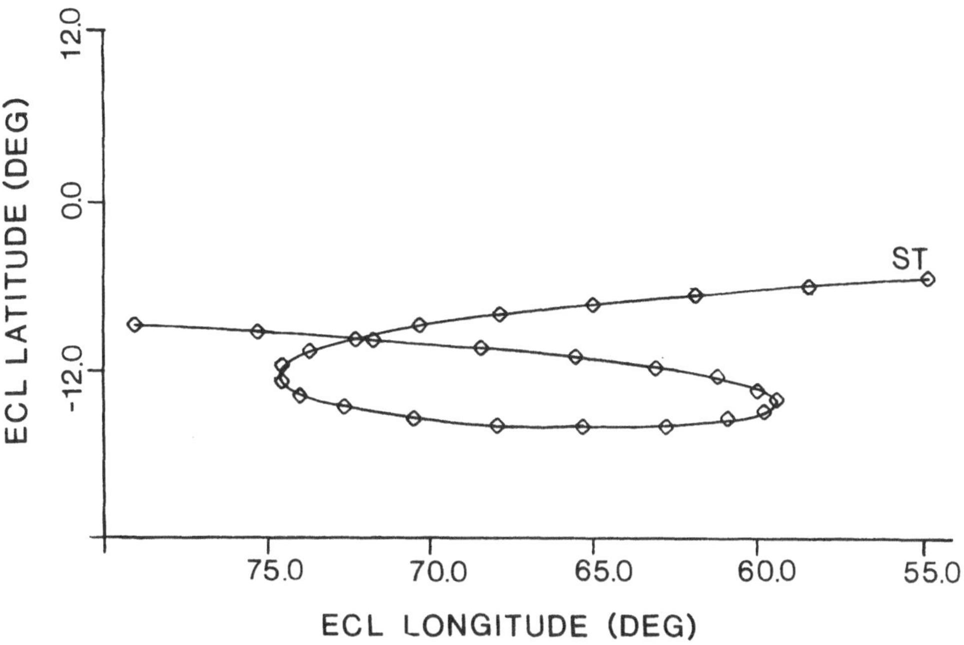

Figure 1. Ephemeris path of (51) Nemausa, with 1980-81 crossing.

and interface).

6. A number of observatories have evinced interest in this project, and some cooperation has been forthcoming in the preliminary orbit refinement program. In particular, an observing program is being initiated south of -10° declination at El Leoncito, Argentina by Dr. Carlos Lopez. Also discussions are underway with Dr. Kaare Aksnes about having observations made in Norway of the brighter objects. We hope that ultimately a significant number of observations will be made at other observatories for this project.

7. Our late colleague, Paul Herget, calculated 10-year ephemerides including perturbations for 22 minor planets and he predicted crossing points. Hence, there is a list of crossing points covering the initial phase prior to 1986, which is being used to set priorities during the current observing program. First epoch

plates at about 30 crossing points have been obtained during the course of the last two years. More recently through Dr. Fricke's interest, Dr. Hans Scholl of the Astronomisches Rechen-Institut kindly computed preliminary ephemerides of all 34 minor planets for the period 1986-2001. Computation of the crossing points revealed 6400 such areas.

8. In conjunction with the international office of the NSF, a cooperative program is being developed with Dr. Jean Kovalevsky, director of F.A.S.T., to use Space Telescope to provide a direct tie between the dynamical system of minor planets and the HIPPARCOS system. A list was prepared of approximately 8000 SAO stars within 20 arcminutes of 6400 minor planet crossing points. This list was submitted to the HIPPARCOS project. They have indicated that all the stars brighter than 9th magnitude in fields with multiple minor planet crossings will be given highest priority and will almost certainly be included in the HIPPARCOS catalogue. These stars amount to about 2000 objects. The other stars have been assigned to the "low priority reservoir," but many will probably be selected because no other stars in their region of the sky are specified by other HIPPARCOS proposals. A similar list has also been prepared of these stars to be observed with the Fine Guidance Sensors of the Space Telescope in order to derive crossing point parameters at the 0.002 arcsecond level for a selected subset of the crossing points.

HIPPARCOS - SPACE TELESCOPE COOPERATION

As members of the Space Telescope Astrometry Team, both Duncombe and Hemenway plan to use the major portion of their "Guaranteed Observing Time" (GOT) to observe these minor planets next to HIPPARCOS stars at crossing points. While these observations may only amount to a few tens during the 2.5 year decreasing percentage GOT, they will establish the usefulness of well placed Space Telescope observations to put stringent constraints on both the solution and systematic effects, by comparing these observations directly with similar observations from the ground, made at the same time. We fully expect to obtain a small but continuing set of crossing point observations with the Fine Guidance sensors of the Space Telescope throughout the 15 year progam.

With regard to the HIPPARCOS task, the relative HIPPARCOS reference system will provide an ideal system against which to reduce the minor planet observations. The HIPPARCOS system will have an overall "solid body" rotation which must be determined. Plans are under way to tie the HIPPARCOS system to extragalactic objects (Hemenway, et al., 1982), (Hemenway and Duncombe, 1984). Other plans are under way to relate the HIPPARCOS system to the FK5. By using observations of the minor planets with respect to HIPPARCOS stars, we expect to provide a dynamical basis for the HIPPARCOS reference system. The project would determine the systematic relationships between an accurate independent dynamical frame, the FK5, an extragalactic frame, and the coordinate system derived from independent radio astronomical observations. Kovalevsky and Mueller have given a cogent and concise overview of the relationship between theoretical, observational, terrestrial, and celestial coordinate systems (Kovalevsky and Mueller, 1981). The effective coordination of an observational program of minor planet dynamics with the international efforts to form a reference coordinate system is perhaps the most important aspect of this project.

CONCLUSIONS

The capability of measuring minor planet motions with respect to a set of background stars to the 0.05 arcsecond level has been demonstrated (Duncombe and Hemenway, 1982). Using an interrupted trail technique which produces point images, a plate to plate consistency of the background star reference frame on minor planet plates of 0.01 arcsecond rms has been obtained. The techniques producing these results are being used with the regular motions of 34 minor planets over large arcs of the sky, to study the systematic effects in the extant Fundamental System. The initial phase of minor planet observation and orbit refinement is in progress. The ultimate aim is to establish a long-term (15-year) project to determine the constants of motion of the FK5 and the HIPPARCOS reference systems with respect to an independently determined dynamical frame.

In embarking on a program of this magnitude and duration, we are encouraged by the example of the Brouwer project and by the support and counsel of many colleagues in the national and international community of astronomers. We are indebted to the late Prof. Paul Herget for his early interest in, and support of, this project.

We gratefully acknowledge support by grants from the National Science Foundation: AST 7916870, AST 8208958, INT 8213072.

REFERENCES

Benedict, G.F. and Shelus, P.J. 1978, in IAU Colloquium #48, Modern Astrometry , ed. Prochazka, F.V. and Tucker, R.H. (: Institute of Astronomy (University Observatory), Vienna, Austria)

Boiko, V.N. 1975, Soviet Astronomy, 19, 261.

Branham, R.L. 1979, Astron. J., 84, 1399, 1402.

Branham, R.L. 1980 Astronomical Papers of the AENA, XXI pt 3.

Brouwer, D. 1935, Astron. J., No. 1022, 57.

Brouwer, D. 1941, in Annals of the New York Academy of Sciences, XLII, 133.

Chiu, L. 1977, Astron. J., 82, 842, 848.

Chiu, L. 1978, Dissertation, University of California at Berkeley.

Duncombe, R.L. 1969, Astron. Papers of the American Ephemeris and Nautical Almanac, XX part II, 139-309.

Duncombe, R.L. and Hemenway, P.D. 1982, Cel. Mech., 26, 207.

Duncombe, R.L. and Hemenway, P.D. 1983, in The Motion of Planets and Natural Satellites, Workshop held in Embu, Brazil Dec. 16-18, 1981, ed. S. Ferraz-Mello and P.E. Nacozy (Sau Paulo: University of Sau Paolo), pp. 215-224.

Dyson, F., 1928, Transactions IAU, 3, 227.

Eichhorn, H., 1985, in IAU Symposium #109, ed. Eichhorn, H. (D. Reidel Publishing Co., Dordrecht, Holland. In press.).

Fricke, W. 1981, in IAU Colloquium #56, Reference Coordinate Systems for Earth Dynamics, ed. E.M. Gaposchkin and B. Kolaczek (: D. Reidel Publishing Co., Dordrecht, Holland), 331.

Fricke, W. 1982, Astron. J., 87, 1338.

Hemenway, P. 1980, Cel. Mech., 22, 89.

Hemenway, Paul D. 1984, Astron. J., 89, 145-150.

Hemenway, P.D., and Duncombe, R.L. 1980, BAAS, 12, 829.

Hemenway, P.D., and Duncombe, R.L. 1981, in IAU Colloquium #56, Reference Coordinate Systems for Earth Dynamics, ed. E.M. Gaposchkin and B. Kolaczek (: D. Reidel Publishing Co, Dordrecht, Holland), 277.

Hemenway, P.D., and Duncombe, R.L. 1985, in Astrometric Techniques, IAU Symposium #109, ed. Eichhorn, H. (Dordrecht, Holland: Reidel), in press.

Hemenway, P.D., Jefferys, W.H., Shelus, P.J., and Duncombe, R.L. 1982, in Proceedings of the International Colloquium on the Scientific Aspects of the HIPPARCOS Space Astrometry Mission, ed. Perryman, M.A.C., and Guyeene, T.D. (Paris: European Space Agency), 61-63.

Jefferys, W.H. 1980, Astron. J., 85, 177.

Kovalevsky, J. 1980, Cel. Mech., 22, 153.

Kovalevsky, J. and Mueller, I.I. 1981, in IAU Colloquium #56, Reference Coordinate Systems for Earth Dynamics, ed. E.M. Gaposchkin and B. Kolaczek (: D. Reidel Publishing Co., Dordrecht, Holland), 375.

Kristensen, L.K. 1980a, Mitteilungen der Astronomischen Gesellschaft, 48.

Kristensen, L.K., Duncombe, R.L. and Hemenway, P.D. 1980b, IAU Circular, 3480.

Lieske, J.H. and Standish, E.M. 1981, in IAU Colloquium #56, Reference Coordinate Systems for Earth Dynamics, ed. Gaposchkin, E.M., and Kolaczek, B. (: D. Reidel Publishing Co., Dordrecht, Holland), 295.

Numerov, 3. 1933, Bull. Inst. Astron. Leningrad, No. 32.

Orelskaya, 1980, Mitteilungen der Astronomischen Gesellschaft, 48, 43.

Pierce, D.A. 1971, Astron. J., 76, 177, pp. 19-20.

Schwarzschild, M. 1967, in Vistas in Astronomy, ed. (: Pergamon Press, Oxford).

Soderheim, S. and Lindegren, L. 1982, Astron. Astrophys., 110,156-162.

Standish, E.M. 1982, BAAS, 14, 580.

Stetson, P.B. 1979, Astron. J., 84, 1056.

Stumpff, P. and Lieske, J.H. 1984, Astron. Astrophys., 130, 211-226.

THE EMPIRICAL INERTIAL SYSTEM DETERMINED IN FK5 BY THE DYNAMICS OF THE PLANETARY SYSTEM

Walter Fricke
Astronomisches Rechen-Institut
Monchhofstrasse 12-14
6900 Heidelberg, Germany, F.R.

The determination of the FK5 equinox has been based on observations of of the Sun and planets at mean epochs from 1900 to 1977 and on lunar occultations. The solution for the equinox correction E(T) and its secular variation $\dot{E}$ indicate the correction $+0^{s}.035$ at 1950.0 to all right ascensions of the FK4 and the correction $+0^{s}.085$ per century to all proper motions in right ascension of the FK4. These corrections will be applied in the construction of the system of the Fifth Fundamental Catalogue, the FK5. Recent observations of the Sun and planets for the improvement of the FK4 equator have not indicated the need for a significant correction. Therefore no change of the equator will be made in the FK5.

Celestial Mechanics **34** (1984) 37. 0008–8714/84.15

OBSERVATIONS OF MINOR PLANETS WITH THE VERY LARGE ARRAY

P. K. Seidelmann, G. H. Kaplan
U.S. Naval Observatory, Washington, D.C. 20390

K. J. Johnston
E. O. Hulburt Center for Space Research
Naval Research Laboratory, Washington, D.C. 20375

C. M. Wade
National Radio Astronomy Observatory [a]
Socorro, New Mexico 87801

ABSTRACT

The weak thermal emission from the largest minor planets can be detected and measured at all points around their orbits at microwave frequencies using the Very Large Array (VLA). Position determinations of astrometric quality have been obtained, and flux measurements have provided size estimates. When enough precise positional observations have been accumulated, the orbits of the minor planets and the Earth can be determined. This will allow the equinox to be located within the radio reference frame, providing a truly fundamental coordinate system for radio source positions. It will also provide a means of relating the optical and radio (quasar) coordinate systems.

[a] The National Radio Astronomy Observatory is operated by Associated Universities, Incorporated, under contract with the National Science Foundation.

Celestial Mechanics **34** (1984) 39–48. 0008–8714/84.15

INTRODUCTION

Historically, optical observations of minor planets have proven useful for the determination of the planetary masses, the orbit of the Earth, and other fundamental astronomical constants. The value of minor planets as "probes" of the solar system results from their very low masses, their rapid motion across the sky, the location of their orbits with respect to the principal planets, the relatively high eccentricities of their orbits, and their small angular sizes as seen from the Earth. The latter property is especially valuable to astrometrists, since it allows for a relatively unambiguous (to order 0.01 arcsecond) association of the centroid of emission with the center of mass.

The Very Large Array (VLA) provides sufficient sensitivity and resolution to detect and measure the thermal emission from the larger minor planets. Thus, astrometric observations of the minor planets in the radio regime are now possible, and we have embarked upon a long-term program of such observations. The VLA is probably the only instrument in the world capable of such observations; the radio fluxes are too weak (of order 0.5 mJy at 6cm) for VLBI techniques. Initial results from our program were described by Johnston, et al. (1982). An update of the progress of our program is given here.

OBJECTIVES OF THE PROGRAM

The VLA minor planet observing program has three major objectives:

(1) Astrometric positions. We have demonstrated that positions of minor planets can be determined from our VLA observations to better than 0.1 arcsecond relative to the radio (quasar) reference frame. The radio observations are not limited to times near opposition (as are optical observations) so that a series of such observations, spanning several years, can considerably improve our knowledge of the orbits of these bodies and of the Earth. A determination of the Earth's orbital elements from these observations is equivalent to locating the dynamical equinox within the radio reference frame, and this is a major goal of our program. Such a determination of the location of the equinox, made entirely from radio observations, would allow the establishment of a truly fundamental reference frame for radio source positions. Furthermore, comparison of minor planet positions determined from radio and optical observations provides information on the relationship between the radio and optical reference frames.

(2) Detection and flux measurement. We are interested in obtaining observations of the largest minor planets at various points in their orbits relative to the Earth, so we have attempted observations at both favorable and unfavorable geometries, with

various degrees of success. The flux received by the VLA from a minor planet is related to its size, distance from the Sun and Earth, albedo, and rotation rate. Assuming that it is a rapidly rotating black body with a known albedo, a measurement of a minor planet's radio flux provides a determination of its diameter. Variations in flux over time scales of hours, if detected, would directly provide the rotation period of the body.

(3) Imaging. Under a combination of favorable circumstances, it may be possible to obtain crude radio images of a few of the largest minor planets. This would provide a direct measurement of diameter. It might be further possible to detect variations in apparent brightness temperature across the image by "binning" scans made on several different days. Such a process, if successful, would yield as a by-product the rotation period of the minor planet being observed, if that information were not otherwise available.

METHOD OF OBSERVING

The observations of the minor planets are interleaved with observations of standard VLA calibration sources. The observational sequence is fifteen minutes of integration on the minor planet followed by five minutes on the calibration source. The apparent positions for each fifteen minute period on the minor planet are calculated directly from the ephemeris for the minor planet. The ephemerides for Ceres, Pallas and Vesta are based on Duncombe (1969), and a special ephemeris for Hygiea computed at the U. S. Naval Observatory is used. Standard VLA software is used for the corrections for the topocentric positions with respect to the phase center of the VLA. Good astrometric calibrators as close as possible to the positions of the minor planets are scheduled based on both the positions of the minor planets and the configuration of the VLA. The "A" configuration, maximum antenna separation of 36 km, may provide sufficient resolution to resolve structures on the largest minor planets when observed at the 2 centimeter wavelength near opposition, if the minor planet is at a high declination, and there are good (i.e., dry and stable) weather conditions. The "B" configuration, maximum antenna spacing of 11 km, provides an adequate combination of signal to noise, astrometric position and resolution. The "C" configuration, maximum antenna spacing 3.4 km, provides reduced resolution but can be used for detection experiments. Characteristics of the different configurations of the VLA are summarized in Table 1.

Since 1981, successful observations have been obtained of 1 Ceres and 2 Pallas at 6 and 2 cm wavelengths, while 4 Vesta and 10 Hygiea have been observed only at 6 cm wavelength.

Table 1. VLA Configurations

Configuration	Maximum Antenna Separation km	Half-Power Beamwidth* (arcseconds) 6 cm	2cm
A	36.3	0.4	0.11
B	11.07	1.1	0.4
C	3.4	3.6	1.2
D	1.0	12.4	4.1

* These approximate synthesized half-power beamwidths are for high declination sources with an aperture untapered, complete weighted and complete filled out to the maximum spacing.

PROBLEMS WITH THE PROGRAM

The feasibility of this observing program has been demonstrated and the initial results presented by Johnston, et al. (1982). Before discussing the subsequent results, it seems appropriate to discuss some of the problems which have been encountered.

A basic requirement of an astrometric program, either for optical or radio techniques, is a single source of ephemerides, a stable list of calibration sources, and pointing control programs and observation reduction programs which are stable, reliable, and accountable. If these conditions are met, observations on one system can be obtained over an extended period of time. Unfortunately, these requirements are not always appreciated by astrophysicists, leading to difficulties when an astrometric program is being pursued on a non-astrometric instrument. The typical observer on the VLA is not interested in positional accuracies of milliarcseconds, as required by our program. The differences in positions between the minor planet and the calibration source are determined by the software system of the VLA control computers. This software provides the capability for making the observations, but it is also a source of problems for an astrometric program. In order to continually improve and upgrade the capabilities of the VLA, it is necessary to make improvements in the hardware and software of the computer system. This continual program of improvement is in direct conflict with the astrometric requirement for a long term, stable observing program.

Our observational program is designed to observe the minor planets at various positions in their orbits as well as from different places in the Earth's orbit. This means that the observations should be made at different times during the year. However, the observing program also requires that the observations be made when the VLA is in the "A" or "B" configuration, or with reduced capabilities when it is in the "C" configuration. The VLA goes through the different configurations in an approximate fifteen month cycle. Thus, there is a systematic pattern to the distribution to the observations which is not optimum for our purposes.

Weather conditions at the observing site also affect the quality of the observations. The presence or absence of humidity is a significant factor in phase stability, affecting the resolution and accuracy of the observations. During the March 25, 1983 observations of Ceres, three inches of snow accumulated on the antennas, lending to antenna gain drop of 30% to 40% and data that are virtually useless. Even relatively minor storms eliminate the option of observing at 2 cm. The efforts to observe surface details on a minor planet require a coincidence of several factors. The VLA needs to be in the "A" configuration at a time of year when the humidity is low. The minor planet needs to be at opposition at a high declination. These combinations must occur simultaneously. To date this situation has not occurred for any of the four large minor planets.

FLUX, TEMPERATURE, AND SIZE

In addition to determining a position for the planets, a flux density is also obtained from an observation. The flux density scale is fixed by observations of 3C286, for which the adopted flux densities are given by Baars, et al. (1977). These flux densities can be converted into an effective brightness temperatures by means of the expression

$$T_B = k \times 10^9 \frac{\lambda^2 D^2 S}{d^2},$$

where k is approximately one if the minor planet's apparent disk is much smaller than the synthesized beam, the wavelength (λ) is in centimeters, the geocentric distance (D) is in astronomical units, the measured total flux density (S) is in Janskys and the assumed circular disk diameter (d) is in km. T_B is the uniform disk brightness temperature in Kelvins computed from the data. If R is the object's heliocentric distance at the time of observation, this brightnesss temperature can be normalized to a brightness temperature (T_{BN}) at a fixed mean heliocentric distance (Rm) by use of the equation

$$T_{BN} = T_B \; (R/Rm)^{1/2}$$

This temperature can be compared to the equilibrium brightness temperature for a rapidly rotating black body at a distance R astronomical units from the Sun, which can be calculated from the expression

Table 2. Microwave measurements of the brightness temperature.

Object	Obs. Date	Ref.	Obs. λ (cm)	D (AU)	R (AU)	Obs. Flux S (Jy)	Assumed Diameter (km)	T_B (K)	T_{BN} (K)	Teq (K)
1 Ceres		1	0.33	2.11	2.98	0.285 ± 0.040	985	147 ± 49	153	165
	5/19/81	4	1.99	2.84	2.56	0.00372 ± 0.00037		127 ± 40	122	
		2	2.85	1.77	2.78	0.0040 ± 0.0018		108 ± 59	108	
		3	3.75	2.20	2.55	0.0024 ± 0.0008		174 ± 78	167	
	5/18/81	4	6.14	2.82	2.56	0.000348 ± 0.000035		112 ± 35	108	
	12/26/81	4	6.14	2.87	2.60	0.000356 ± 0.000036		118 ± 40	114	
	5/9/82									
	3/25/83	5	6.14			0.00134 ± 0.00017		Snow - No Usable Data		
	12/17/83									
2 Pallas	12/29/81	5	6.14	2.16	2.25	0.000278 ± 0.000027	538	175 ± 43	158	164
	3/24/82									
	4/11/83									
	3/2/84									
4 Vesta	1/7/84									
10 Hygiea	2/27/83							Another source in field		
	4/13/83	5	6.14	1.82	2.82	0.000254 ± 0.000070	450	162 ± 61	154	156
	12/10/83									

References

1 Ulich and Conklin (1976)
2 Andrew (1974)
3 Briggs (1973)
4 Johnston et al (1982)
5 This paper

Symbols

D = geocentric distance
R = heliocentric distance
T_B = effective brightness temperature (from S) assuming uniform disk
T_{BN} = normalized brightness temperature computed for Rm = semi-major axis of orbit (See Table 3 for values of Rm)
Teq = equilibrium brightness temperature for isothermal blackbody at distance Rm

$$Teq = 279\ R^{-1/2}(1 - p)^{1/4}$$

where p is the bolometric albedo (Morrison 1974). The microwave observations obtained so far are summarized in Table 2.

One can also invert this process and assume that T_{BN} is equal to Teq, and use the measured value of S to obtain a value for d. A comparison is given in Table 3 between optical data diameters for the minor planets as compiled by Schubart and Matsoń (1979) and the computed diameters based on this program of observations. Our value for the diameters of Pallas and Hygiea are in good agreement with the optical determinations. Our value for the diameter of Ceres is based on several observations and is not consistent with the optical value, but there has not been an occultation determination of the diameter of Ceres. Thus, our determination of the diameter may be more accurate than the optical determinations.

In a later section we will address the determination of asteroid diameters by directly measuring the size of the disk.

OBSERVED POSITIONS

The VLA data from each day's observation of a minor planet are processed through the standard VLA reduction packages and a map is generated, which is then "cleaned". The coordinates of points on the map are effectively determined by the assumed positions of the calibrator sources observed along with the minor planet. Thus the primary sources of systematic error in the resulting minor planet position determinations are errors in the assumed positions of the calibrators. However, only good calibrators are used with positions believed to be accurate to 0.03 arcseconds or better. Other sources of possible systematic error include errors in assumed values of UT1, polar motion, nutation components, baselines, and atmospheric delay. However, since we always choose calibrators which are not more than a few degrees from the minor planet being observed, those effects should contribute only differentially at less than the 0.02 arcsecond level.

Because the atmosphere is the primary source of phase noise, especially at 2 cm, it is the main contributor to the random errors of the position determinations. The minor planet positions are obtained by fitting a circularly symmetric gaussian to the "cleaned" maps. This process is quite sensitive to the signal-to-noise ratio of the map, but should provide positions from our data with random errors less than 0.05 arcseconds.

The data reduction process followed here is similar to that used by Johnston et al. (1984) in determinations of the positions of the radio stars UX Ari and HR1099 relative to nearby quasar calibrators. Those positions show a scatter of 0.02 arcseconds and thus are better than the optical positions of the same stars.

Table 3. Diameters of Asteroids

Asteroid	Heliocentric Distance Rm	Bolometric Albedo p	Teq K	VLA Determination Diameter km	Optical Diameter km	
1 Ceres	2.767	0.06	165	818 ± 82	985 ± 150	(1)
2 Pallas	2.772	0.08	164	562 ± 56	538 ± 50	(2)
4 Vesta	2.361	0.26	168	------------	544 ± 80	(3)
10 Hygiea	3.136	0.027	156	458 ± 145	443	(1)

Method of Optical Diameter

(1) Radiometric and Polarimetric
(2) Occultation and Speckle Interferometry
(3) Radiometric, Polarimetric and Speckle Interferometry

Our positions of Ceres are consistent with the orbit determined from optical data. However, our positions of Pallas are systematically shifted by about 0.6 arcseconds. It will be interesting to continue to "track" Pallas around its orbit with the VLA to discover the source of this discrepancy.

EQUINOX SOLUTIONS

Jackson (1968) and Branham (1979) have used optical minor planet observations for determining the equinox. The same method can be used for the radio frequency observations of the minor planets. The VLA observations could also be used in conjunction with the optical observations. In this case the optical observations would be used to determine some unknowns and the VLA observations used to determine the radio frequency equinox and selected unknowns.

DETAILED OBSERVATIONS OF CERES

When Ceres was at opposition in 1982 and 1983 and the VLA was in the "A" configuration, high resolution 2 cm observations were attempted. At opposition, Ceres disk has a diameter of 0$''\!.$7 while the VLA synthesized beam was 0$''\!.$1. The diameter of the minor planet was measured directly and confirmed the earlier result of Johnston, Seidelmann, and Wade (1982). The observations in 1982 appeared to display a lack of uniform emission across the disk. Observations were repeated in 1983 after the system temperatures of the VLA's 2 cm receivers were improved. However, due to the low declination (-28°) of Ceres at this opposition and the summer (August) atmospheric conditions, a reasonable image of the disk was not obtained. Observations are proposed for the November 1984 opposition which will hopefully result in a good image of the disk.

SUMMARY

The feasibility of observing the minor planets with the VLA has been demonstrated. Positional observations can be obtained with accuracies better than a tenth of an arcsecond. This observing program, like any other astrometric observing program, requires extreme care in the preparation of the observing data and continuous vigilance concerning the methods of observing and the reduction of the data. As a by-product of this observing program, inferences can be made as to the physical size, surface characteristics, and rotational elements of minor planets.

We plan to continue gathering observational data for the four largest minor planets, with the intent of obtaining for each minor planet about ten observations scattered in position with respect to the minor planet's and the Earth's orbit. We will also pursue the possibilities of determining more about the physical characteristics of the minor planets and try to resolve the discrepancy concerning the size of Ceres and the apparent inconsistencies in the position of Pallas.

REFERENCES

Andrew, B. H. (1974) Icarus 22, 454.

Baars, J. W. M., Genzel, R., Pauliny-Toth, I.I.K., and Wetzel A. (1977) Astron. Astrophysics 61, 99.

Branham, R. L. (1979) Astron. Papers of the American Ephemeris 21, pt II, Government Printing Office, Washington, D.C.

Briggs, F. H. (1973) Astrophys. J. 184, 637.

Duncombe, R. L. (1969) Astron. Papers of the American Ephemeris 20, pt. II, Government Printing Office, Washington, D. C.

Jackson, E. S. (1968) Astron. Papers of the American Ephemeris 20, pt. 1, Government Printing Office, Washington, D. C.

Johnston, K. J., Seidelmann, P. K. and Wade, C.M. (1982) Astron. J. 87, 1593-9.

Johnston, K. J., Wade, C. M., Florkowski, D. R., and de Vegt, C. (1984) to be submitted to A. J.

Morrison, D. (1974) Astrophys. J. 194, 203-212.

Schubart, J. and Matson, D. (1979) Asteroids, edited by T. Gehrels, University of Arizona, Tucson, p. 84.

Ulich, B. H. and Conklin, E. K. (1976) Icarus 27, 183.

REVIEW OF CONCEPTS OF STABILITY

Victor Szebehely, Professor
D.D. Cockrell Chair
University of Texas
Austin, Texas, USA

ABSTRACT. Concepts of stability, associated nomenclature and names of originators are reviewed emphasizing some global aspects as well as specific applications to dynamics and to celestial mechanics. Due to the many definitions and interpretations used, details concerning only a few fundamental concepts (Hill's, Liapunov's, Poincaré's stability, etc.) are offered. Short definitions and descriptions are also given for about 50 concepts of stability in the form of a dictionary. Several of the definitions presently in use are not unique, in fact are often contradictory and/or repetitive. Furthermore, the credits given to the originators are not uniform or consistent in the literature and several cross-references seem to be necessary. The list of references given contains some fundamental works and it does not attempt to be complete. The entire length of this paper would have been easily taken up by a more detailed list of references, which list would still not have been "complete".

In the "Introduction" some historical and fundamental aspects of stability are discussed. This is followed by the review of "Major Stability Concepts" in some detail. The "Dictionary of Stability" forms Part 3 and "Conclusions and Future Research" are in Part 4, treating generalizations of present techniques and outlining problems of fundamental importance and of inherent difficulties. Part 5 gives an abbreviated list of references.

1. INTRODUCTION

The subject of stability is excellently suited to a Colloquium dedicated to the memory of Alexander von Humbolt whose fundamental idea was that the successful integration of the various fields of natural sciences will lead to progress in human culture. Indeed, stability is one of the fundamental subject unifying the sciences and serving

Celestial Mechanics **34** (1984) 49–64. 0008–8714/84.15

excellently the purposes of history and philosophy.

Hagihara (1957) anticipated the purpose of this Colloquium when he considered stability a "fascinating and difficult problem of human culture". He formulated the problem of the stability of the solar system as follows: "Will the present configuration of the solar system be preserved for some long interval of time? Will the planets eventually fall into the Sun or will some of the planets recede gradually from the Sun so that they no longer belong to the solar system? What is the interval of time, at the end of which the solar system deviates from the present configuration by a previously assigned small amount?" Then Hagihara continues: "The question has long been an acute problem in celestial mechanics since Laplace, not to say the Egyptian or the Caldean civilization. The term "stability" has often been discussed by various mathematicians and the solution of the problem becomes more and more complicated and difficult to answer as we dig deeper and deeper into it. Present day mathematics hardly enables us to answer this question in a satisfactory manner for the actual solar system. We must limit ourselves here to describing the present status of the efforts toward solving this fascinating but difficult problem of human culture."

Solar system stability problems appear to form a rather small sub-set of stability research. Indeed, Prigogine (1980) points out that the origin of life may be related to successive instabilities (or bifurcations). In thermodynamics and statistical mechanics we study the role of entropy and in modern dynamics we describe chaotic motion and instability by Kolmogorov's entropy.

Moving to even more general ideas, consider the stability of theorems in any field of science. Often theorems are formulated as follows. "If certain conditions are given, then a specific statement is true." We might ask by whom are these conditions given and how well are they fulfilled in any actual application? Small errors and uncertainties in the "given conditions" might destroy the validity of the conclusion if the theorem is unstable to small disturbances. Such theorem-stability might be applicable to the establishment of physical laws which are often based on observations with uncertainties. What errors are allowed (and how do we estimate the actual errors and uncertainties) in order not to destroy the validity of the stated law as the consequence of observations? Clearly, the sensitivity of laws and theorems to observational errors or to the uncertainties of the conditions is critical.

We might go one step further and state that depending on the specific definition of stability used, real systems may always present instabilities when suitably large disturbances are introduced. Once again, the important idea is to find the "proper" disturbance and the "proper" stability condition when a given system or phenomenon is investigated. We must realize that stability depends on the disturbance allowed and on the type of equivalence we want (how much deviation is allowed when the system is still called "stable"). Furthermore, disturbances should be of the relevant type and physical repeatability should be demanded only for striking features. This leads us to the idea of qualitative versus quantitative stability.

We might realize that no physical laws, no theorems, no constants and no initial conditions are exact, consequently, the results of any investigation (philosophical, theoretical, experimental, numerical, etc.) will depend on the stability of the problem investigated (Bellman, 1968 and Szebehely, 1979). The answer might not be correct if the system is sensitive to errors made by our approximations, that is if the system is unstable. On the other hand the theories derived, the physical laws established and in general, our results will be "good approximations" (to what?) if the system is stable. The above discussion will alert the reader to the overriding importance of the concept of stability and of its application to a wide variety of fields in addition to mathematical physics, astronomy, celestial mechanics and engineering.

The ultimate step of course is the realization that no matter what field one pursues, stability ideas always need precise considerations and careful follow-ups. Indeed, to accomplish any results of "validity" the stability of the approach can not be ignored since uncertainties in models, theories, and in various inputs will influence the results.

2. MAJOR STABILITY - CONCEPTS

The reader will not be surprised to find after the preceedings that there are at least 50 terminologies and concepts of stability, regularly used in the literature. In this Section Hill's, Liapunov's, Poincare's and Kolmogorov's ideas are selected to offer some foundations and details. As will be shown later, not all but most of the stability concepts used are associated with these ideas. The advantages and disadvantages of these methods depend of course on the applications, nevertheless, some general ideas and fundamental differences will be outlined.

2.1 Hill's Stability

Hill (1878) proposed the use of the energy-integral to establish bounds of motion for conservative systems with time-independent potentials. In general and simplified terms, the basic idea may be stated as follows. Let

$$v^2 = 2F(q_1,q_2 \dots q_n) - C, \tag{1}$$

where v is the velocity ($v^2 = \dot{q}_i\dot{q}_i$, using the summation convention), F is the potential function, q_i are the coordinates and C is a constant of the motion. Note that Hill's original idea was to establish the boundaries of the planar lunar orbit, using his simplified potential and the associated Jacobian integral. The method might be used for any problem where integral(s) of the motion exist. As long as the degree of freedom is less than three the associated Hill surfaces might be constructed and the possible and forbidden regions of the motion might be established. Note that the considerable advantage of Hill's method is that knowledge of solutions are not needed to establish stability, a significant fact when non-integrable dynamical systems are investigated as is usually the case in practically important cases. On the other hand it is of considerable importance to give the proper interpretations of the results obtained by Hill's method. For instance, in the lunar problem Hill's method shows that the Moon's orbit is bounded around the Earth, using the approximations of the circular, planar restricted problem of three bodies (Sun-Earth-Moon). It is interesting to note that Hill's simplified version of the lunar theory shows higher stability for the Moon than using the model of the complete restricted problem. And if the "restricted" assumption gives way to the use of the general problem of three bodies then the generalization of Hill's method shows instability. These remarks demonstrate that the Moon's stability according to Hill's method shows model-dependence, a not unexpected but at the same time not generally known result.

Many additional comments could be made about Hill's method but for brevity's sake only a recent result is mentioned. A quantitative measure of stability replaces Hill's originally qualitative method (Szebehely 1977), as follows. If the constant of integration of the orbit to be investigated is C_{ac} (for actual) and the critical value (at which bifurcation occurs) of the same constant is C_{cr}, then the measure of stability is defined by $S = (C_{ac} - C_{cr})C_{cr}^{-1}$. (Note that $C_{cr} \geq 3$ in the restricted problem using proper definitions and units.) The physical meaning of this measure is that it shows the distance in phase space between the actual and the critical orbits. If $S < 0$ we have Hill-type instability, suggesting that the orbit may be unstable and, in the case of the lunar orbit, the Moon may leave the Earth and become a planet of the Sun. The dimensionless generalization of the Jacobian constant (C) in the case of the general problem of three bodies (Zare, 1977) is $C = - hc^2 / \bar{m}^5 G^2$, where h is the total energy, c is the angular momentum, $\bar{m}$ is the average mass of the participating bodies and G is the gravitational constant. The stability measure is given by the same formula as before but the limiting surfaces never close completely inspite of the use of the energy-integral and of the momentum integral because of the high degree of freedom (9 vs 3).

2.2 Definitions of Liapunov's and of Poincaré's Stability

Liapunov and Poincaré contributed several fundamental ideas to the field of stability. In this section a comparison of their considerably similar basic ideas is offered and in the next sections Liapunov's characteristic number, Liapunov's function, Poincaré's surface of section and Kolmogorov's representation on tori are discussed.

The two fundamental concepts contributed by Liapunov (1892) and Poincaré (1892) differ in their selection of correspondence between points on the original orbit (the stability of which is to be investigated) and points on the disturbed orbit. Liapunov established "isochronous" correspondence, while Poincare referred to "normal" correspondence. Referring to Figure 1, we show the original orbit (A) and the disturbed orbit (B). The normal correspondence is represented by the deviation $\bar{x}$ and the isochronous relation by $\bar{y}$.

The analytical expression and the definition of Liapunov's stability are shown first.

Given the system of differential equations for the vector $\bar{x}$ as $\dot{\bar{x}} = F(\bar{x},t)$ with a solution $\bar{x} = \psi(t)$ as shown by the curve A on Fig. 1. This solution is stable if given any $\varepsilon > 0$, there exists a $\delta > 0$ such that

$$|\bar{y}_o| = |\bar{\phi}(t_o) - \bar{\psi}(t_o)| < \delta,$$

$$|\bar{y}| = |\bar{\phi}(t) - \bar{\psi}(t)| < \varepsilon.$$

Here $\bar{\phi}(t)$ is the solution obtained when the disturbance is applied, shown as curve B. As seen from the inequalities as well as from Figure

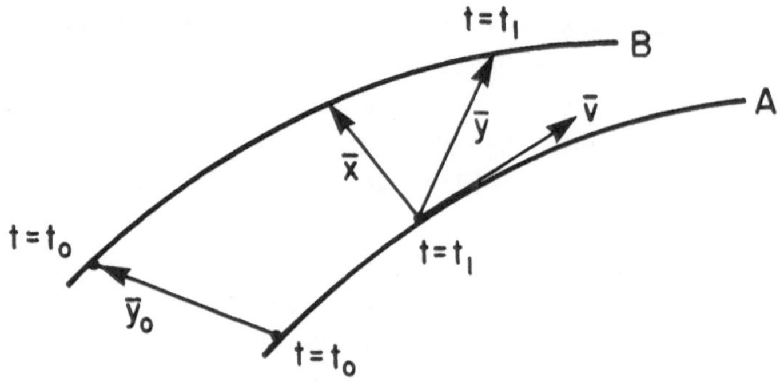

Figure 1. Isochronous and Normal Correspondences

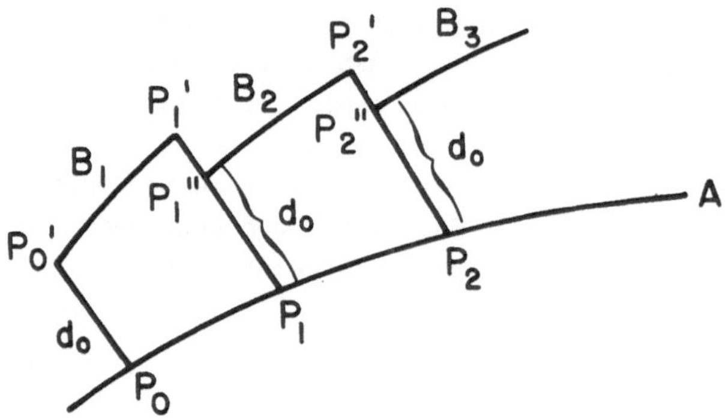

Figure 2. Computation of Liapunov's Characteristic Exponents

1, Liapunov's stability refers to isochronous correspondence.

Using the normal vector $\bar{x}$ instead of $\bar{y}$ we obtain Poincaré's "normal" stability also known as orbital stability.

If either $\bar{x}$ or $\bar{y}$ goes to zero as $t \to \infty$ we speak about asymptotic stability. Note that when the semi-major axis (a) of an elliptic two-body orbit is disturbed (changed) by δ , the associated ε value will become 2a after sufficient number of revolutions. Consequently, elliptic orbits are, according to Liapunov's definition unstable but orbitally stable.

The often controversial idea of influencing stability by transformations should be mentioned here. If stability is an invariant property of a dynamical system, then transformations of the variables should not affect the stability properties. Consider for instance Levi-Civita's (1903) transformation or its generalization, the Kustaanheimo-Stiefel (1971) transformation. The unstable elliptic two-body orbit appears now as the stable solution of a linear differential equation representing a harmonic oscillator. In this differential equation, however, the energy appears as a constant, therefore, no perturbations are possible which would change the energy and this fact offers Liapunov stability. The situation is similar when the equations of motion are formulated using orbital elements.

2.3 Liapunov's Characteristic Numbers and the Liapunov Functions

Liapunov's (1892) characteristic numbers indicate the stability of solutions by means of applying repeated disturbances and numerically integrating the outcome. Figure 2 shows the process. Curve A is once again the original undisturbed solution and the segments B_1, B_2, etc. represent the disturbed solutions.

The change of one of the initial conditions at $t = t_0$ is d_0 so the integration of the disturbed orbit starts at P'_0. When point P'_1 is reached the original disturbance, $d_0 = P_1 P_1''$ is once again introduced and the segment B_2 is obtained. Consequently if $d_i = P_i P'_i$ and $d_0 = P_0 P'_0 = P_1 P_1'' = P_i P_i''$ we have

$$\frac{d_i}{d_0} = \frac{P_i P'_i}{P_0 P'_0}.$$

Liapunov's characteristic number is defined by

$$L = \lim_{n \to \infty} L_n$$

where $L_n = \frac{1}{nt} \sum_{n=1}^{\infty} \ln \frac{d_i}{d_o}$

If $L > 0$ we have instability or chaotic motion, if $L = 0$ we have quasi-periodic motion and if $L < 0$ we have asymptotic stability or attraction.

Note that the solution must be known for the application of this approach as well as repeated accurate numerical integrations are needed. This might be considered a disadvantage of this method when compared with Hill's approach. On the other hand Liapunov's characteristic number will certainly describe the details of the stability aspects of a given solution for various disturbances while Hill's method will offer only possible boundedness without consideration of the specific orbits.

Liapunov's theorem (also known as his direct method) does not require the knowledge of a solution and offers the stability information in a region of the phase-space by means of establishing the existence of Liapunov's function. The L-function associated with the differential equation $\dot{x}_i = X_i(x_j)$ is $V(x_j)$ if

$$\dot{V} = \frac{dV}{dt} = X_i \frac{\partial V}{\partial x_i} \leq 0 \quad .$$

The existence of the L-function offers stability, in fact we have asymptotic stability if $-\dot{V}$ is positive definite. If $\dot{V}$ is positive definite, we have instability.

2.4 Poincaré's Surface of Section

One of the important method in modern dynamics is the numerical use of Poincaré's (1892) surface of section technique which reveals the phase space behavior of two-degrees of freedom dynamical systems. From this, important conclusions might be drawn concerning stability. The technique became widely used when high-speed computations reached sufficient level concerning accuracy and speed. While it is applicable to higher than two-degrees of freedom dynamical systems, most users limit it to the study of such systems. The basic idea might be described by considering the system $\ddot{x} = F(x,y,\dot{x},\dot{y},t)$ and $\ddot{y} = G(x,y,\dot{x},\dot{y},t)$.

After numerical integration we prepare a two-dimensional representation by plotting the series of points $P(\dot{y},y)$ corresponding to $x = 0$ and $\dot{x} > 0$. In other words, the four-dimensional flow in the x, $\dot{x}$, y, $\dot{y}$ phase-space is represented by the points of intersections of the trajectory with the x = 0 plane. This might also be looked at as a point-transformation in the y, $\dot{y}$ plane, the transformation resulting in the intersections. If the transformation has an invariant point, this will correspond to a periodic orbit, while smooth curves in the y, $\dot{y}$ plane represent quasi-periodic solutions. If the points are covering the plane we speak about random or chaotic motion. Application to stability now consists of establishing such plots and observing the possible change of the qualitative nature of the solutions as disturbances are applied.

2.5 Kolmogorov's Tori

Kolmogorov's original idea proposed in 1954 is now known with various proofs, several modifications and improvements as the Kolmogorov-Arnold-Moser (KAM) theorem (Arnold 1963, Moser 1966). The basic idea is that an unperturbed two-body motion can be represented as a flow on a torus. Will the motion stay on a distorted torus when perturbations, resonances, etc. are applied or will the torus completely disappear? Will the flow cover the torus completely (corresponding to quasi-periodic orbits with irrational commensurabilities), will the flow close after a finite number of revolutions (corresponding to periodic orbits) or will the resonances combined with the perturbations result in the elimination of the torus, corresponding to chaotic motion? The KAM theory offers conditions for pertubations and commensurabilities, which if satisfied will preserve the torus. Once again stability is associated with qualitative changes of the trajectories. When some orbits in multi-dimensional space drift between KAM tori we speak about Arnold's diffusion. The result is emerging that non-integrable systems have non-zero measure of unstable regions of the phase space in general.

3. DICTIONARY OF STABILITY

In this section short definitions are given of various concepts and terminologies used in stability research. The definitions will not be given by mathematical relations for the sake of brevity, especially, since the basic concepts discussed in Section 2 will be frequently referred to. We will use the abbreviation st for stability and will try to avoid other, possibly confusing abbreviations. Only few of the fundamental concepts of linear stability analysis are included. Note that some of the definitions are repetitive, some contradictory, nevertheless, they represent the present state of art.

asymptotic st: effect of disturbance as $t \to \infty$; convergent st; extreme st

characteristic roots: solutions of the characteristic equation; eigenvalues of the matrix representing system of linear equations

characteristic st: st of a system without perturbations

complete st: asymptotic st in the large; st to all orders; practical st; formal st

conditional st: st of a subset of solutions

convergent st: see asymptotic st

discrete st: st conditions satisfied for certain discrete times

dangerous region: region of instability

extreme st: see asymptotic st

first order st: see linear st

formal st: see complete st

future st: st for $0 \leq t \leq +\infty$

hierarchical st: no orbit crossings in planetary systems; no change of the order of the Jacobian distances

Hill st: solutions bounded in the configuration space

infinitesimal st: see linear st

isochronous st: see text

L-st: Liapunov st; strong st; see isochronous st

Lagrange st: st of equilibrium solutions; all solutions stay in bounded regions of the phase space

Laplace st: all solutions stay in bounded regions of the phase space; no collisions and no escapes for many-body problems

Liapunov st: see isochronous st;

linear st: st of linear or linearized systems;

Lindblad st: st of stars at edges of flat nebulae

Malkin st: st under persistent perturbations; modified L-st

negative st: st for $0 \geqq t \to -\infty$

neutral st: st of equilibrium when other equilibria are near by

non-linear st: st of non-linear systems

normal st: see text

numerical st: the st of the method used for numerical integration

orbital st: see text

permanent st: see future st

planetary system st: no secular terms in the semi major axes

Poincaré st: see text

Poisson's st: system returns infinitely often to positions arbitrary near to its original position

positive st: see future st

practical st: magnitude of disturbances related to characteristic roots; uniformly bounded solution for finite time; effect of initial conditions and disturbances;

S-st: temporary st for $0 \leqq t \leqq S$

safe region: region of st

secular st: no secular terms appear in the solution

semi-permanent st: temporary stability for long time

straight-forward st: effect of initial conditions

structural st: effect of change in model, chage of constants or change of form of the differential equations

temporary st: see st for finite time

topological st: topological equivalence of the set of disturbed and original solution curves

trigonometric st: no real parts of the characteristic roots and no commensurability for linear systems unilateral st: asymptotic st without oscillation; only negative real parts of the characteristic roots

vertical st: st in the direction normal to the plane of the orbits of the primaries in the restricted problem of three bodies

weak st: two different kinds of motion in the same neighborhood in the phase space.

4. CONCLUSIONS AND FUTURE RESEARCH

It is hoped that the reader at the end of this review arrives at the conclusion that stability research is one of the most challenging undertaking for the human intellect. Indeed, it covers much wider fields than orbit mechanics and it extends from the predictions of the future behavior of all phenomena to the establishment and validity of physical laws, theories and theorems. Instabilities might be responsible for the initiation of our existence or for its termination, as mentioned in the introduction.

Regarding future stability research, it should not be surprizing to know that its direction and success are highly unpredictable. Even with precise definitions of stability and assuming that our present state is well known, the outcome should be rather uncertain, according to the basic principles mentioned in this paper. Nevertheless, it might be expected that as new engineering systems are developed or new physical concepts emerge, they will be subjected to stability analysis. It is now clear that no single stability technique or definition is to be applied for different phenomena. Flexibility is expected and recommended concerning the selection of the method of analysis to best suit the specific system and the special dominant questions of importance. The basic concepts (Hill's, Kolmogorov's Liapunov's, Poincaré's, etc.) will dominate stability research for the time being but new concepts should emerge, dictated by the practical requirements of various new systems.

The behavior of systems in the phase-space might show qualitative changes as the system-parameters change (structural stability) or as the initial conditions vary (straight forward stability). The essentially new idea is emerging according to which the qualitative changes are to be emphasized since small changes not effecting the general behavior of the system might be ignored. The change from well ordered and well predictable periodic or quasi-periodic motion to chaotic behavior is to be predicted or to evaluated. Such changes are not expected for linear systems, therefore, interest is directed to systems governed by non-linear laws.

A recent example, subject to further research might be shortly described here to illustrate the point. The motion about the triangular equilibrium points of the restricted problem of three bodies is considered stable if the value of the mass- parameter (μ) is sufficiently small. This condition is usually satisfied in the solar system considering various combinations of planets and their satellites. If the particle is not at the equilibrium point initially, i.e. if the initial conditions deviate by a "small" amount in the phase space, the motion is known as "libration" around the equilibrium point. As the initial conditions are varied the limit of the librational region is reached and "circulation" sets in. (The phenomenon shows some analogy to the motion of the spherical pendulum.) Further increase of the deviations results in chaotic motion, escape, capture, etc. The product of the limiting deviations in the configuration space (Δq_i) and in the momentum space (Δp_i), according to some of our preliminary results (Szebehely, 1979) give a constant order of magnitude, showing a strictly formal analogy to Heisenberg's uncertainty relation. In other words,

$$\sum_{i=1}^{3} \Delta q_i \Delta p_i = K = \text{constant},$$

regardless of the direction along which the deviations were applied. Large limiting initial deviations in the position are associated with small limiting initial deviations in the momentum in any given direction (and vice versa). The value of K gives the combination of the limiting deviations which results in non-librational or chaotic motion. If the value of the product is less than K, the motion is libration (periodic or quasi-periodic). If the value of the product is larger than K, the motion is circulation, capture, escape or chaotic. The motion changes qualitatively as the product exceeds the value of K. In this way the behavior depending on the initial conditions is evaluated. Since the critical value of K is expected to change with the mass-parameter, the dependence of the behavior on the system-parameter, μ is to be evaluated. In other words, straight forward and structural stability are evaluated corresponding to change of the character of the motion.

The above example might indicate the future direction of stability research of dynamical systems. Stability might be associated with change of the values of the initial conditions or with change of values of the system-parameters. The region of phase space for which the type of motion is the same represent a stable region. The region where the motion changes qualitatively might be termed unstable.

Acknowledgements

This paper could not have been prepared without the contributions and cooperation of my colleagues, co-workers and students who influenced my work on stability for the last thirty years. Support received from the National Science Foundation and from the University of Texas to prepare this paper and to attend the Alexander von Humbolt Conference is gratefully acknowledged.

REFERENCES

The following list concentrates on the fundamental and basic reference articles in addition to recent books which the reader might find useful. Note that some of the text book references of this list are not mentioned in the present paper and they cover many aspects of stability of rather general interest.

Arnold, V.I., Dokl. Akad. Nauk, SSR 137, 255, 1961; 138, 13, 1961; 142, 758, 1962; 145, 487, 1962. Also Usp. Mat. Nauk, 18, 13, 1963; 18, 91, 1963.

Bellman, R., "Some Vistas of Modern Mathematics", Univ. of Kentucky Press, 1968.

Birkhoff, G.D., "Collected Mathematical Papers", American Math. Soc., N.Y., 1950.

Brillouin, L., "Scientific Uncertainty and Information", Academic Press, N.Y., 1964.

Chandrasekhar, S., "Principles of Stellar Dynamics", Dover Publ., N.Y., 1960.

Hagihara, V., "Stability in Celestial Mechanics", Kasai Publ., Tokyo, 1957.

Helleman, R.H.G., in "Fundamental Problems in Statistical Mechanics", E. Cohen, Editor, p. 165, North Holland Publ., Amsterdam, 1980.

Hill, G., "Collected Mathematical Works", Carnegie Inst., Washington, D.C., 1905-1907.

Horton, C.W., L.E. Reichl and V. Szebehely (Editors), "Long-Time Prediction in Dynamics", Wiley Publ., 1963.

Jefferys, W.H. and V. Szebehely, Comments of Astrophysics, 8, 9, 1978.

Jorna, S. (Editor), "Topics in Nonlinear Dynamics", American Inst. of Physics, N.Y., 1978.

Kolmogorov, A.N., Proc. Intern. Cong. Math., Amsterdam 1954, 1, 315, North Holland Publ. Amsterdam, 1957 and Dokl. Akad. Nauk, SSR 98, 527, 1954.

LaSalle, J. and S. Lefschetz, "Stability by Liapunov's Direct Method with Applications", Academic Press, 1961.

Levi-Civita, T., Ann. Math. [3] 9 1, 1903.

Liapunov, A.A., Communications Math. Soc. Krakow, 2, 1, 1892.

Malkin, I.G. "Theory of Stability of Motion", State Publ. House, Moscow-Leningrad, 1952.

Moser, J., SIAM Review, 8, 145, 1966.

Moser, J., "Lectures on Hamiltonian Systems", Memoirs, American Math. Soc., 81, 1, 1968.

Moser, J., "Stable and Random Motion in Dynamical Systems", Princeton Univ., Press, Princeton, N.J., 1973.

Pars, L.A., "A Treatise on Analytical Dynamics", Heineman Press, London, 1965.

Poincaré, H., "Les Methodes Nouvelles de la Mecanique Celeste," Gauthier-Villars, Paris, 1892-1899.

Prigogine, I., "From Being to Becoming", W.H. Freeman and Company, San Francisco, 1980.

Prigogine, I. and R. Herman, "Kinetic Theory of Vehicular Traffic", American Elsivier Publishing Company, Inc., New York, 1971.

Roy, A.E., "Orbital Motion", Adam Hilger Ltd. Publ., Bristol, 1978.

Siegel, C. and J. Moser, Lectures on Celestial Mechanics, Springer, N.Y., 1971.

Stiefel, E.L. and G. Scheifele, "Linear and Regular Celestial Mechanics", Springer, N.Y., 1971.

Szebehely, V., Celestial Mechanics, 4, 116, 1971; 9, 359, 1974; 15, 107, 1977; 18, 383 and 391, 1978; 22, 7, 1980; 23, 3, 1981.

Szebehely, V., "Theory of Orbits", Academic Press, N.Y., 1967.

Szebehely, V., Proc. Nat. Acad. Sc., 75, 5743, 1978.

Szebehely, V. (Editor), "Instabilities in Dynamical Systems", D. Reidel Publ., Holland, 1979.

Ulam, S.M., "Problems in Modern Mathematics", Interscience Publishers, New York, 1980.

Whittaker, E.T., "A Treatise on the Analytical Dynamics of Particles and Rigid Bodies", Cambridge Univ. Press, London, 1904.

Wintner, A., "Analytical Foundations of Celestial Mechanics", Princeton Univ. Press, Princeton, 1947.

Zare, K., Celestial Mechanics, 16, 35, 1977.

THREE-BODY PROBLEM

Christian Marchal
D.E.S.
ONERA 92320
Chatillon, France

Junzo Yoshida
Dept. of Physics
Kyoto Sangyo University
Kamizamo, Kita-Ku, Kyoto 603
Japan

Sun Yi-Sui
Dept. of Astronomy
Nanjing University
Nanjing, China

ABSTRACT. In the first part of this paper [Marchal, Yoshida, Sun Yi-Sui 1985] we have analyzed three-body systems satisfying the condition $r \leqq kR$ where k is a suitable constant, r the mutual distance of the two masses of the "binary" and R the distance between the center of mass of the binary and the "third mass".

That condition $r \leqq kR$ puts limits on the acceleration of the third mass and these limits allow us to determine the corresponding "escape velocities".

In this second part we look for initial conditions under which the inequality $r \leqq kR$ will remain forever satisfied and we develop the corresponding tests of escape and their applications.

This leads to a major improvement of the knowledge of the nature of three-body motions especially in the vicinity of triple close approaches.

The region of bounded motions is much smaller than was generally expected and numerical computations of particular solutions show that we approach very near to the true limit.

1. INTRODUCTION

The usual tests of escape are efficient when at least two of the mutual distances are large, but they are generally useless when the mutual distances are small and they give an inaccurate picture of the region of phase space where occur bounded motions.

Celestial Mechanics **34** (1984) 65–93. 0008–8714/84.15

Fortunately when the distance of two of the three bodies is small in comparison to their distance to the third body, the motion of that third body is a slowly perturbed Keplerian motion that can accurately be analyzed and that leads to some easy tests of escape. For these tests we need a "condition of isolation" of the third body over long periods, a condition that is obtained for sufficiently large values of the Sundman function and that is very often satisfied even for very small mutual distances.

We will thus arrive to a very efficient test of escape especially for triple close approaches.

2. SUMMARY OF THE FIRST PART [Marchal, Yoshida, Sun Yi-Sui 1985]

2.1 The acceleration of the third body

Let us consider a three-body system and its Jacobi decomposition (figure 1), with the "binary" m_1, m_2 ($m_1 \geq m_2 > 0$) and the "third point mass" m_3.

The vector $\vec{R}$ equal to $(\overrightarrow{0_{1.2}, m_3})$ has an acceleration function of $\vec{R}$, $\vec{r}$ and the three masses m_1, m_2, m_3:

$$d^2\vec{R}/dt^2 = -\mu(\alpha\overrightarrow{r_{13}}\, r_{13}^{-3} + \beta\overrightarrow{r_{2.3}} \cdot r_{23}^{-3}) \quad (1)$$

with:

$$\mu = GM = \text{gravitational constant} \quad (2)$$

$$G = \text{constant of the law of universal attraction} \quad (3)$$

$$M = m_1 + m_2 + m_3 = \text{total mass} \quad (4)$$

$$\alpha = m_1/(m_1 + m_2) \geq \frac{1}{2} \; ; \quad \beta = m_2/(m_1 + m_2) = 1 - \alpha \quad (5)$$

$$\overrightarrow{r_{13}} = \vec{R} + \beta\vec{r} \; ; \quad \overrightarrow{r_{23}} = \vec{R} - \alpha\vec{r} \quad (6)$$

If $\vec{R}$ and the three masses are given but if our only information on $\vec{r}$ is $r \leq kR$ (k being a suitable constant smaller than α^{-1}) then $d^2\vec{R}/dt^2$ is unknown but always falls into the prolate ellipsoid of figure 2.

The point A corresponds to the collinear limit case:

$$\vec{r} = k\vec{R} \quad (7)$$

$$d^2\vec{R}/dt^2 = \overrightarrow{OA} = -\mu\vec{R}(1 + \varepsilon)/R^3 \quad (8)$$

$$\varepsilon = \frac{\alpha}{(1 + k\beta)^2} + \frac{\beta}{(1 - k\alpha)^2} - 1 \geq 0 \quad (9)$$

ε is generally small (for instance $k \leq 0.5$ implies $\varepsilon \leq k^2$).

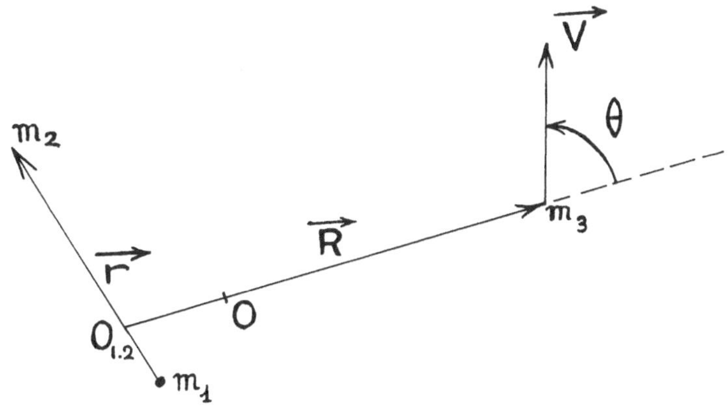

Figure 1. Geometry of the three-body problem.

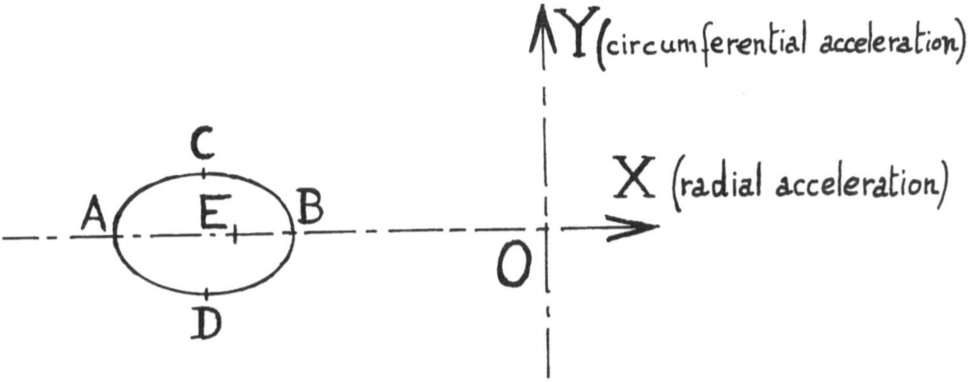

Figure 2. Domain of $\frac{d^2\overline{R}}{dt^2}$ for given values of $\overline{R}$ and the three masses.

The prolate ellipsoid of figure 2 is a function of μ/R^2 and ε only, its major axis AB is along the $-\vec{R}$ direction and its two minor axes, as CD , have a length of 2/3 of AB .

2.2 The escape velocities of the third body

Let us assume that in the three-body system of interest the lengths r and R verify:

$$\left.\begin{array}{l} r \leqq kR \text{ for all future time} \\ (k \text{ is a suitable constant smaller than } \alpha^{-1}) \end{array}\right\} \qquad (10)$$

In these conditions the limitations on $d^2\vec{R}/dt^2$ presented in figure 2 allow to have a rough picture of the future motion and to look for a test of escape of the third body.

The results are presented in the figure 3 in terms of the parameters ϕ and Ψ proportional to the radial velocity $V\cos\theta$ and the circumferential velocity $V\sin\theta$:

$$\phi = V\cos\theta \cdot \sqrt{R/\mu} \;;\quad \Psi = V\sin\theta \cdot \sqrt{R/\mu} \geqq 0 \qquad (11)$$

The limit curve is a function of ε only (ε being given in (9)) it starts at A_ε , where $\phi = \sqrt{2 + 2\varepsilon}$ and $\Psi = 0$, and its differential equation is:

$$(2\phi^2 + 4\Psi^2 - 4 - \varepsilon)\frac{d\Psi}{d\phi} + 2\phi\Psi + \varepsilon\left[4 + 9\left(\frac{d\Psi}{d\phi}\right)^2\right]^{\frac{1}{2}} = 0 \qquad (12)$$

If the initial velocity corresponds to a ϕ , Ψ point outside the limit curve of figure 3 the third body will escape when $t \to +\infty$ and we can thus define an "escape velocity" V_{E+} function of μ , R , ε and the angle θ equal to $(\vec{R}, \vec{V})$:

$$V_{E+} = f(\varepsilon, \theta)\cdot\sqrt{\mu/R} \;;\quad \phi = f\cos\theta \;;\quad \Psi = f\sin\theta \qquad (13)$$

and thus for any μ , R , ε and θ :

$$\left.\begin{array}{l} (10) \text{ and initially } V \geqq V_{E+} \text{ implies } m_3 \text{ and } R \\ \text{escape to infinity when } t \to +\infty . \end{array}\right\} \qquad (14)$$

That escape will forever be with:

$$V \geqq V_{E+} \;;\quad V^2 - V_{E+}^2 \text{ non-decreasing ;}$$

$$\frac{d^2(R^2)}{dt^2} \geqq 2\left(V^2 - V_{E+}^2 + \frac{\mu(1+\varepsilon)}{R}\right) > 0 \qquad (15)$$

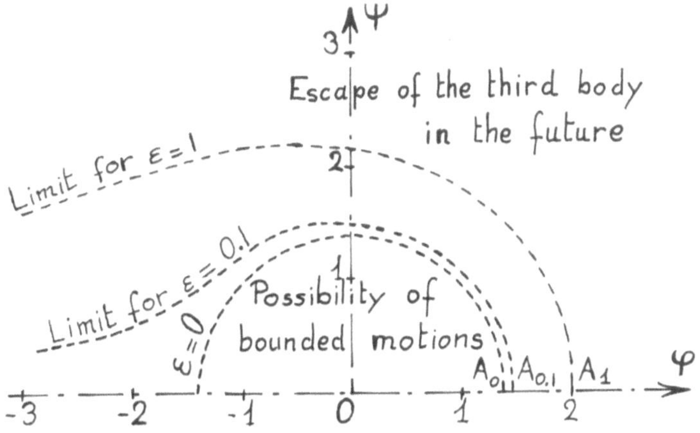

Figure 3. Conditions for escape of the third body.

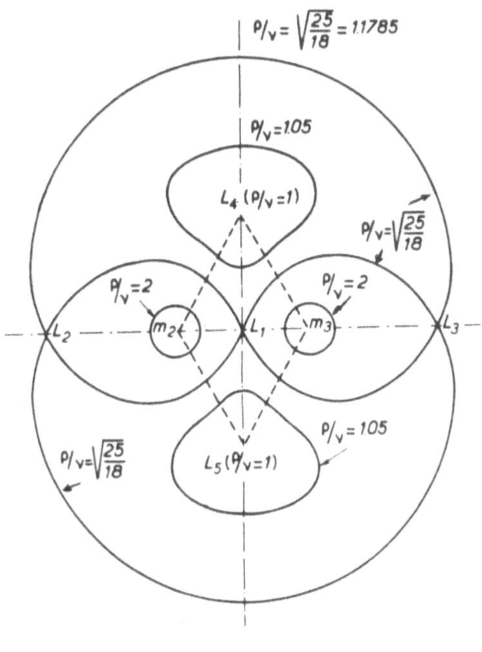

Figure 4. Level curves of the function ρ/υ for three equal masses.

hence R will have no local maximum and at most one local minimum in the future.

Symmetrically, when $t \to -\infty$, we will define the escape velocity V_{E-} by:

$$V_{E-}(\mu, R, \varepsilon, \theta) = f(\varepsilon, \pi-\theta)\cdot\sqrt{\mu/R} = V_{E+}(\mu, R, \varepsilon, \pi-\theta) \quad (16)$$

The function $f(\varepsilon, \theta)$ is defined for $\varepsilon \geqq 0$; $0 \leqq \theta < \pi$, however the integration of (12) is complex. We can nevertheless write the following:

$$\partial f/\partial\varepsilon \geqq 0 \ ; \quad \partial f/\partial\theta \geqq 0 \ ; \quad f \geqq \sqrt{(2 + 2\varepsilon)} \quad (17)$$

$$\{\varepsilon \text{ and/or } \theta = 0\} \text{ implies } f = \sqrt{(2 + 2\varepsilon)} \quad (18)$$

$$\{\theta \to \pi \text{ and } \varepsilon > 0\} \text{ implies } f \cong \left[\frac{\varepsilon}{2(\pi-\theta)} \operatorname{Log}\left(\frac{1}{\pi-\theta}\right)\right] \quad (19)$$

As simple upper and lower bounds of f we can use:

$$1 + [1 + 4(\varepsilon + \varepsilon^2)g(\theta)]^{\frac{1}{2}} \leqq \{f(\varepsilon, \theta)\}^2 \leqq 2 + 2\varepsilon \cdot g(\theta) \quad (20)$$

and even:

$$f(\varepsilon, \theta) \leqq \underbrace{\inf}_{\theta_2 \in (\theta - \frac{\pi}{2} ; \frac{\pi}{2})} \{[2 + 2\varepsilon \cdot g(\theta_2)]^{\frac{1}{2}}/\cos(\theta_2 - \theta)\} \quad (21)$$

the increasing function $g(\theta)$ being defined by:

$$g(\theta) = \frac{1}{4} + \frac{1}{2\sin^2\theta}\int_{\cos\theta}^{1} dx\sqrt{(4 + 5x^2)} \quad (22)$$

that is:

$$g(\theta) = \frac{1}{4} + \frac{3 - \cos\theta\sqrt{(4 + 5\cos^2\theta)}}{4\sin^2\theta} - \frac{\sqrt{5}}{5\sin^2\theta} \operatorname{Log}\left\{\frac{\cos\theta + \sqrt{(0.8 + \cos^2\theta)}}{1 + \sqrt{1.8}}\right\} \quad (23)$$

hence:

$$g(0) = 1 \ ; \quad g\left(\frac{\pi}{2}\right) = 1.43041 \ ; \quad g(\pi) = \infty \quad (24)$$

The bounds given in (20), (21) are good for small θ but they become weak for large θ.

3. SECOND PART - GENERALITIES AND NOTATIONS

The previous section and especially the conditions (14) emphasizes the interest of initial conditions satisfying (10) that is leading to $r \leqq kR$ for all future time; however it may seem difficult to obtain information valid for all future time.

A first possibility is given by the Hill-type stability [see the next section or, for instance, Marchal and Bozis 1982]. That stability is a generalization to the general three-body problem of the Hill stability of the circular restricted three-body problem [Szebehely 1967]. When the product c^2h is sufficiently negative (with $\vec{c}$ = angular momentum and h = energy) the zone of possible motion in phase space is divided into three disconnected parts. The motion remains forever confined to one of these parts and one of the three bodies remains forever "isolated", the corresponding ratio r/R remains forever bounded with a least upper bound smaller than α^{-1} and even than one.

However the Hill-type stability is rather rare and on the other hand the following analysis will lead to much better results.

We will need the following notations:

A) Integrals of motion (in the axes of the center of mass)

$$\vec{c} = \text{angular momentum} = \sum_{j=1}^{3} m_j \vec{r}_j \times \vec{V}_j \; ; \quad \vec{V}_j = \frac{d\vec{r}_j}{dt} \tag{25}$$

$$h = \text{energy} = \frac{1}{2}\sum_{j=1}^{3} m_j V_j^2 - G\left[\frac{m_1m_2}{r_{12}} + \frac{m_1m_3}{r_{13}} + \frac{m_2m_3}{r_{23}}\right] \tag{26}$$

B) With the Jacobi notations these integrals can be written:

$$\vec{c} = m\vec{r} \times \vec{v} + \mathcal{M}\vec{R} \times \vec{V} \tag{27}$$

$$h = \frac{1}{2}(mv^2 + \mathcal{M}V^2) - G\left[\frac{m_1m_2}{r} + \frac{m_1m_3}{r_{13}} + \frac{m_2m_3}{r_{23}}\right] \tag{28}$$

with m , $\mathcal{M}$, $\vec{v}$ being defined by:

$$m = m_1m_2/(m_1 + m_2) \; ; \quad \mathcal{M} = m_3(m_1 + m_2)/M \; ; \quad \vec{v} = d\vec{r}/dt \tag{29}$$

C) Let us now define the following fixed lengths a and p :

$$M^* = m_1m_2 + m_1m_3 + m_2m_3 \tag{30}$$

$$a = \text{"generalized semi-major axis"} = -GM^*/2h \tag{31}$$

$$p = \text{"generalized semi-latus rectum"} = Mc^2/G(M^*)^2 \tag{32}$$

a and p are indeed the semi-major axis and the semi-latus rectum of the relative orbit of the primaries when the smallest mass is negligible (restricted case).

D) We will use the following variable lengths ρ and ν :

ρ is the "mean quadratic distance":

$$M^*\rho^2 = m_1m_2r_{12}^2 + m_1m_3r_{13}^2 + m_2m_3r_{23}^2 \tag{33}$$

ν is the "mean harmonic distance":

$$\frac{M^*}{\nu} = \frac{m_1m_2}{r_{12}} + \frac{m_1m_3}{r_{13}} + \frac{m_2m_3}{r_{23}} \tag{34}$$

The harmonic mean of positive numbers is always smaller than or equal to their quadratic mean, hence always:

$$\inf\{r_{12};\ r_{13};\ r_{23}\} \leq \nu \leq \rho \leq \sup\{r_{12};\ r_{13};\ r_{23}\} \tag{35}$$

The equality $\nu = \rho$ is obtained either for a two-body case ($\nu = \rho = r_{12}$) or for a restricted case ($\nu = \rho$ = distance between the two primaries) or when all r_{ij} are equal (equilateral triangle).

ρ and ν are related to the semi-moment of inertia I and to the potential U :

$$2I = m_1r_1^2 + m_2r_2^2 + m_3r_3^2 = mr^2 + \mathcal{M}R^2 = M^*\rho^2/M \tag{36}$$

$$U = G\left[\frac{m_1m_2}{r_{12}} + \frac{m_1m_3}{r_{13}} + \frac{m_2m_3}{r_{23}}\right] = \frac{GM^*}{\nu}$$

E) Finally we will use:

$$\rho' = d\rho/dt \tag{37}$$

$$j = \frac{p}{2\rho} + \frac{\rho}{2a} + \frac{\rho\rho'^2}{2\mu} = \text{Sundman function} \tag{38}$$

With these notations the usual relations become:

Lagrange-Jacobi identity ($d^2I/dt^2 = U + 2h$):

$$d^2(\rho^2)/dt^2 = 2\mu\left(\frac{1}{\nu} - \frac{1}{a}\right) \tag{39}$$

Sundman inequality ($4I(U + h) \geq c^2 + (dI/dt)^2$):

$$\frac{\rho}{\nu} \geq j \tag{40}$$

The equations (37) - (39) imply:

$$\frac{dj}{dt} = \frac{\rho'}{\rho}\left(\frac{\rho}{\nu} - j\right) \; ; \quad \frac{d[\rho(j-1)]}{dt} = \rho'\left(\frac{\rho}{\nu} - 1\right) \tag{41}$$

hence (40) and (35) imply that ρ , j and $\rho(j - 1)$ always vary in the same direction.

In a two-body case and in a restricted case ρ and ν are identical and (39) gives directly the evolution of the mutual distance in a two-body motion. In a general three-body case $\rho \geq \nu$ and the two-body evolution appears as a limit. The inequality (40) also allows to obtain long term information on the evolution of ρ and j , informations that can be extended to the n-body problem [Marchal 1975].

4. THE RATIO ρ/ν AND THE HILL-TYPE STABILITY

The ratio ρ/ν is defined by (30), (33), (34), it is a function of the masses and the mutual distances but it is independent of the scale of the triangle of the three masses, hence it is possible to draw a map of the ρ/ν = constant curves in terms of the position of m_1 with respect to m_2 and m_3 (see the figure 4 in the case of three equal masses).

These curves generalize the Hill's curves of the circular restricted three-body problem. The ratio ρ/ν is infinite at m_2 , at m_3 and at infinity, it is minimum (and equal to one) at the triangular Lagrangian point L_4 and L_5 and finally it has three saddle points at the collinear Lagrangian points L_1 , L_2 and L_3 .

In the figure 4 the three masses are equal and the three collinear Lagrangian points are at the same level: $\rho/\nu = \sqrt{\frac{25}{18}} = 1.1785...$

If the three masses are unequal let us relabel them in ascending order:

$$0 < m_A \leq m_B \leq m_C \tag{42}$$

If we give to the collinear Lagrangian points the subscript of the mass that is between the two other masses (as they are in figure 4) we always obtain:

$$1 < \left(\frac{\rho}{\nu}\right)L_C \leq \left(\frac{\rho}{\nu}\right)L_B \leq \left(\frac{\rho}{\nu}\right)L_A \leq \sqrt{\frac{343}{243}} = 1.1881... \tag{43}$$

The upper bound is obtained for $m_B = m_C = 1.6\ m_A$.

Let us now consider the equation (38) and the Sundman inequality (40):

$$\frac{\rho}{\nu} \geq j = \frac{p}{2\rho} + \frac{\rho}{2a} + \frac{\rho\rho'^2}{2\mu} \tag{44}$$

In a negative energy case the generalized semi-major axis a is positive and, for given p and a, the minimum of the right member is obtained when $\rho' = 0$ and $\rho = \sqrt{(ap)}$, that minimum is $\sqrt{(p/a)}$ hence:

$$h < 0 \text{ implies } \frac{\rho}{\nu} \geq \sqrt{(p/a)} \tag{45}$$

The ratio p/a is a function of the masses and the integrals of motion, it is proportional to c^2h :

$$p/a = -2c^2hM/G^2(M^*)^3 \tag{46}$$

and we obtain the following:

If p/a is larger than one the three bodies cannot approach the triangular Lagrangian configurations and there are two forbidden regions in figure 4 around L_4 and L_5.

If $\sqrt{p/a}$ is larger than $(\rho/\nu)L_A$ the three collinear Lagrangian configurations themselves become forbidden and the system has the Hill-type stability. In the figure 4 the zone of possible motion is divided into three disconnected parts and m_1 is forever either near m_2 or near m_3 or very far away: there is a small binary that the isolated body can neither approach nor disrupt. (For instance if $m_1 = m_2 = m_3$ and if $p/a = 4$ the limit is given by the $\rho/\nu = 2$ curves of figure 4 and the corresponding ratio r/R remains forever less than or equal to 0.19115...)

With (42) and (43) we can notice that, if the isolated mass is the smallest, the Hill-type stability occurs as soon as $\sqrt{p/a} > (\frac{\rho}{\nu})L_B$.

5. THE RATIOS r/R AND ρ/ν . THE CONDITIONS OF ISOLATION.

In the previous section we have essentially considered the ratio ρ/ν while in the first part our main ratio was r/R and we need the relation between these two ratios. Hence let us consider three given masses and a given ratio r/R less than α^{-1}, and let us look for possible values of ρ/ν.

ρ, given in (33) and (36), is a function of r and R :

$$M^*\rho^2/M = mr^2 + \mathcal{M}R^2 \tag{47}$$

ν is not a function of r and R, however:

$$\left.\begin{aligned} \frac{M^*}{\nu} &= \frac{m_1m_2}{r_{1.2}} + \frac{m_1m_3}{r_{1.3}} + \frac{m_2m_3}{r_{2.3}} \\ r_{1.2} &= r \\ m_1r_{1.3}^2 + m_2r_{2.3}^2 &= mr^2 + (m_1 + m_2)R^2 \end{aligned}\right\} \tag{48}$$

For given r and R the maximum of ν is obtained in the isosceles case ($r_{1.3} = r_{2.3}$) and the minimum in the collinear case with m_2 between m_1 and m_3, hence if we put :

$$r/R = \lambda \; ; \quad \lambda < \alpha^{-1} \tag{49}$$

we obtain:

$$\left.\begin{aligned} & f(\lambda) \leq \rho/\nu \leq F(\lambda) \\ & f(\lambda) = (m_3 + m)^{-3/2} \cdot (\frac{m_3}{\lambda^2} + M\alpha\beta)^{1/2} \cdot (m + \frac{m_3\lambda}{\sqrt{1 + \lambda^2\alpha\beta}}) \\ & F(\lambda) = (m_3 + m)^{-3/2} \cdot (\frac{m_3}{\lambda^2} + M\alpha\beta^{1/2} \cdot (m + \frac{m_3\lambda\alpha}{1+\lambda\beta} + \frac{m_3\lambda\beta}{1-\lambda\alpha}) \end{aligned}\right\} \tag{50}$$

The curves $f(\lambda)$ and $F(\lambda)$ are presented in the figure 5, they always give the same picture with one and only one minimum of $f(\lambda)$ and $F(\lambda)$ in the $(0, \alpha^{-1})$ interval.

The minimum of $f(\lambda)$ is obtained at E and corresponds to the equilateral triangles (Lagrange central configuration).

The minimum of $F(\lambda)$ is obtained at C and corresponds to the collinear central configuration (or Euler central configuration) with m_2 between m_1 and m_3.

Let us now consider a large value of the Sundman function j such as j_0 (figure 5). We know that if $r/R < \alpha^{-1}$ the point $(r/R \; ; \rho/v)$ will fall into the shaded region of the figure 5, we also know that $\rho/\nu \geq j$ (Sundman inequality (40)), hence $j = j_0$ implies either $r/R \leq \lambda_H$ or $r/R \geq \lambda_K$. In the former case we will consider that m_3 is "isolated."

That definition or isolation ($j > F_2$; $r/R \leq \lambda_3$) can even be extended to ($j \geq F_2$; $r/R \leq \lambda_3$) indeed the case $j = F_2$; $r/R = \lambda_3$ cannot be crossed, it only corresponds to Euler motions with a collinear central configuration and with forever $j = F_2$; $r/R = \lambda_3$. Hence if initially the mass m_3 is isolated it will remain isolated as long as $j \geq F_2$ and the ratio r/R will remain less than or equal to $\lambda(j)$ with $F[\lambda(j)] = j$ and $\lambda(j) \leq \lambda_3$.

The interest of that notion of isolation comes from the figure 4 and the equations (37) - (41). If initially j is large and ρ' is positive one of the three bodies is isolated and j will remain non-decreasing as long as ρ is non-decreasing, hence, because of the expression of j itself, we will obtain large intervals of time with a large j and an isolated body.

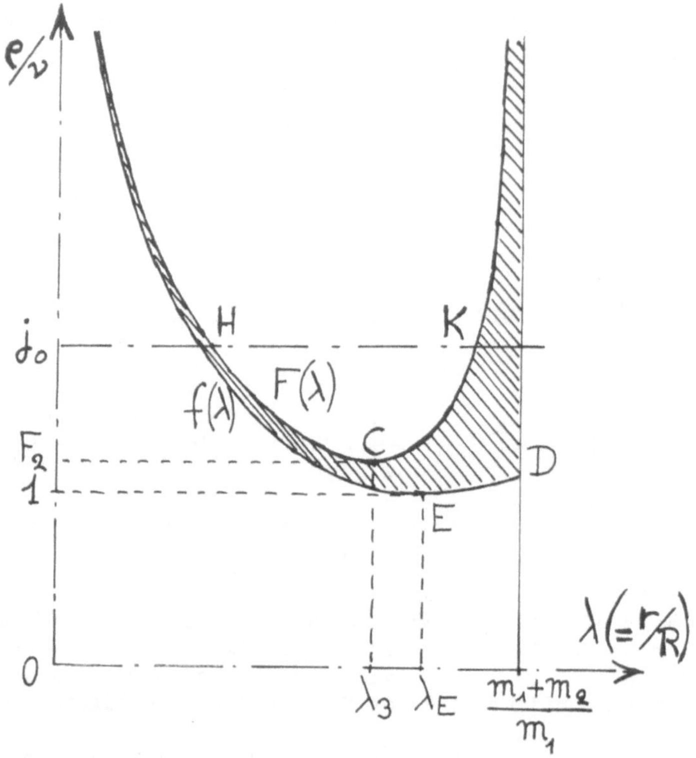

Figure 5. The relationship between r/R and ρ/υ.

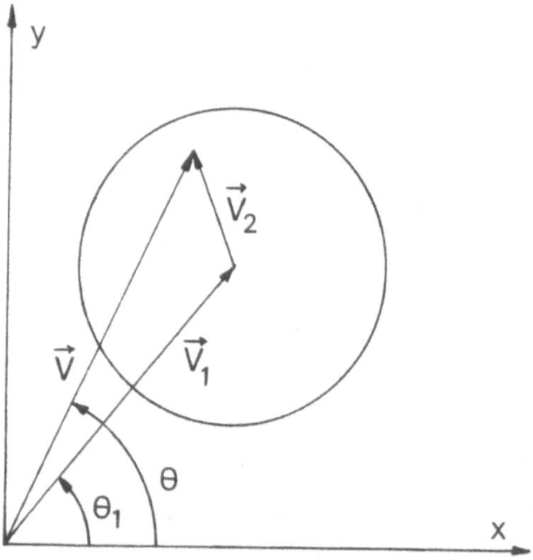

Figure 6. The decomposition of $\overline{V}$.

6. THE CASES OF ISOLATION

The figure 4 shows that when the three masses are equal one of them is necessarily isolated as soon as $j \geqq \sqrt{25/18}$ (since $\rho/\nu \geqq j$).

In the case of unequal masses we obtain a discussion similar to that of the Hill type stability with j instead of $\sqrt{(p/a)}$

The ordinate F_2 of the $F(\lambda)$ curve of figure 5 is equal to $(\rho/\nu)L_2$, i.e. to the value of ρ/ν for the collinear central configuration with m_2 between the two other masses.

Let us relabel the three masses in ascending order as in (42), (43):

$$\left.\begin{aligned} & 0 < m_A \leqq m_B \leqq m_C \\ & \left(\frac{\rho}{\nu}\right) L_K = F_k \ ; \quad k = \{A, B, C\} \\ & 1 < F_C \leqq F_B \leqq F_A \leqq \sqrt{\frac{343}{243}} = 1.1881... \end{aligned}\right\} \qquad (51)$$

F_k is thus the value of ρ/ν for the collinear central configuration with m_k between the two other masses, it is also the value of j for the corresponding Euler motion and with obvious λ_A, λ_B, λ_C and R_A, R_B, R_C we obtain:

$$\left.\begin{aligned} & m_A \text{ is isolated iff } j \geqq F_B \text{ and } r_{BC}/R_A \leqq \lambda_A \\ & m_B \text{ is isolated iff } j \geqq F_A \text{ and } r_{AC}/R_B \leqq \lambda_B \\ & m_C \text{ is isolated iff } j \geqq F_A \text{ and } r_{AB}/R_C \leqq \lambda_C \end{aligned}\right\} \qquad (52)$$

Note the asymmetry: F_A appears twice and F_C doesn't appear (because m_C is never the smallest mass of the binary).

F_A, F_B, F_C and λ_A, λ_B, λ_C are given by the collinear central configurations and/or by the minimum of the corresponding function $F(\lambda)$ of (50). They are complicated functions of the mass ratios but λ_B and λ_C have a simple relation since they correspond to the same central configuration:

$$\left(\frac{1}{\lambda_B} - \frac{m_C}{m_A + m_C}\right)\left(\frac{1}{\lambda_C} - \frac{m_B}{m_A + m_B}\right) = 1 \ ;$$
$$0 < \lambda_C \leqq \frac{M + m_A}{2M} \leqq \lambda_B \leqq \lambda_A < 1 \qquad (53)$$

If $j \geqq F_A$ one of the three masses is necessarily isolated and the discussion (52) allows to recognize the isolated mass, however we can already write the following simple rules:

$$\left.\begin{array}{ll} \text{A)} & \text{The mass opposed to the largest side of the triangle of the three masses is never isolated.} \\ \text{B)} & \text{If } r_{BC} \text{ is the smallest side of the triangle and if } j \geq F_B \text{ the mass } m_A \text{ is isolated.} \\ \text{C)} & \text{If } r_{AC} \text{ is the smallest side of the triangle and if } j \geq F_A \text{ the mass } m_B \text{ is isolated.} \end{array}\right\} \quad (54)$$

The analysis (52) remains necessary in the two following remaining undecided cases:

A) $F_B \leq j < F_A$ and r_{BC} is the second largest side of the triangle. Question: is m_A isolated or not?

B) $F_A \leq j$ and r_{AB} is the smallest side of the triangle. Question: Which, among m_C and the furthest mass from m_C, is the isolated mass? (It is never m_C if $2r_{AB} > \sup\{r_{AC} ; r_{BC}\}$).

Note an exceptional singularity: two masses are isolated at the same time if and only if the motion is a Euler motion with the smallest mass in the middle. The two largest masses are then isolated with forever $j = F_A$; $r_{AC}/R_B = \lambda_B$; $r_{AB}/R_C = \lambda_C$.

The expression of the Sundman function j is given in (38):

$$j = \frac{p}{2\rho} + \frac{\rho}{2a} + \frac{\rho\rho'^2}{2\mu} \qquad (55)$$

hence we obtain a large j and the isolation of one of the three bodies in the following cases:

A) When $p \neq 0$ (non-zero angular momentum) and ρ is small (triple close approach).

B) When ρ is large and a positive (negative energy).

C) For sufficiently large $|\rho'|$.

7. THE TESTS OF ESCAPE

7.1 Cases of positive or zero energy

If the energy integral h is positive or zero many results are already known and our analysis can only improve the details.

A) Systems of non-negative energy have always an escape at both ends, indeed from (31), (35) and (39):

$$\left.\begin{array}{l} 0 \leq \nu \leq \rho \; ; \quad -\dfrac{1}{a} = \dfrac{2h}{GM^*} \geq 0 \\[2ex] d^2(\rho^2)/dt^2 = 2\mu\left(\dfrac{1}{\nu} - \dfrac{1}{a}\right) \geq \dfrac{2\mu}{\nu} \geq \dfrac{2\mu}{\rho} > 0 \end{array}\right\} \qquad (56)$$

The function $\rho^2(t)$ is concave up and ρ has one and only one minimum: ρ_m at some time t_m. Furthermore $d^2(\rho^2)/dt^2 \geq 2\mu/\rho$ implies for all time $2(\rho - \rho_m)(\rho + 2\rho_m)^2 \geq 9\mu(t - t_m)^2$, hence ρ goes to infinity for both $t \to -\infty$ and $t \to +\infty$ at least as $[4.5\ \mu t^2]^{1/3}$, (and even at least as $|t| \cdot \sqrt{(-\mu/a)}$ if $h > 0$ since then $d^2(\rho^2)/dt^2 \geq -2\mu/a > 0$).

With (40), (41) the Sundman function j varies in the direction of ρ and has one and only one minimum, j_m, at the same time t_m.

After t_m, i.e. when $\rho' \geq 0$, as soon as j goes up above the limit of isolation one of the three masses is and will remain forever isolated and since ρ goes to infinity the isolated mass will escape to infinity. This escape is always hyperbolic with the only exception of the parabolic Euler motion for which $h = 0$; $\rho/\nu \equiv j \equiv j_o = F_A$ or F_B.

The final nature of the motion will only depend on the evolution of the binary m_1, m_2 that is on the sign of its final energy $h_{1.2}$:

$$h_{1.2} = h - \mathcal{M}R_\infty'^2/2 \tag{57}$$

The limits on R'_∞ given in the section 5.D. of the first part [Marchal, Yoshida, Sun Yi-Sui 198] allow generally an early determination of the final nature of the motion.

7.2 Tests of escape for systems of negative energy

For escape tests of these systems we will need the escape velocities of figure 3 and, according to (14), we will look for initial conditions leading to $r \leq kR$ for all future time.

All our tests of escape of the mass m_3 will satisfy at some initial time t_o and for a suitable scalar k:

$$\begin{cases} r_o/R_o \leq k \leq \lambda_3 & (58) \\ V_o \geq V_{oE+}(\mu, R_o, \varepsilon, \theta_o) & (59) \end{cases}$$

ε being given by (9) and V_{oE+} being the corresponding escape velocity.

The demonstration of the test will be the demonstration of $r \leq kR$ for all future time.

First test - A test for Hill-stable systems.

The mass m_3 and the length R will escape to infinity when $t \to +\infty$ if exists a scalar k such that at some initial time t_o.

$$r_o/R_o \leq k \leq \lambda_3 \tag{60}$$

$$\sqrt{(p/a)} \geq F(k) \tag{61}$$

$$V_o \geq V_{oE+}(\mu, R_o, \varepsilon, \theta_o) \tag{62}$$

The function $F(k)$ is defined in (50).

Since for systems of negative energy the Sundman function j always remains larger than or equal to $\sqrt{(p/a)}$ we will have forever $\rho/\nu \geq j \geq F(k) \geq F(\lambda_3)$, the mass m_3 will remain forever isolated, the system has the Hill-type stability and the figure 5 shows that $r \leq kR$ for all time.

Second test

Let us now consider a complicated test that will be used for the demonstration of the simpler tests met afterwards.

The mass m_3 and the length R will escape to infinity when $t \to +\infty$ if exists a scalar k such that at some initial time t_o :

$$\begin{cases} r_o/R_o \leq k \leq \lambda_3 & (63)\\ R'_o \geq o & (64)\\ V_o \geq V_{oE+}(\mu, R_o, \varepsilon, \theta_o) & (65)\\ R_o \geq A_k & (66) \end{cases}$$

We will assume that $\sqrt{(p/a)} \leq F(k)$, if not the simple test (60) - (62) can be used, and A_k is the length defined by:

$$A_k = [aF(k) + \sqrt{[a^2 \cdot F(k)^2 - ap]}] \cdot [M^*]^{\frac{1}{2}} \cdot [M(\mathcal{M} + mk^2)]^{-\frac{1}{2}} \tag{67}$$

Let us demonstrate this second test.

The propositions (14) and (10) show that we only need to verify $r \leq kR$ for all future times, hence let us do the demonstration by contradiction and let us assume that exists a bounded time t_f such that:

$$\left\{\begin{array}{l} t_o \leq t \leq t_f \ \text{ imply } \ r/R \leq k \\ r/R > k \ \text{ immediately after } \ t_f \end{array}\right\} \tag{68}$$

The conditions (15) lead to:

$$t_o \leq t \leq t_f \Rightarrow \begin{cases} V \geq V_{E+}(\mu, R, \varepsilon, \theta) & (69)\\ d^2(R^2)/dt^2 \geq 2\mu(1 + \varepsilon)/R > 0 & (70)\\ R' \geq o \ ; \ R \geq R_o \geq A_k & (71) \end{cases}$$

The relation between r , R and ρ given in (36) can be written:

$$M^* \frac{\rho^2}{R^2} = M\mathcal{M} + \mathcal{M}m \frac{r^2}{R^2} \tag{72}$$

hence at t_f , since $r_f = kR_f$:

$$\rho f = R_f[M(M + mk^2)]^{\frac{1}{2}} \cdot [M^*]^{-\frac{1}{2}} \quad (73)$$

and, since $R_f \geqq R_o \geqq A_k$:

$$\rho_f \geqq aF(k) + \sqrt{(a^2F(k)^2 - ap)} \quad (74)$$

which implies:

$$\frac{p}{2\rho_f} + \frac{\rho_f}{2a} \geqq F(k) \quad (75)$$

and:

$$j_f = \frac{p}{2\rho_f} + \frac{\rho_f}{2a} + \frac{\rho_f \rho_f'^2}{2\mu} \geqq F(k) + \frac{\rho_f \rho_f'^2}{2\mu} \quad (76)$$

On the other hand, with the figure 5 and the relation (40):

$$r_f/R_f = k \text{ implies: } F(k) \geqq \rho_f/\nu_f \geqq j_f \quad (77)$$

The conditions (76) and (77) lead to:

$$F(k) \geqq \frac{\rho_f}{\nu_f} \geqq j_f \geqq F(k) + \frac{\rho_f \rho_f'^2}{2\mu} \quad (78)$$

hence:

$$\rho_f' = 0 ; \quad F(k) = \rho_f/\nu_f = j_f \quad (79)$$

The Lagrange-Jacobi identity gives at t_f :

$$(\rho^2)_f'' = 2\mu(\frac{1}{\nu_f} - \frac{1}{a}) = 2\mu(\frac{F(k)}{\rho_f} - \frac{1}{a}) \quad (80)$$

that is, with (74):

$$(\rho^2)_f'' \leqq 0 \quad (81)$$

Thus at t_f , with (70), (71), (79), (81):

$$R_f' \geqq 0 ; \quad (R^2)_f'' > 0 ; \quad \rho_f' = 0 ; \quad (\rho^2)_f'' \leqq 0 \quad (82)$$

hence, even if $R_f' = 0$, we obtain $\rho^2/R^2 < \rho_f^2/R_f^2$ immediately after t_f and also, with (72), $r^2/R^2 < r_f^2/R_f^2 = k^2$ which is contradictory to the hypothesis (68).

We thus arrive to the following conclusion: after t_o the ratio r/R will remain forever less than or equal to k, the mass m_3 and the length R will escape to infinity and the second test (63) - (66) is demonstrated.

Note that for this second test k can also be chosen in $[\lambda_3, \alpha^{-1}]$.

Third test.

The mass m_3 and the length R will escape to infinity when $t \to +\infty$ if exists a scalar k such that at some initial time t_o :

$$\left.\begin{array}{l} r_o/R_o \leq k \leq \lambda_3 \\ j_o \geq F(k) \end{array}\right\} \text{These conditions imply the isolation of } m_3 \tag{83}$$

$$R'_o \geq 0 \tag{84}$$

$$\rho'_o \geq 0 \tag{85}$$

$$V_o \geq V_{oE+}(\mu, R_o, \varepsilon, \theta_o) \tag{86}$$

As in the first tests ε is related to k, α, β by (9) and $V_{oE+}(\mu, R_o, \varepsilon, \theta_o)$ is the corresponding escape velocity.

Demonstration of the third test.

According to (14) and (10), as for the first tests, we only need to verify $r \leq kR$ for all future times, it is obvious if $\sqrt{(p/a)} \geq F(k)$ (first test), if not we obtain the following.

A) As long as ρ is non-decreasing the Sundman function j is non-decreasing and larger than or equal to $F(k)$. Since $\rho/\nu \geq j$ the figure 5 and the section 5 imply that m_3 remains isolated and satisfies $r \leq kR$.

B) If ρ is forever non-decreasing the question is solved and it remains to analyze the situation if ρ reaches a local maximum ρ_M at some future time t_M. We will demonstrate that at t_M the conditions (63) - (66) of the second test are satisfied.

In the interval $[t_o, t_M]$ we have already $\rho' \geq 0$ and $r \leq kR$ and we have also from the conditions (15), (40) and (41):

$$V \geq V_{E+}(\mu, R, \varepsilon, \theta) \tag{87}$$

$$d^2(R^2)/dt^2 \geq 2\mu(1 + \varepsilon)/R > 0 \tag{88}$$

$$d_j/dt = \frac{\rho'}{\rho}\left(\frac{\rho}{\nu} - j\right) \geq 0 \tag{89}$$

hence at t_M:

$$R_M R'_M \geq R_o R'_o \geq 0 \tag{90}$$

$$R_M \geq R_o > 0 \tag{91}$$

$$j_M \geq j_o \geq F(k) \tag{92}$$

Thus at t_M the conditions (63), (64), (65) of the second test are satisfied and it remains to verify the condition (66) that is here $R_M \geq A_k$.

We have assumed that at t_M the mean quadratic distance ρ has a local maximum, hence:

$$\rho'_M = 0 \; ; \quad j_M = \frac{p}{2\rho_M} + \frac{\rho_M}{2a} \tag{93}$$

and from the Lagrange-Jacobi identity (39):

$$(\rho^2)''_M = 2\mu\left(\frac{1}{\nu_M} - \frac{1}{a}\right) \leq 0 \; ; \quad \text{that is} \quad \nu_M \geq a \tag{94}$$

Since $\rho/\nu \geq j$:

$$\rho_M \geq \nu_M j_M \geq a j_M \tag{95}$$

and, with (93):

$$\rho_M = a j_M + \sqrt{(a^2 j_M^2 - ap)} \tag{96}$$

on the other hand

$$\left.\begin{aligned} M^* \rho^2 &= M(\mathcal{M}R^2 + mr^2) \\ r_M &\leq kR_M \end{aligned}\right\} \quad \text{imply:} \quad R_M \geq A_M \tag{97}$$

with, ρ_m being given by (96):

$$A_M = [a j_M + \sqrt{(a^2 j_M^2 - ap)}] \cdot [M^*]^{\frac{1}{2}}[M(\mathcal{M} + mk^2)]^{-\frac{1}{2}} \tag{98}$$

The comparison between A_m and A_k is the comparison between j_m and $F(k)$ and then, with (92) and (97):

$$F(k) \leq j_M \quad \text{and} \quad A_k \leq A_M \leq R_M \tag{99}$$

Hence finally we found that at the t_M all the conditions (63) - (66) of the second test are satisfied. Thus in all cases the ratio r/R will remain forever less than or equal to k, the mass m_3 and the length R will escape to infinity and the third test is demonstrated.

Fourth test.

The mass m_3 and the length R will escape to infinity when $t \to +\infty$ if exists a scalar k such that at some initial time t_o :

$$\left[\begin{array}{l} \left.\begin{array}{l} r_o/R_o \leq k \leq_M \lambda_3 \\ \dfrac{p}{2\rho_o} + \dfrac{\rho_o}{2a} \geq F(k) \end{array}\right\} \\ R'_o \geq 0 \\ V_o \geq V_{oE+}(\mu, R_o, \varepsilon, \theta_o) \end{array}\right.$$

These conditions imply

$j_o \geq F(k)$ and the isolation of m_3 (100)

(101)

(102)

Demonstration.

If $\sqrt{(p/a)} \geq F(k)$ and/or $\rho'_o \geq 0$ the conditions of the first and/or the third test are satisfied.

If $\sqrt{(p/a)} < F(k)$ the condition $\dfrac{p}{2\rho_o} + \dfrac{\rho_o}{2a} \geq F(k)$ implies: either

$$\begin{array}{l} \rho_o \geq aF(k) + \sqrt{(a^2F(k)^2 - ap)} \quad \text{or} \\ \rho_o \leq aF(k) - \sqrt{(a^2F(k)^2 - ap)} \end{array} \qquad (103)$$

In the former case we found immediately $R_o \geq A_k$ in a way similar to (96), (97), (98) and the test is a part of the second test.

The only really new case is the case when:

$$\sqrt{(p/a)} < F(k) \ ; \ \rho'_o < 0 \ ; \ \rho_o \leq aF(k) - \sqrt{(a^2F(k)^2 - ap)} \qquad (104)$$

We found then:

$$\left.\begin{array}{l} \dfrac{\rho_o}{\nu_o} \geq j_o > \dfrac{p}{2\rho_o} + \dfrac{\rho_o}{2a} \geq F(k) \\ \rho_o \leq \sqrt{(ap)} < aF(k) \end{array}\right\} \quad \nu_o \leq \frac{\rho_o}{j_o} < \frac{\sqrt{ap}}{F(k)} < a \qquad (105)$$

As long as ρ is decreasing the sum $\dfrac{p}{2\rho} + \dfrac{\rho}{2a}$ is increasing and the larger quantity j remains larger than $F(k)$, hence the mean harmonic distance ν (that satisfies $\nu \leq \rho/j$) remains less than a and even less than $\sqrt{ap}/F(k)$. Since $(\rho^2)'' = 2\mu(\frac{1}{\nu} - \frac{1}{a})$ the curve $\rho^2(t)$ is concave up and ρ will reach a minimum ρ_m at some future time t_m.

Let us examine the conditions at t_m.

During $[t_o, t_m]$ the Sundman function j remains larger than $F(k)$, the mass m_3 remains isolated, the ratio r/R remains smaller than k and then:

$$\{t_o \leq t \leq t_m\} \text{ imply } V \geq V_{E+} \ ; \ (R^2)'' > 0 \ ; \ R' \geq 0 \qquad (106)$$

At t_m we obtain also $\rho_m' = 0$ and all the conditions (83) – (86) of the third test of escape are satisfied. Thus, even in the last case (104), the conditions (100) – (102) of the fourth test always lead to an escape of m_3 with forever $r \leq kR$.

Third and fourth tests.
The union of the third and fourth tests give a more efficient test that can be written as follows.
The mass m_3 and the length R will escape to infinity when $t \to +\infty$ if exists a scalar k such that at some initial time t_0 :

$$
\begin{cases}
r_0/R_0 \leq k \leq \lambda_3 & \qquad\qquad m_3 \text{ is} \qquad (107) \\[2ex]
\dfrac{p}{2\rho_0} + \dfrac{\rho_0}{2a} + \dfrac{\rho_0\rho_0'}{2\mu} \cdot \sup\{0;\ \rho_0'\} \geq F(k) & \qquad\qquad \text{isolated} \quad (108) \\[2ex]
R_0' \geq 0 & \qquad\qquad\qquad\qquad\qquad (109) \\[2ex]
V_0 \geq V_{0E+}(\mu,\ R_0,\ \epsilon,\ \theta_0) & \qquad\qquad\qquad\qquad\qquad (110)
\end{cases}
$$

The ratio r/R will remain forever less than or equal to k and we can note that the best value of k for this test is the smallest possible giving an equality in (108): it doesn't destroy (107) and gives the smallest possible ϵ and V_{0E+}.

Fifth test.
Let us draw back the condition (109) in the above test, we will obtain:
The mass m_3 and the length R will escape to infinity when $t \to +\infty$ and/or when $t \to -\infty$ if exists a scalar k such that at some initial time t_0 :

$$
\begin{cases}
r_0/R_0 \leq k \leq \lambda_3 & \qquad\qquad\qquad\qquad\qquad (111) \\[2ex]
\dfrac{p}{2\rho_0} + \dfrac{\rho_0}{2a} + \dfrac{\rho_0\rho_0'}{2\mu} \cdot \sup\{0;\ \rho_0'\} \geq F(k) & \qquad\qquad\qquad (112) \\[2ex]
V_0 \geq V_{0E+}(\mu,\ R_0,\ \epsilon,\ \theta_0) & \qquad\qquad\qquad\qquad\qquad (113)
\end{cases}
$$

Demonstration.
If $R_0' \geq 0$ we found again the conditions (107) – (110) and m_3 escapes when $t \to +\infty$.
If $R_0' \leq 0$ the figure 1 implies $\theta_0 \geq 90°$ and the relations (16) and (17) lead to:

$$
V_{0E+}(\mu,\ R_0,\ \epsilon,\ \theta_0) \geq V_{0E-}(\mu,\ R_0,\ \epsilon,\ \theta_0) \qquad (114)
$$

Hence, as soon as $\dfrac{p}{2\rho_0} + \dfrac{\rho_0}{2a} \geq F(k)$, the conditions symmetrical to that of the test (100) - (102) are satisfied and m_3 escapes when $t \to -\infty$.

It remains to verify the fifth test when $R'_o < 0$ and

$\frac{p}{2\rho_o} + \frac{\rho_o}{2a} < F(k)$ which imply $\rho'_o > 0$ and $V_o \geq V_{oE+} \geq V_{oE-}$.

In that final case the conditions (83) - (86) of the third test are satisfied with the only exception of the condition (84) that is $R'_o \geq 0$. Fortunately this condition (84) takes no part in the demonstration until the analysis at the time t_M. If then $R'_M \geq 0$ the demonstration is valid and m_3 escapes when $t \to +\infty$. On the contrary if $R'_M \leq 0$ we find again $r_M \leq kR_M$; $R_M \geq A_M \geq A_k$ and also $V_M \geq V_{ME+} \geq V_{ME-}$, the conditions symmetrical to that of the second test are satisfied and m_3 escapes when $t \to -\infty$.

Hence in all cases of the test (111) - (113) we find at least one escape of m_3 and this fifth test is demonstrated.

Some auxiliary conditions allow to determine on which side m_3 is escaping but they are generally complicated, for instance we can put:

$$\left.\begin{aligned} S &= \frac{p}{2\rho_o} + \frac{\rho_o}{2a} + \frac{\rho_o\rho'_o}{2\mu} \cdot \sup\{0;\ \rho'_o\} \\ A_o &= [as + \sqrt{(a^2s^2 - ap)}] \cdot [M^*]^{\frac{1}{2}} \cdot [M(M + mk^2)]^{-\frac{1}{2}} \end{aligned}\right\} \quad (115)$$

and we obtain, whith the conditions (111) - (113):

The mass m_3 escapes when $t \to +\infty$ if $R'_o \geq 0$ and/or $R_o \leq A_o$ and/or $\sqrt{p/a} \geq F(k)$ (first test) and/or $\{m \leq 2m_3\ ;\ \rho \leq as\}$.

If $R_o \leq A_o$ it implies $A_o \leq A_M$ and the demonstration of $R'_M \geq 0$ at $t = t_M$ is easy.

The mass m_3 escapes when $t \to -\infty$ if $\{R'_o \leq 0;\ \frac{p}{2\rho_o} + \frac{\rho_o}{2a} \geq F(k)\}$ and/or $\{R'_o \leq 0\ ;\ R_o \geq A_k\}$ (second test).

Since $s \geq F(k)$ and $A_o \geq A_k$ we know at least one direction of escape.

7.3 Summary of the section 7 on the tests of escape

If we neglect the rarely useful first test and the complicated informations related to the second test and to (115), we can express the information of this section in the most efficient following form.

Let us assume that at some initial time t_o the mass m_3 is isolated (i.e. that $j_o \geq F_2$; $r_o/R_o \leq \lambda_3$), we have then two possible most efficient values of k^2.

A) k_1 defined by:

$$j_o = F(k_1) \; ; \quad k_1 \leq \lambda_3 \qquad (116)$$

B) k_2 defined by:

$$\frac{p}{2\rho_o} + \frac{\rho_o}{2a} = F(k_2) \; ; \qquad k_2 \leq \lambda_3 \qquad (117)$$

The functions j and f are defined in (38) and (50)

k_2 exists iff $\frac{p}{2\rho_o} + \frac{\rho_o}{2a} \geq F_2 = F(\lambda_3)$; if it exists it verifies $k_1 \leq k_2 \leq \lambda_3$ with $k_1 = k_2$ when $\rho' = 0$.

The isolation of m_3 implies $r_o/R_o \leq k_1$.

The mass m_3 and the length R will escape to infinity if the following conditions are satisfied.

If the energy integral h is positive or zero the mass m_3 and the length R will escape to infinity at least when $t \to (\text{sign } \rho_o') \cdot \infty$ and on both sides if k_2 exists. These escapes will forever be with an isolated mass m_3 and with $r/R \leq k_1$ on the side of (sign ρ_o'), $r/R \leq k_2$ on the opposite side.

If the integral h is negative our tests require an initial escape velocity and we thus need the scalars ε_1 and ε_2 related to k_1 and k_2 by (9).

In the case when $R_o'\rho_o' \geq 0$ the escape condition is:

$$V_o \geq \inf\{V_{oE+}(\mu, R_o, \varepsilon_1, \theta_o) \; ; \; V_{oE-}(\mu, R_o, \varepsilon_1, \theta_o)\} \qquad (118)$$

The mass m_3 and the length R will then escape to infinity when $t \to [\text{sign}(R_o' + \rho_o')]\infty$, and on both sides if $R_o' = \rho_o' = 0$.

This or these escapes will be forever with $r/R \leq k_1$.

If $R_o'\rho_o' < 0$ we can try the two following tests.

If:

$$V_o \geq V_{oE(\text{sign } R_o')}(\mu, R_o, \varepsilon_2, \theta_o) \qquad (119)$$

the mass m_3 and the length R escape to infinity when $t \to (\text{sign } R_o')\infty$, with forever $r/R \leq k_2$. If:

$$V_o \geq V_{oE(\text{sign } R_o')}(\mu, R_o, \varepsilon_1, \theta_o) \qquad (120)$$

we have an escape of m_3 with forever $r/R \leq k_1$ on the side determined by rules generalizing those given after (115).

Thus we can write that an isolated body escapes to infinity if at some time it has an escape velocity.

These escapes verify always $d^2(R^2/dt^2 > 2\mu/R$ and are almost always hyperbolic (R/t goes to a bounded and non-zero R_∞'), the only exception being that of Euler parabolic motions (for which $h = 0$; $\rho/\nu \equiv j \equiv j_0 = F_2$; $r/R \equiv \lambda_3$).

It seems that the escape condition (118) remains valid when $R'_0\rho'_0 < 0$ (with an escape of m_3 for $t \to (\text{sign } R'_0)\infty$, with forever $r/R \leq k_1$). If that conjecture was true it would be stronger than (119) and (120), however we have found neither a counter-example nor a demonstration.

8. APPLICATION - ANALYSIS IN THE ρ , ρ' PLANE

Let us consider again the figure 1, the velocity vector $\vec{V}$, its radial component $x = V\cos\theta$ and its circumferential component $y = V\sin\theta$.

If we are especially interested into the parameters ρ and ρ' we can decompose $\vec{V}$ into $\vec{V}_1 + \vec{V}_2$ (figure 6), with:

$$\vec{V}_1 = (x_1, y_1) \; ; \; \begin{cases} x_1 = R\rho'/\rho \\ y_1 = R\sqrt{\mu p}/\rho\rho' \end{cases} \tag{121}$$

For given masses, integrals of motion, $\vec{r}$, $\vec{R}$, ρ and ρ' the velocity $\vec{V}_1$ is at the center of the zone of possible $\vec{V}$, zone delimited by:

$$\vec{V} = \vec{V}_1 + \vec{V}_2 \; ; \; \|\vec{V}_2\| \leq r \sqrt{\left[\frac{2\mu m}{M\rho^3}\left(\frac{\rho}{\nu} - j\right)\right]} \tag{122}$$

The equality in (122) corresponds to plane motions while all three-dimensional motions give there a strict inequality.

If we neglect ν we obtain with (50):

$$r/R = \lambda \; ; \; \|\vec{V}_2\| \leq r \sqrt{\left[\frac{2\mu m}{M\rho^3}\{F(\lambda) - j\}\right]} \tag{123}$$

Let us assume now that the mass m_3 is isolated ($j \geq F_2$; $r/R \leq \lambda_3$ and hence, with the figure 5, $r/R \leq k_1 \leq \lambda_3$ with $F(k_1) = j$) . We can look for the conditions under which the knowledge of the constants m_1 , m_2 , m_3 , a , p , μ and the variables ρ, ρ' is sufficient for the satisfaction of our tests of escape.

For instance we can use the very efficient escape conditions (116), (118), (120) that is:

$$V \geq V_E\,(\mu, R, \varepsilon_1, \theta)$$

$$V_E = V_{E(\text{sign }\rho')}; \quad \text{if } \rho' \neq 0$$

with (124)

$$V_E = \inf(V_{Et}, V_{E-}); \quad \text{if } \rho' = 0$$

and

$$\varepsilon_1 = \frac{\alpha}{(1 + k_1\beta)^2} + \frac{\beta}{1 - k_1\alpha)^2} - 1 \; ; \quad k_1 \leq \lambda_3 \; ; \quad F(k_1) = j$$

For a given value of the ratio $r/R = \lambda$ it is sufficient to verify that the circle of the figure 6 is entirely in the escape zone.

If $\rho' > 0$ the angle θ_1 of the figure 6 belongs to $[0; \frac{\pi}{2}]$ and we must verify $V \geq V_{E+}$, however with (13) and (21) we found:

$$V_{Et} = f(\varepsilon_1, \theta) \cdot \sqrt{\frac{\mu}{R}} \leq \underbrace{\inf}_{\theta_2 \in \{\theta - \frac{\pi}{2} \; ; \; \frac{\pi}{2}\}} \frac{\sqrt{\{2\mu[1 + \varepsilon_1 g(\theta_2)]/R\}}}{\cos(\theta - \theta_2)} \tag{125}$$

this implies, with $\theta_2 = \theta_1$:

$$V_{E+}(\mu, R, \varepsilon_1, \theta) \cdot \cos(\theta - \theta_1) \leq \{2\mu[(1 + \varepsilon_1 g(\theta_1)]/R\}^{\frac{1}{2}} \tag{126}$$

but for all points of the circle of the figure 6:

$$V \cdot \cos(\theta - \theta_1) \geq |\vec{V}_1| - |\vec{V}_2| \geq \frac{R}{\rho^2} \sqrt{(\mu p + \rho^2 \rho'^2)} - $$

$$- r \sqrt{[\frac{2\mu m}{M p^3} \{F(\lambda) - j\}]} \tag{127}$$

hence, if $\rho' > 0$ and for a given ratio $r/R = \lambda$, we found the following sufficient condition of escape:

$$\{2\mu[1 + \varepsilon_1 g(\theta_1)]/R\}^{\frac{1}{2}} \leq \frac{R}{\rho^2} (\mu p + \rho^2 \rho'^2)^{\frac{1}{2}} -$$

$$- r \{\frac{2\mu m}{M p^3} (F(\lambda) - j)\}^{\frac{1}{2}} \tag{128}$$

which can be written:

$$\{\frac{p}{2\rho} + \frac{\rho\rho'^2}{2\mu}\}^{\frac{1}{2}} \geq \lambda\{\frac{m}{M} [F(\lambda) - j]\}^{\frac{1}{2}} + \{\frac{\rho^3}{R^3} [1 + \varepsilon_1 g(\theta_1)]\}^{\frac{1}{2}} \tag{129}$$

If $\rho' \leq 0$ we obtain a similar expression with $g(\pi - \theta_1)$

instead of $g(\theta_1)$.

However λ is not given in terms of ρ and ρ' and we must verify (129) for all possible λ in the available range $[0; k_1]$. On the other hand from $r/R = \lambda$ and (72):

$$\frac{\rho}{R} = \{\frac{M}{M^*}(M + m\lambda^2)\}^{\frac{1}{2}} \tag{130}$$

Hence we finally arrive to the following sufficient condition of escape of the isolated mass m_3 :

$$\left.\begin{aligned} \{\frac{p}{2\rho} + \frac{\rho\rho'^2}{2\mu}\}^{\frac{1}{2}} \geq \sup_{\lambda\in[0;\, k_1]} \{\lambda[\frac{m}{M}(F(\lambda) - j)]^{\frac{1}{2}} + \\ + [\frac{M(M + m\lambda^2)}{M^*}]^{\frac{1}{4}} \cdot [1 + \varepsilon_1 g(\omega_1)]^{\frac{1}{2}}\} \\ \text{with:} \quad \omega_1 = \inf\{\theta_1;\ \pi - \theta_1\};\quad j = F(k_1) \end{aligned}\right\} \tag{131}$$

The angle ω_1 has a very small influence because ε_1 is generally small and because $g(\omega_1)$ only varies between 1 and 1.43041 when ω_1 takes all possible values between 0 and $\pi/2$.

If we neglect ω_1 , that is if we take in (131) the largest possible $g(\omega_1) = 1.43041$, we arrive to the figure 7 for the case of three equal masses.

This figure 7 is drawn in terms of the two most efficient parameters, the ratio ρ/a and the sum $\frac{p}{2\rho} + \frac{\rho\rho'^2}{2\mu}$. These two parameters give easily the Sundman function j equal to $\frac{\rho}{2a} + \frac{p}{2\rho} + \frac{\rho\rho'^2}{2\mu}$.

Let us note the two following main results.

A) The elliptic Lagrange and Euler motions approach very near to the limit of escape especiallly for $\rho/a = 0$ where we found $\frac{p}{2\rho} + \frac{\rho\rho'^2}{2\mu}$ equal to 1 for Lagrange motions, equal to 1.1785 for Euler motions and equal to 1.1879 for the full curve.

B) The figure 7 shows the wide importance of unbounded motions, importance that can be emphasized if we remember that all systems with a positive or zero energy integral h have unbounded motions and if we note that all bounded motions must always remain below the full curve of figure 7 but also cannot remain in the vicinity of the origin; indeed with any initial (ρ_o, t_o) we always obtain [Marchal 1975]:

$$t_1 = t_o \pm \frac{T}{2} \text{ implies } \rho(t_o) + \rho(t_1) \geq 2a$$

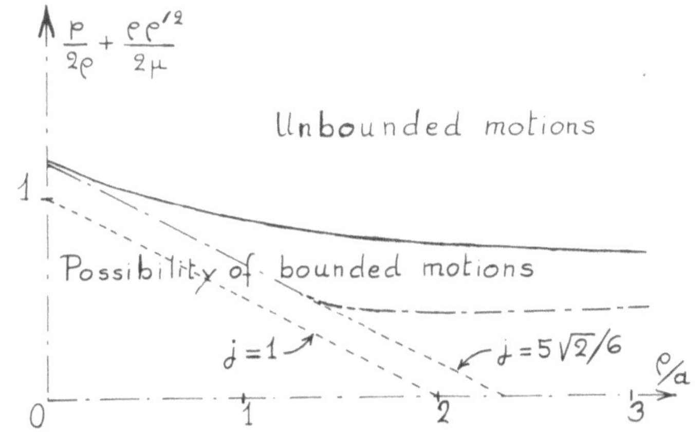

Figure 7. Sufficient condition for escape of m_3 for the case of three equal masses.

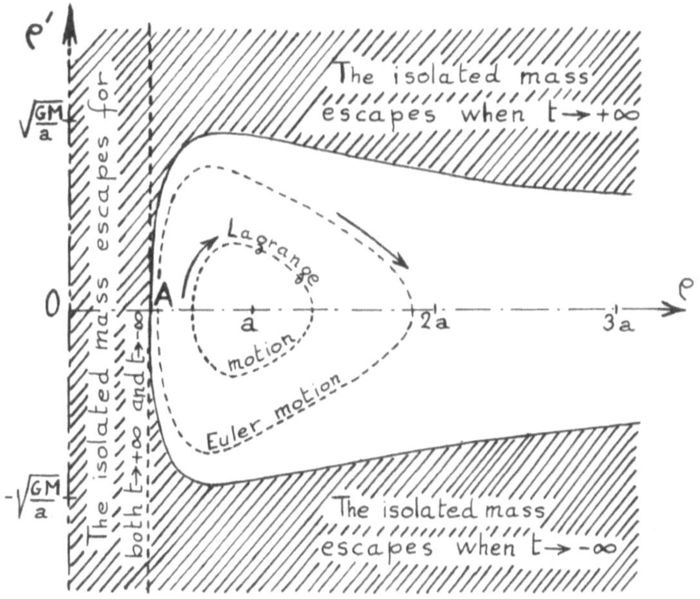

Figure 8. Sufficient condition for escape of m_3 for the case of three equal masses with $\rho/a = 8/9$.

$$t_2 = t_o \pm \frac{T}{4}\left(\frac{p}{\rho_o} + \frac{\rho_o}{a}\right) \quad \text{implies} \quad \rho(t_o) \cdot \rho(t_2) \geq ap \qquad (132)$$

with $T = 2\pi\sqrt{(a^3/\mu)}$ = period of the elliptic Lagrange motions

Let us consider now the more restricted but also more detailed analysis related to the figure 8.

That figure corresponds to a system with three equal masses and a ratio p/a equal to 8/9 (for instance: $G = m_1 = m_2 = m_3 = 1$; $c = 2$; $h = -1$).

The shaded horeshoe region only corresponds to unbounded orbits, its limit is a little better than the full curve of figure 7 because the informations related to the angle ω_1 can now be used. This limit is symmetrical with respect to the ρ-axis and for large ρ we obtain $\rho'^2 \cong \frac{\mu}{\rho}\sqrt{\frac{32}{37}}$.

All bounded orbits remain forever in the central non-shaded region, as do the Lagrange and Euler motions with central configurations (dotted curves).

The non-shaded region is infinite along the ρ-axis but ends at A on the left, hence bounded orbits cannot approach triple collisions, their ρ will forever remain larger than $\rho(A) = 0.436802a$. The same is true for all orbits which at any time have no isolated body (orbits entering the region encircled the dotted curve of Euler motion) or even for all orbits which at some time have a Sundman function j smaller than $j(A) = 1.235897$.

Conversely all orbits that reach the region where $\rho \leqq \rho(A)$ have an hyperbolic-elliptic escape at both ends (without exchange), their ρ has only one minimum, their Sundman function j is always larger than or equal to $j(A)$ and their escaping mass is for all time isolated.

Note that the elliptic Euler motion approaches very near to the limit and it seems that bounded motions can be met in almost all the non-shaded area.

A comparison with other studies can be found in [Marchal, Yoshida, Sun-Yi-Sui 1983, figure 9] and it seems that we approach very near to the limit of escape.

9. CONCLUSIONS

The new tests of escape developed in the two parts of this study bring a very large amount of information especially for three-body system of negative energy and we can write that an isolated body escapes to infinity if at some time it has an escape velocity.

The application of these tests to system of three equal masses has led to the smallest present delimitation of the region of bounded motions and these results can be extended easily to systems of unequal masses.

It would be interesting to compute numerically some bounded orbits, for instance periodic orbits, in the vicinity of our limits in order to know how near we approach to the true limit.

REFERENCES

Marchal, C.: 1975, 26th Congress of the International Astronautical Federation (Lisbon), ONERA T.P. 1975-77.

Marchal C., Bozis, G.: 1982, Celes. Mech. 26, 311.

Marchal C., Yoshida J., Sun Yi-Sui: 1983 36th Congress of the International Astronautical Federation (Budapest), IAF 83-319 (to appear in Acta Astronautica, 1984).

Marchal C., Yoshida J., Sun Yi-Sui: 1984, Celes. Mech. 33, 193.

Szebehely V.: 1967, Theory of Orbits, Academic Press, New York.

THE LYAPUNOV CHARACTERISTIC EXPONENTS - APPLICATIONS TO CELESTIAL MECHANICS.

Cl. Froeschlé
Observatoire de Nice
B.P. 139
06003 Nice Cedex
France

ABSTRACT. After a presentation of Lyapunov characteristic exponents (LCE) we recall their basic properties and numerical methods of computation. We review some numerical computations which are concerned with LCEs and mainly related to celestial mechanics problems.

1. INTRODUCTION

In recent years much work, not only numerical, has been devoted to the investigation of the ergodic properties of classical dynamical systems (1). However, while the extreme cases of integrable and of ergodic systems are at least partially understood in a mathematical context (1,2), the situation is different for many models of physical interest to which Celestial Mechanics Problems obviously belong. Of course the now classical KAM theory has been the great break-through to prove the persistence in some regions of the phase space of uniform integrals (23). Another important break-through for understanding the behaviour of the so-called chaotic regions is the theory of Lyapunov exponents (hereafter called LCE) which measure the mean exponential rate of divergence of nearby trajectories. The use of such exponents dates back to Lyapunov (7), but in a form adapted to the theory of dynamical systems and to ergodic theory it was only in 1968 that Oseledec (30) published his non-commutative Ergodic Theorem which provides a general and simple way to compute all the LCEs. The first numerical characterisation of the stochasticity of a phase space trajectory in terms of the divergence of nearby trajectories was introduced by Hénon and Heiles (24) and then further studied by Chirikov (8), Ford (10), Froeschlé (12), Froeschlé and Scheidecker (13,14) who in order to give a precise quantitative definition of exponential divergence and thus of stochasticity have been led to considering the spectral properties of a linear operator. Studying directly the behaviour of the eigen-values of linear tangential mapping for discrete dynamical system they computed LCEs without mentioning the terms (Comme Monsieur Jourdain qui faisait de la prose sans le savoir).

The connection between LCEs and the preceding numerical works has

Celestial Mechanics **34** (1984) 95–115. 0008–8714/84.15

been given and popularised by Benettin et al. (6) who give also a simple procedure to compute all LCEs.

In this paper, we shall frequently refer to the work of Benettin et al. (6) especially for Parts 2 and 3 .

In Section 2 following (6) we display the basic features of LCEs. After the introduction of the basic notations and terminology we introduce LCEs through divergence of nearby orbits and thus the spectral properties of a linear operator. The simple example of a periodic orbit will be first explored and then generalized. Theoretical results of LCEs are presented with a special emphasis on Hamiltonian systems. Last but not least we give through Pesin's formula the connection between Kolmogorov entropy and LCEs.

In § 3 we describe the numerical techniques to compute LCEs with a special emphasis on the Benettin et al. one.

In § 4 we present a review of some numerical computations of LCEs performed in the context of Celestial Mechanics in order to answer some specific questions.

2.- LYAPUNOV CHARACTERISTIC EXPONENTS

2.1 Divergence of nearby orbits and LCEs

It is well known that nearby trajectories of integrable systems diverge linearly. On the other hand many numerical experiments show clearly that the so-called stochastic region is characterised by exponential-like divergence of such trajectories. To give a precise quantitative definition of exponential divergence and thus of stochasticity one is naturally led to consider the spectral properties of a linear operator. Let us fix some basic notations and the terminology.

In the framework of ergodic theory, a classical dynamical system is given by a collection (M, μ, ϕ^t) where :

- M is an n dimensional compact smooth Riemannian manifold
- μ is a probability measure on M
- ϕ^t a measure - preserving flow, i.e. a one parameter group of diffeomorphisms $\phi^t : M \to M$ with composition law

$$\phi^{t+t'} = \phi^t \circ \phi^{t'}$$

In addition ϕ^t is measure preserving i.e. $\mu(\phi^{-t}(A)) = \mu(A)$ for any measurable set $A \subset M$.

Let us consider a point $x \in M$ and a nearby point y and denote by d(,) the distance on M induced by the Riemannian metric. More precisely if y(s) is any regular curve on M such that y(0) = x; consider the expression

$$\psi(x,y,s,t) = d(\phi^t(y(s)), \phi^t(x)) / d(y(s),x)$$

If s is small, one is naturally led to introduce the tangent vector

$$w = \left.\frac{\partial y}{\partial s}\right|_{s=0} \qquad T_xM$$

where T_xM is the tangent space to M at x. Hence we have

$$d(y(s), x) = s \,||\, w \,||\, + o(s)$$

with
$$\lim_{s\to 0} \frac{o(s)}{s} = 0$$

$||\quad||$ being the norm on T_xM induced by the metric on M.
In the same way we have

$$d(\phi^t(y(s)), \phi^t(x)) = s||D\,\phi^t_x\, w|| + o(s)$$

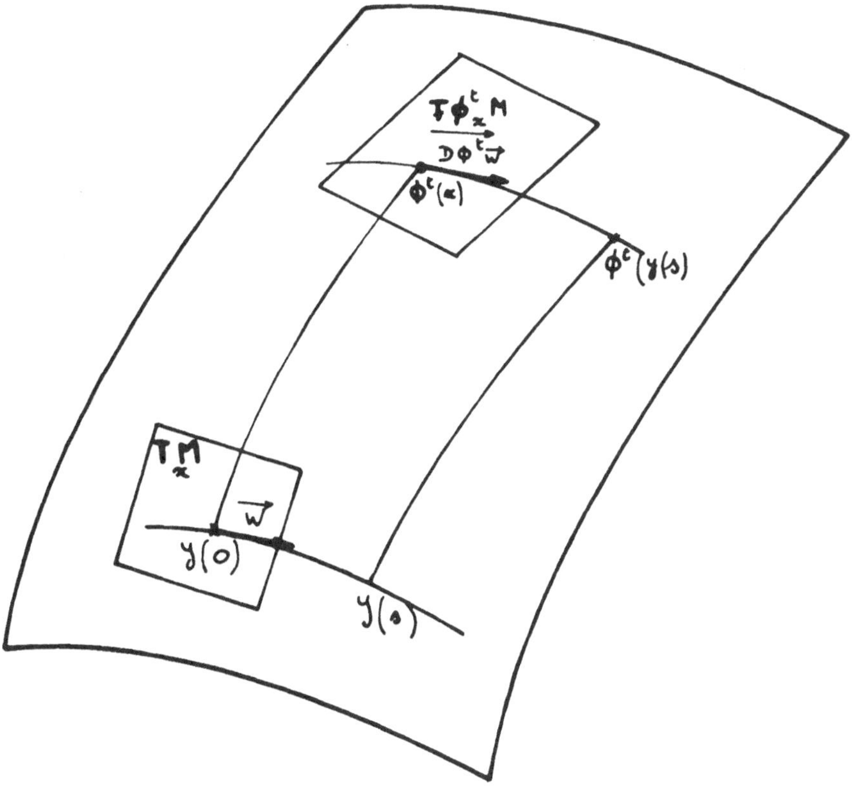

Fig. 1 : Divergence of nearby orbits.

$D\,\phi_x^t$ is the linear tangent mapping which maps T_xM into $T_{\phi^t(x)}M$.

The composite mapping theorem reads

$$D\,\phi_x^{t+t'} = D\,\phi^{t'}_{\phi^t(x)} \circ D\,\phi_x^t \tag{1}$$

The limit of $\psi\,(x,y,s,t)$ when $s \to 0$ is given by $||D\,\phi_x^t\,w||\,/\,||\,w\,||$ and we then say that we have exponential divergence of nearby orbits if the limit

$$\lim_{t\to\infty} \frac{1}{t}\,\ell n\,\frac{||D\,\phi_x^t\,w||}{||\,w\,||} = \chi(x,w) \tag{2}$$

exists and is positive. This number $\chi(x,w)$ is called the Lyapunov characteristic exponent of the flow ϕ^t relative to x and w. (M being compact all equivalent norms give the same limit).

In the particular case of a periodic orbit of period , $D\,\phi_x$ is a mapping of T_xM onto itself. Suppose there are n independant eigenvectors $e_1,\dots,e_n$ with eigenvalues $\lambda_1,\dots,\lambda_n$ such that $|\lambda_1|>|\lambda_2|> \dots >|\lambda_n|$. Then

$$||D\,\phi_n^{k\tau}\,(e_i)||\,/\,||e_i|| = e^{\chi_i\,k\,\tau}\,,\quad i = 1,\dots,n$$

where $$\chi_i = \frac{\ell n|\lambda_i|}{\tau}$$

it follows immediately that

$$\chi(x,e_i) = \chi_i$$

Futhermore if, given k such that $1 < k < n$, then for a vector

$$w = a_k\,e_k + \dots + a_n\,e_n \text{ with } a_k \neq 0$$

the k-th component dominates and

$$\chi(x,w) = \chi_k(x) \tag{3}$$

Therefore we will take into consideration the following sequence of linear subspaces $L_1 \supset L_2 \supset \dots \supset L_{n+1}$ of T_xM, defined by

$$L_1 \equiv [e_1,\dots,e_n] = T_xM$$

$$L_2 \equiv [e_2,\dots,e_n]$$

.

.

$\cdot$
$\cdot$

$$L_n \equiv [e_n]$$

$$L_{n+1} \equiv [0]$$

where $[e_k,\ldots,e_n]$ denotes the linear space spanned by the vectors $e_k,\ldots,e_n$. Such a structure of linear space with decreasing dimension each containing the following one is called a filtration. The preceding result can then be written

$$\chi(x,w) = \chi_i(x) \text{ if } w \subset L_i \setminus L_{i+1}$$

(i.e. $w \in L_i$ but $w \notin L_{i+1}$).

The problem is to generalize this construction to non-periodic orbits. To use concepts as eigenvectors and eigenvalues is difficult since $T_{\phi_x^t}M$ is generally different from T_xM. However if one is able to identify naturally the tangent space at different points of M (i.e. in the case of a mapping of R^n) then $D\phi_x^t$ becomes a linear operator on the n-dimensional Euclidian space and its asymptotic spectral properties can be studied.

One idea is to study directly the behaviour of the eigenvalues $\lambda_1^t,\ldots,\lambda_n^t$ of $D\phi_x^t$ in order to see whether $\lim_{t\to\infty} \ln|\lambda_i^t|/t$ exists. This idea is at the basis of numerical computations of Refs. (13,14,15). In the general case the structure of a filtration which arises from the very definition of LCEs, allows one to overcome the difficulty. Actually such a structure follows from a slightly wider definition of LCEs

$$\chi(x,w) = \lim_{t\to\infty} \sup \frac{1}{t} \ln||D\phi_x^t w|| \tag{4}$$

However Oseledec (30) has shown that under weak smoothness condition $\lim_{t\to\infty}\sup$ can be replaced by $\lim_{t\to\infty}$

From the definition we have

$$\chi(x,w+w') < \max\{\chi(x,w),\chi(x,w')\} \tag{5}$$

and $\chi(x,cw) = \chi(x,w)$

for any w and w' belonging to T_xM and $c \neq 0$.

From these relations it is clear that for any real α the set of vector $\{w \in T_xM \; ; \; \chi(x,w) < \alpha\}$ is a linear subspace of T_xM (by convention we put $\chi(x,0) = -\infty$). Therefore as w spans T_xM, the numbers $\chi(x,w)$ take at most n (n = dim M) distinct values $\nu_1,\ldots,\nu_s$ with $1 < s < n$ (s and the ν_i generally depend on x).

We suppose that $\nu_1 > \nu_2 > \ldots > \nu_s$ and we define the linear subspaces $L_i(x)$ by

$$L_i(x) = \{w \in T_x M \ ; \ \chi(x,w) \leqslant \nu_i\} \quad 1 \leqslant i < s$$

and $L_{s+1}(x) = \{0\}$

We have thus a filtration

$$T_x M = L_1(x) \supset L_2(x) \supset \ldots \supset L_{s+1}(x) = \{0\}$$

with the fundamental property

$$\chi(x,w) = \nu_i \quad \text{if } w \quad L_i(x) \setminus L_{i+1}(x)$$

The number $k_i(x) = \dim L_i(x) - \dim L_{i+1}(x)$ is by definition the multiplicity of ν_i and we are thus able, taking for each i, k_i linearly independent vectors in $L_i(x) \setminus L_{i+1}(x)$, to construct a basis $e_1, \ldots, e_n$ of $T_x M$. The corresponding LCEs $\chi_1(x) = \chi(x,e_1)$, ..., $\chi_n(x) = \chi(x,e_n)$ ordered by $\chi_1(x) > \ldots > \chi_n(x)$ constitute the spectrum of the LCEs at x.

It is clear that Eq. (3) holds and if one takes at random a vector w in $T_x M$ then $\chi(x,w) = \chi_1(x)$ i.e. the maximal LCE at x, as it is illustrated on Fig. 2 in the particular case n = 2.

$$\chi(x,w) = \chi_1(x)$$

only if $w' \in L_2(x)$ one has : $\chi(x,w') = \chi_2(w)$.

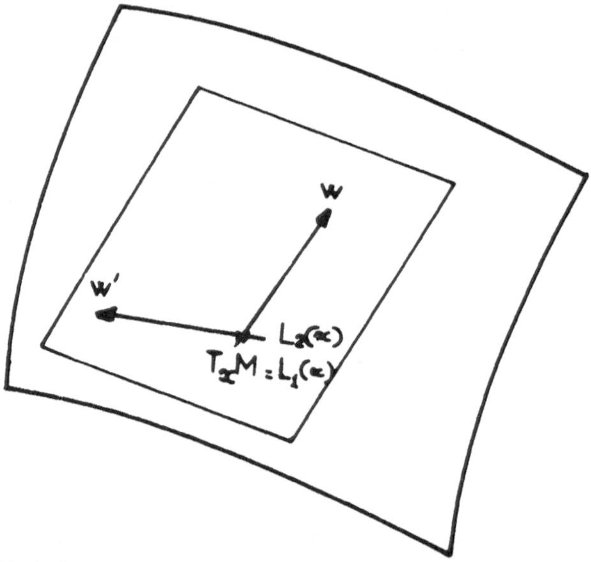

Fig. 2 : Definition of LCEs in the case n = 2 with the corresponding filtration : $L_1(x) = T_x M$ is a 2-dimensional plane and $L_2(x)$ is a line.

2.2. Theoretical results on LCEs

2.2.1. For almost all $x \in M$ and all non zero vectors $w \in T_x M$ the limit

$$\lim_{t\to\infty} t^{-1} \ln ||D\phi_x^t (w)|| = \chi(x,w) \tag{6}$$

exists and is finite (M being compact all equivalent norms give the same limits). From the preceding subsection we have :

$$\lim_{t\to\infty} t^{-1} \ln ||D\phi_x^t (e_i)|| = \chi_i(x) \tag{7}$$

Then for any vector $w = \sum_{i=k}^{n} a_i e_i$ with $a_k \neq 0$

one has :

$$\chi(x,w) = \chi_k(x) \tag{8}$$

Thus for almost all vectors $w \quad T_x M \in \chi(x,w) = \chi_1(x)$

2.2.2. LCEs of order p . Let $(w^1,\ldots,w^p)$ be a system of p linearly independent vectors of $T_x M (1 < p < n)$, and $V^p(w^1,\ldots,w^p)$ the p-dimensional volume of the corresponding parallelepiped (as in § 2.2.1. the metric is irrelevant).
The limit

$$\lim_{t\to\infty} t^{-1} \ln V^p (D\phi_x^t (w^1),\ldots,D\phi_x^t (w^p)) = \chi^p(x,w^1,\ldots,w^p) \tag{9}$$

exists and is finite for almost all $x \in M$. The χ^p are called LCEs of order p (previously defined LCEs were of order 1).

2.2.3. For almost all vectors $w^1,\ldots,w^p$ belonging to $T_x M$ we have

$$\chi^p(x,w^1,\ldots,w^p) = \sum_{i=1}^{p} \chi_i(x) \quad p = 1,\ldots,n \tag{10}$$

This relation, see Ref. (6), is at the basis of the computational technique recalled in Section 3.

2.3. The symplectic and Hamiltonian Cases

Let us consider an Hamiltonian system with N degrees of freedom; then M is defined by the surface of constant energy $H = h$ and $n = \dim(M) = 2N-1$. M is still supposed to be compact.
Then we have in addition

$$\chi_i(x) = -\chi_{n-i+1}(x) \qquad i = 1,\ldots,n \tag{11}$$

The spectrum of LCEs at point x is then

$$\{\chi_1(x),\ldots,\chi_{N-1}(x),\ 0,\ -\chi_{N-1}(x),\ldots,\ -\chi_1(x)\}$$

we have one LCE equal to zero which corresponds to the LCE of a vector tangent to the trajectory.

Usually one does not work on the restriction of ϕ^t to H = h but on the whole 2N dimensional phase space. In such a case the manifold M is taken as the interior of the surface of constant energy h and is of dimension 2N. Theorems analogous to those given in section 2.2. hold of course with n replaced by $\hat{n}= n+1 = 2N$. Indicating the LCEs by $\hat{\chi}_1,\ldots,\hat{\chi}_{2N}$, we have the following spectrum

$$\hat{\chi}_1(x),\ldots,\hat{\chi}_{N-1}(x), 0,\ 0, -\hat{\chi}_{N-1}(x),\ldots,\ -\hat{\chi}_1(x)$$

Moreover one has $\hat{\chi}_i(x) = \chi_i(x)$ for $i = 1,\ldots,N-1$

If in addition to the energy integral there exists p additional uniform (isolating) integrals then 2p LCEs vanish. Therefore LCEs can be used to determine the dimension of manifold, embedded in phase space, on which the orbit lies. Let us notice that for non Hamiltonian systems the LCEs allow us to define the Lyapunov dimension of an attractor (9).

REMARK 1 : The fundamental formula (11) holds as well for symplectic diffeomorphisms which are the counter part of Hamiltonian systems for discrete dynamical systems.

REMARK 2 : The relation :

$$\sum_{i=1}^{n} \chi_i(x) = 0 \tag{12}$$

is trivially verified in the case of Hamiltonian systems by using Liouville theorem and Eq. (10).

2.4. LCEs as indicators of ergodicity

The LCEs $\chi_i(x)$ $(i = 1,\ldots,n)$ are integrals of motion i.e. for almost all x we have

$$\chi_i(\phi^t(x)) = \chi_i(x)$$

which is a direct consequence of the definition of the LCEs. Indeed $\chi(x,w)$ depends only on the orbit and on the direction of w i.e.

$$\chi(\phi^t(x),\ D\phi^t(cw)) = \chi(x,w) \text{ for any } t \text{ and } c \neq 0.$$

As a consequence if the dynamical system is ergodic, the functions $\chi_i(x)$ are almost everywhere constant on M. Conversely, if it occurs that some of the $\chi_i(x)$ are not almost everywhere constant on M then the system is certainly not ergodic.

2.5. LCEs and Kolmogorov entropy

We first recall the definition of metric entropy.

Let us consider a given measurable partition of M denoted by $P = \{A_1,\ldots,A_n\}$. We let each cell A_i evolves backwards under the flow and obtain the partition $\phi^{-t}(P) = \{\phi^{-t}(A_1),\ldots,\phi^{-t}(A_n)\}$.

For any two measurable partitions $P = \{A_1,\ldots,A_\ell\}$ and $Q = \{B_1,\ldots,B_m\}$ we define their composition PVQ as the partition $\{A_i \cap B_j\}$ $\begin{matrix} 1 < i < \ell \\ 1 < j < m \end{matrix}$

Let us restrict t to the natural numbers. For any initial partition P and any t we consider the partition

$$P_t = PV\,\phi^{-1}(P)\ V \ldots V\ \phi^{-t}(P)$$

In order to exhibit the exponential decrease of the measure μ of an element C_i of P_t, we consider the quantity :

$$h(P) = -\lim_{t\to\infty} \frac{1}{t} \sum_{i=1}^{k_t} \mu(C_i)\ \ell n\ (\mu(C_i)).$$

where k_t denotes the the number of elements C_i of P_t. The quantity $h(P)$ has the significance of a mean exponential rate.

The Kolmogorov entropy of the dynamical system (M, μ, ϕ^t) is defined as

$$h = \sup h(P)$$

where the sup is taken over all measurable partition of M.

Since the Kolmogorov entropy is positive only when the average measure of an element of P_t decreases exponentially (going backwards in time) it is not surprising to find that entropy is related to exponential divergence of nearby trajectories (going forwards in time). Pesin's formula gives the precise connection between Kolmogorov entropy and LCEs which states that under suitable smoothness conditions one has :

$$h = \int_M \left[\sum_{i=1}^{N} \chi_i(x)\right] d\mu(x) \tag{13}$$

where the sum $\sum_{i=1}^{N}$ is extended over all positive LCEs. In particular, if the system is ergodic $h = \sum_{i=1}^{N} \chi_i$ where χ_i is the (almost everywhere constant) value of $\chi_i(x)$.

3. NUMERICAL TECHNIQUES TO COMPUTE LCEs

Let us consider the typical case in which the flow ϕ^t is defined by an autonomous first order system

$$\dot{x} = f(x) \tag{14}$$

with $x = (x_1,\ldots,x_n)$ U open set of R^n. We identify all tangent spaces with R^n through the coordinates $x_1,\ldots,x_n$. A tangent vector w evolves satisfying the variational equation

$$\dot{w} = A(x(t))w \tag{15}$$

where $x(t) = \phi^t(x)$ is a solution of Eq. (14) and A(x) is the Jacobian matrix $A(x) = (\frac{\partial f}{\partial x})$. In principle the LCEs of any order k could be obtained by choosing randomly k vectors in R^n, integrating Eqs. (14) and (15). Then taking $k = 1,\ldots,n$ Eqs. (9) and (10) give the LCEs. But practically this is not possible because in general in the stochastic region, the vectors become too large and the angles between their directions too small to allow a numerical computation of volumes. The following procedures overcome these difficulties :

3.1. Calculation of the largest LCE

The computation of the maximum exponent χ_1 has been used extensively as a test of stochasticity (1,8,10,12).

Naive application and integrations of Eqs. (2), (14) and (15) lead after a sufficiently large time to a computer overflow as the norm of w increases exponentially with t. The difficulty is overcome making use of the linearity of Eq. (15).

Having taken an initial vector w of norm 1, one renormalizes the evolved vectors at arbitrary time $j\tau$ $(j = 1, 2, \ldots)$, then we put :

$$\chi(x,\ell) = \frac{1}{\ell\tau}\sum_{j=1}^{\ell} \ln \alpha_j$$ where α_j is the

renormalizing factor at time $j\tau$. Then we obtain

$$\lim_{\ell\to\infty} \chi(x,\ell) = \chi_1(x)$$

Actually integration of Eq. (15) is not necessary. As shown on Fig. 3 let us consider two orbits starting at P_o and P'_o. Denote by d the distance dist $(P_o\ P'_o)$. After a time τ, P_o is in P_1 and P'_o in P'_1. If d_1 = dist $(P_1\ P'_o)$ then by an homothesis of center P_1 and of rate d/d_1 we get two new starting points $P_1\ P''_1$ such that dist $(P_1\ P''_1)=d$ and iterate the processus as suggested on Fig. 3. It is shown in (3) that the quantity

$$\gamma(P_o,\ d,\ \tau,\ n) = \frac{1}{n\ \tau}\sum_{i=1}^{n} \ln \frac{d_i}{d}$$

goes to $\chi_1(P_o)$ as n tends to infinity.
In the case of a mapping F :

$$X_{n+1} = F(X_n) \tag{16}$$

the problem reduces to compute the natural logarithm of eigenvalues of the Jacobian matrix related to the following equations :

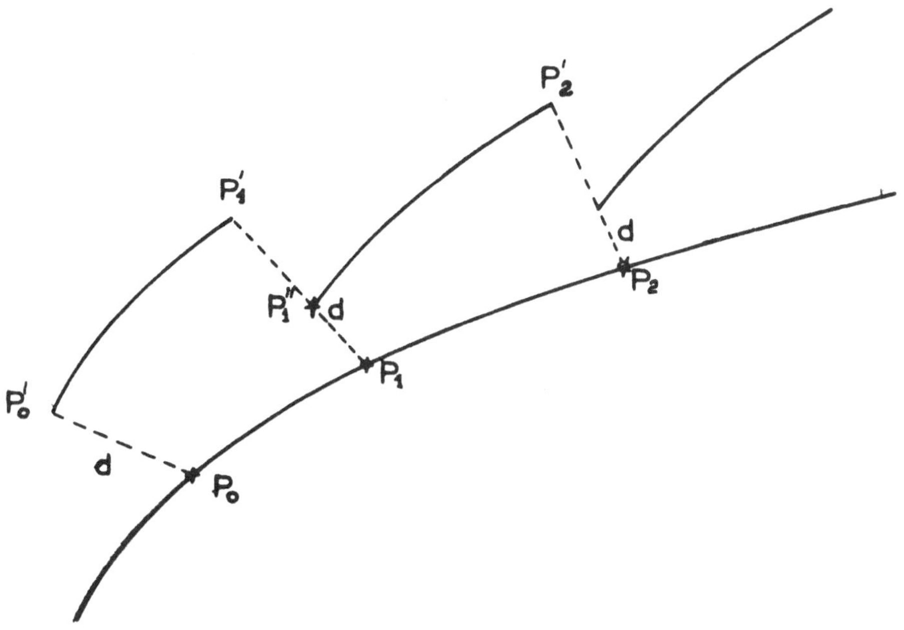

Fig. 3 : Numerical computation of the maximal Lyapunov characteristic exponent.

$$X_n = F^n(X_o) \tag{17}$$

$$Y_n = [DF^n(X_o)]\ Y_o \tag{18}$$

where $DF^n(X_o)$ is obtained through the composite mapping theorem

$$DF^n(X_o) = DF(X_{n-1}) \circ DF(X_o)$$

As in the previous case, overflow arises for elements of the Jacobian matrix. This difficulty is overcome by the same trick (12) : one takes into account the linearity of Eq. (18) and divides at regular intervals the Jacobian matrix by a given scaling factor.

3.2. Computation of all LCEs

Using the definition of LCEs of order p and relation (10) Benettin et al. (6) gave a simple procedure to compute all the LCEs of order one. In this case a further difficulty occurs, due to the fact that when at least two vectors are involved the angles between their directions may become too small for numerical computations. Benettin et al. (6) overcome this difficulty by extending the trick described in subsection (3.1.). They notice first, that an invertible linear mapping as $D\phi^t_x$ maps p-dimensional linear subspaces onto linear subspaces of the same dimension; and that the "rate of growth" of any p-dimensional volume under the action of any such linear mapping is an intrinsic quantity of the subspace involved. Thus given $w^1,\dots w^p$ orthonormal, one can replace at regular time intervals τ the p evolved tangent vectors by a set of new orthonormal vectors, using the Gram-Schmidt procedure, spanning the same p-dimensional subspace.

Let us denote $v^i_o = w^i, \quad i = 1,\dots,p$

One defines and computes recursively

$$\tilde{v}^i_k = D\phi^{\tau}_{\phi(k-1)\tau\,(x)}(v^i_{k-1})$$

$$\alpha^i_k = ||(\tilde{v}^i_k)_{\perp}||,$$

$$v^i_k = (\tilde{v}^i_k)\;/\alpha^i_k$$

where $(\tilde{v}^i_k)_{\perp}$ is the component of $\tilde{v}^i_k$ orthogonal to all the (already orthonormal) v^j_k with $j < i$ i.e.,

$$(\tilde{v}^i_k)_{\perp} = \tilde{v}^i_k\ ,\ i = 1,$$

$$(\tilde{v}^i_k)_{\perp} = \tilde{v}^i_k - \sum_{j=1}^{k-1} < v^j_k\ ,\ \tilde{v}^i_k > v^j_k \qquad i > 1$$

where < > is the Euclidean scalar product.
Then using Eq. (10) for $p = 1,\dots,n$ and the linearity of $D\phi^t_x$ one has

$$\gamma_i(x,t) = \frac{1}{N\tau}\sum_{k=1}^{N} \ln \alpha^i_k$$

with $t = N\tau$

and $\chi_i(x) = \lim_{N\to\infty} \gamma_i(x,N)$

Even for mappings where other method (15) allows the computation of eigenvalues of Eq. (18) (again an iterative procedure is necessary to overcome numerical difficulties) the Benettin et al. method appears to be the simplest one to handle.
Last but not least a major difficulty still exists (for any method) in numerically evaluating the LCEs: there is no a priori condition for determining the number of iterations that must be used. Thus other techniques like surface of section plots may be useful to clarify the basic mechanism of the chaotic motion in each case. However for problems with more than three degrees of freedom LCEs computations are almost the only tools for estimating stochasticity.

4. NUMERICAL COMPUTATIONS OF LCEs IN THE CONTEXT OF CELESTIAL MECHANICS.

Many numerical experiments have been performed to compute LCEs especially in the context of Physics (4,5,8,13,15,26). Besides the problem of measuring the stochasticity of orbits and Kolmogorov entropy, LCEs are also interesting to measure diffusion speed in the phase space and to give an estimate of the most basic property of an attractor, i.e. its dimension (9). Of course in the case of conservative dynamical systems there are no attractors but the fundamental problem is to determine the dimension of the manifold on which the motion takes place in phase space, i.e. the number of isolating integrals. Of course the numerical works described in the following chapters do not pretend to exhaustivity but just give examples how LCEs have been used for studying Hamiltonian systems, especially in celestial mechanics. Non conservative systems will not be treated in this paper.

4.1. LECs as tools for fundamental problems

In a previous paper Froeschlé (11) studied a sample of orbits parametrized by the initial osculating semi-major axis $a = 0.235749(d_o)^{1/2}$ (d_o being a free parameter) of the three dimensional three body problem (11). Using three different methods, stereoscopic view, slice cutting and local fitting, Froeschlé (11) tried to determine if other isolating integrals exist besides the so-called Jacobi integral, i.e. the dimension of the manifold on which the motion takes place in the six dimensional phase space. Using the LCEs, R. Gonczi and C. Froeschlé (22) studied the same problem.

Fig. (4) shows, in logarithmic scales the typical behaviour of the three positive γ_i (P_0, t) as a function of time for an initial point P_0 lying in an integrable region. As the system is Hamiltonian the other three are opposite to the first three ones. The three γ_i (P_0,t) appear to be linearly decreasing functions of time varying roughly as t^{-1} and the largest one shows very small fluctuations with respect to the expected linear behaviour. The dotted line shows the variations of $\gamma(P_0, d, \tau, n)$ with $t = n\tau$, $\tau = 5$ units of time for two initially close orbits starting at P_0 and P'_0 such that :

$d = \text{dist}(P_0\, P'_0) = 10^{-3}$. The agreement between $\gamma_1(P_0, t)$ and $\gamma(P_0, d, \tau, n)$ is quite good.

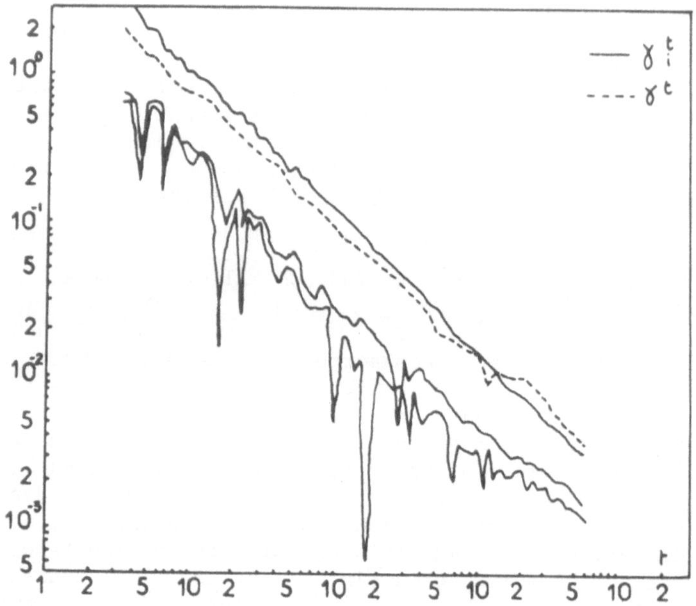

Fig. 4 : Variation of the $\gamma_i^t(P_0)$ (whose limits are identified with the LCEs) as functions of the time t. The initial point P_0 lies in an integrable region $d_0 = 0.15$). The dotted line shows the variation of $\gamma^t(P_0)$ whose limit as $t \to \infty$ is the largest LCE computed by the method of the divergence of nearby orbits.

All the other integrable cases, $d_0 < 0.25$ show exactly the same behaviour.

The same curves for a point P_0 in the stochastic region (namely with $d_0 = 0.3$) are shown in Fig. 5. We found that $\gamma_i(P_0, t\ max)$ satisfies $\gamma_i(P_0, t\ max) = -\gamma_{7-i}(P_0, t\ max)$ up to the third digit, suggesting that the approach to the limit $t \to \infty$ is good. Again $\gamma_3(P_0, t)$ seems to tend linearly to zero, in agreement with the existence of the Jacobi integral. We note rather large fluctuations of this curve (at least on a logarithmic scale). However, the two other positive exponents seem to have strictly positive limits, and a good estimation of these limits can be found already at t = 1000. Therefore, the systematic exploration as a function of the parameter d_0 will be performed only until $t_{max} = 1000$.

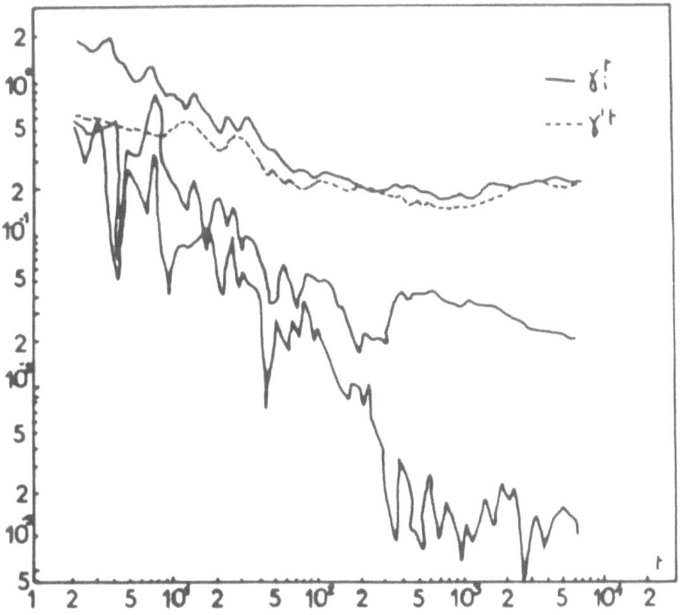

Fig. 5 : Same curves as in Fig. 4 but for an initial point P'_0 in a stochastic region ($d_0 = 0.3$).

Again in Fig. (5) the computation of the largest LCE using nearby orbits (dotted line) is in good agreement with the previous results as well as the same calculation with the regularized orbits.

In Fig. (6) the different values of γ_i (P, 1000) are shown when d_0 takes values from 0.1 to 0.5. When P is in the integrable region ($d_0 < 0.25$) we see that the three numbers remain small. After that value we observe a threshold around $d_0 = 0.25$ where two isolating integrals disappear : this is expressed by a sharp increase of γ_1 and γ_2. After that these two numbers remain constant. This suggest that the considered orbits belong to a connected stochastic zone.

The LCEs appear to be good estimators of stochasticity and allow one to perform not only a qualitative but also a quantitative study of a given set of orbits whose qualitative behaviour has already been studied by Froeschlé (11). Concerning the number of isolating integrals the quasi simultaneous disappearance of two of them is observed and thus confirm the results previously obtained by Froeschlé. Already with three degrees of freedom the method appears very interesting compared with other methods. Of course the method can be and has been applied for systems with more than three degrees of freedom (5).

Studying the planar restricted three body problem, W.M. Jefferys and Zhao-Hua Yi (28) have computed LCEs with various mass ratios μ and Jacobi constant C for various cases of satellite and asteroidal motions. Defining as "stability region" any region of the phase space such that every orbit within the region has all LCEs equal to zero, they have found that there exists a maximum value C* of C depending on μ such that no unstable orbit are found when C > C*. Of course for such problems with two degrees of freedom the method of surface of section can be used to produce plots of the stability region (27). Hence the computation of LCEs is less essential.

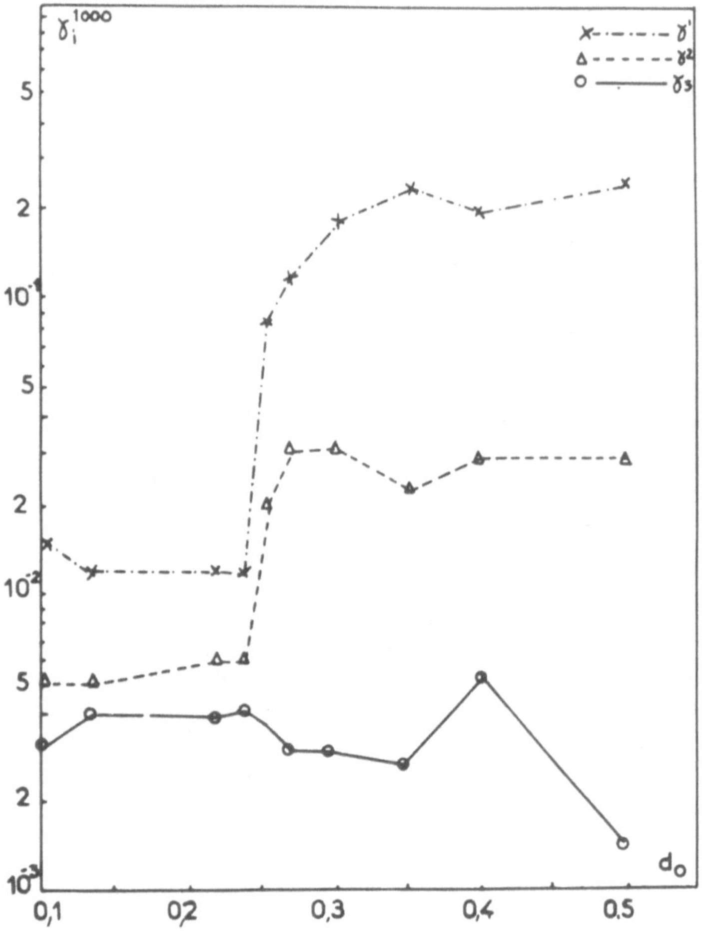

Fig 6 : Variation of $\gamma_i^{1000}(P_0)$ (whose values estimate the LCEs as a function of d_0.

G. Benettin et al. (5) have studied the variation of the Kolmogorov entropy of a one-dimensional self-graviting system consisting of N plane parallel sheets with uniform density when N increases. This model which is of astrophysical interest (It can be considered as the simplest simulation of a system of stars. The stars are stratified into plane parallel layers and all parameters vary only in one dimension), had been previously studied by Froeschlé and Scheidecker (16). They found using as a criterium of stochasticity linear or exponential divergence of nearby orbits that, with increasing N, the relative measure μ_S of the stochastic region tends rapidly to one.

The system is described by the Hamiltonian

$$H(x,u) = \frac{1}{2} m \sum_{i=1}^{N} u_i^2 + 2\pi G m^2 \sum_{\substack{i,j=1 \\ i > j}}^{N} |x_j - x_i|,$$

From the particular form of the potential it follows that the motion on any surface of constant energy reproduces on scale the motion on any other surface of constant energy. Namely if M_E is the surface of constant energy, which turns out to be compact, and if ϕ^t denote the Hamiltonian flow, the mapping $\psi_\lambda : M_E \to M_{\lambda^2 E}$ given by $\psi_\lambda (x,u) = (\lambda^2 x, \lambda u)$ satisfies: $\phi^{\lambda t} \circ \psi_\lambda = \psi_\lambda \circ \phi^t$.
futhermore it can be shown that the Kolmogorov entropy h(E) is related to $h(\lambda^2 E)$ by : $h(E) = \lambda\, h(\lambda^2 E)$.

Table I shows the variation of Kolmogorov entropy h with the number N of sheets. At fixed specific energy $\frac{E}{N}$ Kolmogorov entropy turns to be a linear function of N. Other systems instead seem to behave completely differently. We wrote that our model has two very particular features: (a) the energy dependence is trivial and (b) the so-called "connectance" is maximal i.e. each particle interacts with equal strength with any other particle.

This property is perhaps responsible for the increasing stochasticity with N, but the very regular behaviour which has been found is an open problem.

TABLE I

N	μ_S	h
2	Integrable	0
3	0.04	0.0008
4	0.86	0.034
5	0.99	0.072
6	1	0.108
7	1	0.13
8	1	0.18
10	1	0.24

4.2. Application to asteroidal motion

Many papers have been devoted to the study of asteroidal orbits in the so-called Kirkwood gaps. Using the methods of surface of section, Giffen (21) and later Froeschlé and Scholl (17,18) have studied more particularly a stable but "ergodic" orbit in the 2/1 Kirkwood gap on the basis of Schubart's (31) averaged Sun - Jupiter - Asteroid model which corresponds to an Hamiltonian system with two degrees of freedom. In this paper an orbit is called stable if its semi-major axis oscillates slightly around a constant mean value. In particular it has been numerically shown (18) that the stable but ergodic orbit under study is trapped inside a bottle shaped space which is circumscribed by a first integral of motion. The top of the bottle being closed by invariant curves (see Fig. (7)) which prevents the "ergodic" orbit from leaving the gap, and thus gives to the orbit the appearance of being stable. For both an integrable orbit and an ergodic one, computations of the maximum LCE are performed. Up to t = 8000 years it is found the surprising result that for both cases the maximum LCE seems to decrease linearly, whereas it is normally expected only in the integrable case. After t = 8000 years the "ergodic" case seems to separate from the integrable case and might approach a limit which is very small compared to limits obtained in Fig. (6).

Many explanations could be given for the low value of $\gamma_1(P_0, T)$ with T = 20 000 years for this quasi-ergodic orbit.

4.2.1. This orbit is trapped in a narrow torus which prevents a wide spread in the phase space. Such cases are well known from the study of two-dimensional area preserving mappings (24).

4.2.2. The orbit is integrable and lies on a manifold with many sheets. In this case the approach of $\gamma_1(P_0, T)$ to zero would need a much longer period of time than our calculations covered.

4.2.3. R.S. Mackay et al (29) have shown that even in region without invariant closed curves there are still invariant Cantor sets, named Cantori, which appear to form major obstacles to transport. Such a phenomenon appears also very clearly through the numerical experiments performed on a 4-dimensional symplectic mapping by Froeschlé et al (19). They have shown sudden jumps by factor of 10 of $\gamma_1(P_0, N)$ for the number of iterations N such that N > N critical (N critical ≅ 10^5).

These computations illustrate a major difficulty in numerically evaluating the LCEs : there is no a priori condition to determine the number of iterations N that should be used. Especially with systems with more than two degrees of freedom where the weak mechanism of Arnold diffusion may connect small and large "seas" (1,4,14,26).

An other reason why the value of γ_1 found by Froeschlé and Scholl is so small may be due to the exclusion of high frequency terms in the Hamiltonian of the restricted three bodies problem by an averaging procedure.

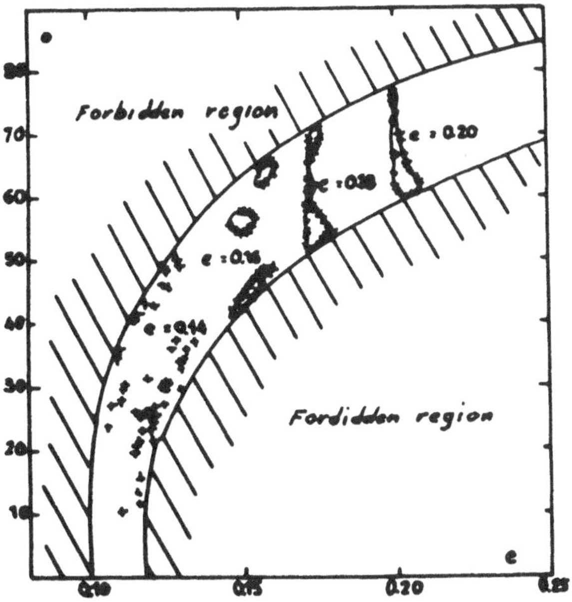

Fig. 7 : Two-dimensional surface of section at the semi-major axis a = 3.27 A.U. on the energy surface $\overline{H}$ = C for different orbits of the average hamiltonian $\overline{H}$ (a,e,σ,ω).

Using both the planar-elliptic problem and a mapping derived by Chirikov (8) procedure, Wisdom (32) has computed maximum characteristic exponents for trajectories near the 3/1 commensurability. This procedure consists essentially to replace the high frequency term by others originated by Fourier expansion of δ-functions. He found about the same value of γ_1 for the "ergodic" orbits but performing a systematic exploration in the 3/1 gap (mainly with the model mapping) the rate of "ergodic" cases appears to be much more larger than the rate found by Froeschlé and Scholl(18). Of course the models are different (an underestimate could arise from averaging procedure and conversely an overestimate from the mapping model) and the time span is much more larger in Wisdom exploration. As a consequence of his results Wisdom has found that the outer boundary of the chaotic zone coincides with the boundary of the 3/1 Kirkwood gap in the actual distribution of asteroids within the errors of the asteroid orbital elements. Problems remain for ejecting mechanism (mars crosser etc...) as stochasticity does not necessarily means escape outside of the gap. For a recent review concerning problems about the dynamical structure of the asteroidal belt see (20).

ACKNOWLEDGEMENTS

It is a pleasure to thank Dr. Ch. Froeschlé and Dr. R. Gonczi for a critical reading of the manuscript. I would like also to express my appreciation to J. Colin, G. Pen and A. Willemse for typing, drawing and photographic works.

REFERENCES

(1) Chaotic behaviour of deterministic systems, Cours des Houches XXXVI, North Holland, 1981.
(2) Arnold V.I., Mathematical methods of classical Mechanics. Translation Springer Verlag, Heidelberg, 1978.
(3) Benettin G., Galgani L. and Strelcyn J.M., 'Kolmogorov Entropy and numerical experiments', Phys. Rev. A 14, p. 2338-2345, 1976.
(4) Benettin G., Strelcyn J.M., 'Numerical experiments on the free motion of a point mass moving in a plane convex region; stochastic transition and Entropy', Phys. Rev. 17, p. 773-785, 1978.
(5) Benettin G., Froeschlé C., Scheidecker J.P., 'Kolmogorov Entropy of Dynamical systems with increasing number of degrees of freedom', Phys. Rev. A 19, p. 2454-2460, 1979.
(6) Benettin G., Galgani L., Giorgilli A., Strelcyn J.M., 'Lyapunov characteristic exponents for smooth Dynamical systems; a method for computing all of them'. Part 1 : Theory; Part 2 : Numerical application, Meccanica March 1980.
(7) Cesari L., Asymptotic Behaviour and Stability Problems in Ordinary differential equations, Springer Verlag Berlin, 1959.
(8) Chirikov B.V., 'An universal instability of many dimensional oscillator systems', Phys. Rep.52, p. 263, 1979.
(9) Farmer J.D., Otte and Yorke J., 'The dimension of chaotic attractors' Physica 7D, p. 153-180, 1983.
(10) Ford J., Fundamental problems in Statistical Mechanics, ed. E.G.D. Cohen, Vol III (North-Holland, Amsterdam) p. 215-255, 1975.
(11) Froeschlé C.,'Numerical Study of Dynamical systems with three degrees of Freedom,'I Graphical Displays of four-dimensional sections',Astron. Astrophys.4, p. 115-128, 1970.,'II Numerical displays of four- dimensional sections', Astron. Astrophys. 5, p. 177-183, 1970.
(12) Froeschlé C., 'A numerical Study of the Stochasticity of Dynamical Systems with two degrees of freedom', Astron. Astrophys. 9, p. 15-23, 1970.
(13) Froeschlé C. and Scheidecker J.P., 'Numerical Study of the Stochasticity of Dynamical Systems with more than two degrees of freedom', Astron. Astrophys.22, p. 431, 1973.
(14) Froeschlé C. and Scheidecker J.P., 'On the disappearance of isolating integrals in systems with more than two degrees of freedom', Astrophys. and Space Sc.25, p. 273, 1973.

(15) Froeschlé C. and Scheidecker J.P., 'Numerical study of the stochasticity of Dynamical System with more than two degrees of freedom', J. Comp. Phys. Vol. 11, n° 3, 1973
(16) Froeschlé C., Scheidecker J.P., 'Stochasticity of dynamical systems with increasing number of degrees of freedom', Phys. Rev. A 12, p. 2137-2143, 1975.
(17) Froeschlé C. and Scholl H., 'On the dynamical topology of the Kirkwood gaps', Astron. Astrophys., 48, p. 389-393, 1976.
(18) Froeschlé C. and Scholl H., 'The stochasticity of pecular orbits in the 2/1 Kirkwood gap', Astron.Astrophs. 93, 62-66, 1981.
(19) Froeschlé C., Gonczi R., 'Lyapunov characteristic numbers and Kolmogorov entropy of a four dimensional Mapping', Nuovo Cimento, vol 55 B, n° 1, p. 59-69, 1980.
(20) Froeschlé C. and Scholl H., 'The dynamical structure of the asteroidal belt (review paper)' Proceeding of Uppsala meeting, p. 115-125, 1983.
(21) Giffen R., 'A study of commensurable motion in the asteroid belt', Astron. Astrophys. 23, p. 387-403, 1973.
(22) Gonczi R., Froeschlé C., 'The Lyapunov characteristic exponents as indicators of stochasticity in the restricted Three-body Problem', Cel. Mech. 25, p. 271-280, 1981.
(23) Guikenheimer J., Moser J., Newhouse S., Dynamical Systems, CIME lectures Birkhauser, 1978.
(24) Hénon M. and Heiles C., 'The applicatibility of the third integral of motion, some numerical experiments', Astron. Journal 69, p. 73-79, 1964.
(25) Hénon M.,'Exploration numérique du problème restreint', I Ann. Astr. 28, p. 499, 1965. II Ann. Astr. 28, p. 992, 1965. III Bull. Astr. Paris, 1, p. 57, 1966. IV Bull. Astr. Paris, 1, fasc. 2, p. 49, 1966.
(26) Hénon M. and Wisdom J., 'The Benettin-Strelcyn oval Billiard revisited' Physica 8D, p. 157-169, 1983.
(27) Jefferys W.H., 'An atlas of surfaces of section for the restricted problem of the three bodies', Publications of the department of Astronomy of the University of Texas at Austin, ser. II., 3, 6, 1971.
(28) Jefferys W.H., and Zhao-Hua Yi, 'Stability in the restricted problem of three bodies with Lyapunov characteristic numbers', Cel. Mech. 30, p. 85-95, 1983.
(29) MacKay R.S., Meiss J.D. and Percival I.C., 'Transport in Hamiltonian system', preprint.
(30) Oseledec V.I., 'A multiplicative ergodic theorem. The Lyapunov characteristic numbers of Dynamical systems (in Russian)', Trudy Mosk. Mat. Obsc. 19, p. 179-210, 1968. English translation in Trans. Mosc. Math. Soc. 19, p. 197, 1968.
(31) Schubart J., 'Long-period effects in nearly commensurable cases of the restricted three-body problem in Smithsonian Astrophys'. Obs. Spec. Rep. n° 149, 1964.
(32) Wisdom J. 'Chaotic behaviour and the origin of 3/1 Kirkwood gap', Icarus 56, p. 51-74, 1983.
(33) Szebehely V. Theory of orbits , Academic Press, 1977.

KOLMOGOROV ENTROPY AS A MEASURE OF DISORDER IN SOME NON-INTEGRABLE HAMILTONIAN SYSTEMS

R. Gonczi[x+], Ch. Froeschlé[x], Cl. Froeschlé[x]

x+ Laboratoire de Physique Théorique, Faculté des Sciences, Parc Valrose - 06034 Nice Cedex (France)
x Observatoire de Nice, B.P. 139 - 06003 Nice Cedex (France)

ABSTRACT. This paper is devoted to study the stochastic behaviour of some Hamiltonian systems with closed velocity curves. We investigate Hamiltonians already studied by Ali and Somorjai (1). These authors, by discussing Poincaré's surfaces of section for several energy values, gave a qualitative evaluation of the stochasticity of the systems.

Here we present a quantitative study of this stochastic behaviour. For each energy we compute the Lyapunov characteristic exponents of fifty orbits chosen at random, in order to calculate the Kolmogorov entropy by Pesin's formula. Our results are in agreement with those of Ali and Somorjai : the disorder does not increase monotonically with increasing energy. However, we find that the largest entropy does not necessarily correspond to the maximum of the stochastic volume. The Kolmogorov entropy thus appears to be a good measure of the degree of disorder of dynamical systems.

1. INTRODUCTION

Since the pioneer work of Hénon and Heiles (6) many numerical studies of non integrable Hamiltonian systems have already been done (4). In particular it was shown that the degree of stochasticity does not always increase monotonically with the energy : Ali and Somorjai (1) extensively studied a system of two one-dimensional (non linear) oscillators harmonically coupled . Using the Poincaré's surface for different values of energy E, the authors showed that the stochasticity increases with E until a critical value E_c ; then for still larger energies the stochasticity slowly decreases and again ordered motion appears. Qualitative results of the stochastic behaviour were presented by plotting the Poincaré's surfaces of section.

In the present paper we study the same dynamical systems, but try to give a more quantitative evaluation of the degree of stochasticity. Therefore for each system we calculate the Lyapunov characteristic exponents of many orbits at several energies. We then evaluate the Kolmogorov entropy, not only of the whole phase space, but also of the "stochastic" region only.

Celestial Mechanics **34** (1984) 117–124. 0008–8714/84.15

In section 2 we briefly recall the model and the results previously obtained by Ali and Somorjai (1). Section 3 is devoted to present our calculations and in section 4 we discuss our results and compare them to the previous ones.

2. THE MODEL AND PREVIOUS RESULTS

Ali and Somorjai (1) investigated dynamical systems with closed velocity curves. They considered a system of two one-dimensional (non linear) oscillators harmonically coupled. The Hamiltonian is given by :

$$H = \frac{1}{2}(\dot{x}^2 + \dot{y}^2) + V_1(x) + V_2(y) + xy$$

where in case A :

$$V_1(x) = (\frac{x^2}{2} - 1)^2 + \frac{x^2}{2}$$

$$V_2(y) = \frac{1}{2}y^2$$

and in case B :

$$V_1(x) = (\frac{x^2}{2} - 1)^2 + \frac{x^2}{2}$$

$$V_2(y) = (\frac{y^2}{2} - 1)^2 + \frac{y^2}{2}$$

They solved the equations of motion and computed 15-25 different orbits for several values of the energy E. As surfaces of section they have taken both the plane y, $\dot{y}$ ($x = 0$, $\dot{x} > 0$) and the plane x, $\dot{x}$ ($y=0$, $\dot{y} > 0$). By analysing qualitatively these surfaces of section for each value of E, the authors studied the evolution of the degree of stochasticity of the system. They concluded that this degree of stochasticity does not increase monotonically : the maximum ratio of irregular orbits is obtained for $E = 4.2$ in case A and for $E = 3.5$ in case B. In both cases, regular orbits reappear for larger energies. In case B for instance, the whole phase space looks again integrable for $E = 5000$. The authors concluded that the relative contribution of the coupling term to the total energy of the system decreases with increasing energy.

3. COMPUTATION OF LYAPUNOV CHARACTERISTIC EXPONENTS (L.C.E.) AND KOLMOGOROV ENTROPY

In this section we give a quantitative evaluation of the stochastic behaviour of the Hamiltonian systems studied by Ali and Somorjai.

3.1. Choice of initial conditions

Given an upper bound H_o of the energy, we determine the minimum and maximum values of x, $\dot{x}$, y, $\dot{y}$ using the peculiar form of the Hamiltonian, i.e. the zero velocity curves are closed. In case A we get the box :

$$x_{max} = - x_{min} = \sqrt{2(1+\sqrt{H_o})}$$

$$y_{max} = - y_{min} = \sqrt{2(1+\sqrt{H_o})}\ (2H_o - 1)$$

$$\dot{x}_{max} = \dot{y}_{max} = - \dot{x}_{min} = -\dot{y}_{min} = \sqrt{2H_o - 2}$$

In case B :

$$x_{max} = y_{max} = - x_{min} = -y_{min} = \sqrt{1 + \sqrt{4H_o - 7}}$$

$$\dot{x}_{max} = \dot{y}_{max} = - \dot{x}_{min} = -y_{min} = \sqrt{2H_o}$$

Then in such a box we choose at random many sets of initial conditions $(x_o, y_o, \dot{x}_o, \dot{y}_o)$ and among them we select 50 orbits which energy E_o is close to a fixed value E, That is : $E \leqslant E_o \leqslant E + \Delta E$ with $\frac{\Delta E}{E} \cong 0.5\%$. Each orbit is then numerically integrated using a Burlish - Stoer (3) method, up to a time t_{max} determined as explained in section 3.2.

3.2. Computation of the L.C.E.'s

Following the general computational method of Benettin et al (2) we calculate the L.C.E.'s of each orbit.

It is known that for an Hamiltonian system with N degrees of freedom, an orbit starting at point P is associated with 2N numbers called L.C.E.'s :

$$\chi_1(P) \geqslant \chi_2(P) \geqslant \dots \geqslant \chi_{2N}(P)$$

where by definition :

$$\chi_i(P) = \lim_{t\to\infty} \gamma_i^t(P)$$

(for the definition of γ_i see (5)).
These numbers verify :

$$\chi_i(P) = - \chi_{2N-i+1}(P).$$

Furthermore, due to the existence of the energy integral, two L.C.E.'s vanish; then in our case (N=2) the computation of only the largest L.C.E. is relevant. Nevertheless the other L.C.E.'s have also been calculated for testing the accuracy of the computational method, using

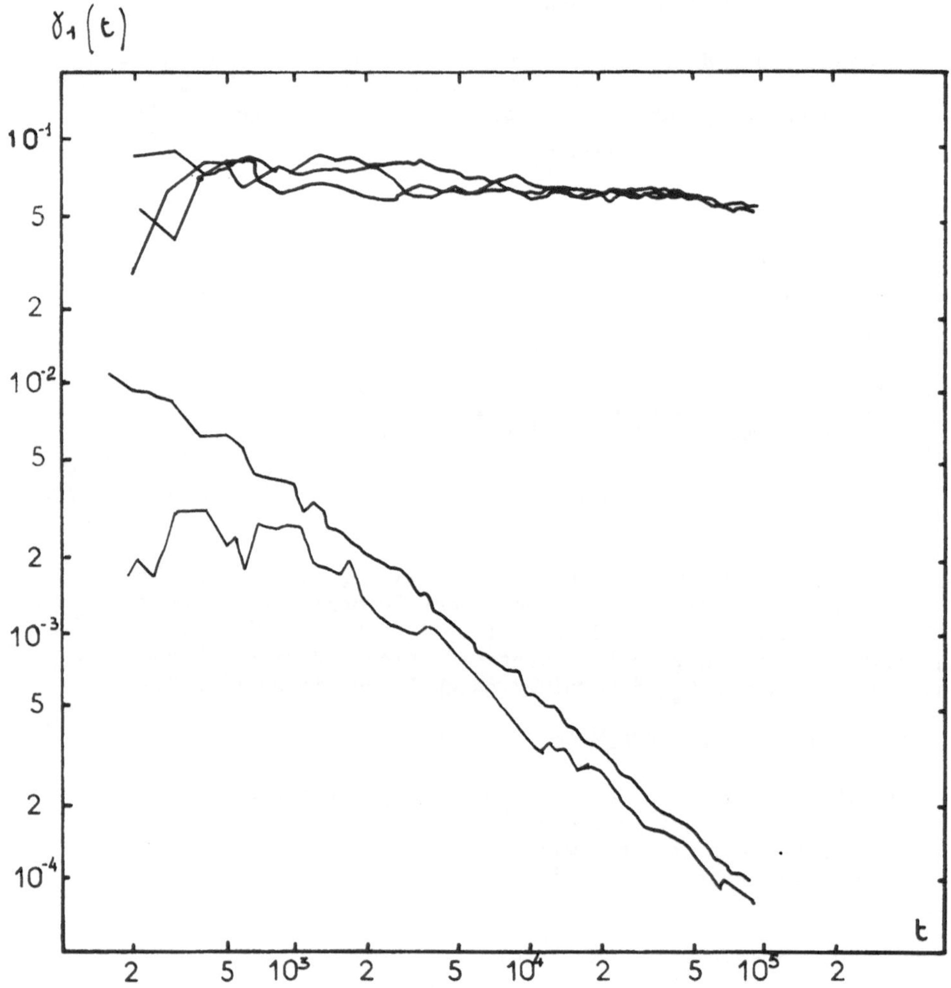

Fig. 1 : Variation of $\gamma_1(t)$ (whose limit is identified with the largest LCN) for 5 randomly chosen orbits.

the above relation.

Figure 1 shows the behaviour of the $\gamma_1^t(P)$ for five different orbits. Clearly for 3 of them γ_1^t appears to converge to a strictly positive value, while the 2 others display the characteristic behaviour of integrable orbits. Already at $t \cong 10^3$ one can easily discriminate between the two kinds of orbital behaviour. Furthermore at this time good approximations of the limits are obtained : L.C.E.'s of stochastic orbits are of opposite sign and constant up to the third digit, and those corresponding to integrable orbits are also close to zero with the same accuracy. Finally, as we integrate a lot of orbits, we make a compromise with economical constraints and decide to stop all integrations at $t_{max} = 1200$ units of time.

3.3. Kolmogorov entropy

Pesin's formula (7) gives the relation between the L.C.E.'s and the Kolmogorov entropy of a dynamical system. Let $\rho(P)$ denote the sum of all positive L.C.E.'s at point P. In our case one has :

$$\rho(P) = \chi_1(P)$$

The formula states that the total entropy is given by :

$$\alpha = \int_M \rho(P)\, d\mu$$

where M denotes the phase space, μ a normalized measure on it and P a point of M.

The entropy is mainly determined by the points P of the stochastic region since in the integrable one the L.C.E.'s vanish. We have thus estimated α by :

$$\alpha \cong \frac{1}{n} \sum_{i=1}^{n} \chi_1(P_i)$$

Where n = 50 is the number of random points P_i. The entropy of the stochastic region has also been estimated by the same formula, where only the stochastic orbits are taken into account.

4. DISCUSSION OF RESULTS

Figs.2a and 2b show, for both cases A and B, the Kolmogorov entropy α and the percentage q of irregular orbits as functions of the energy E.

The percentage q gives a measure of the relative volume of the stochastic zone, which is closely connected to the stochastic surfaces analysed by Ali and Somorjai. Therefore it is not surprising to find in both cases the same qualitative behaviour : very low value of q at small energies ($E \lesssim 1.$), sharp increase until a maximum value, then for larger energies ($E \gtrsim 100.$) reappearance of ordered motion since q clearly decreases.

Concerning the entropy α, the curves show in both cases a bump for

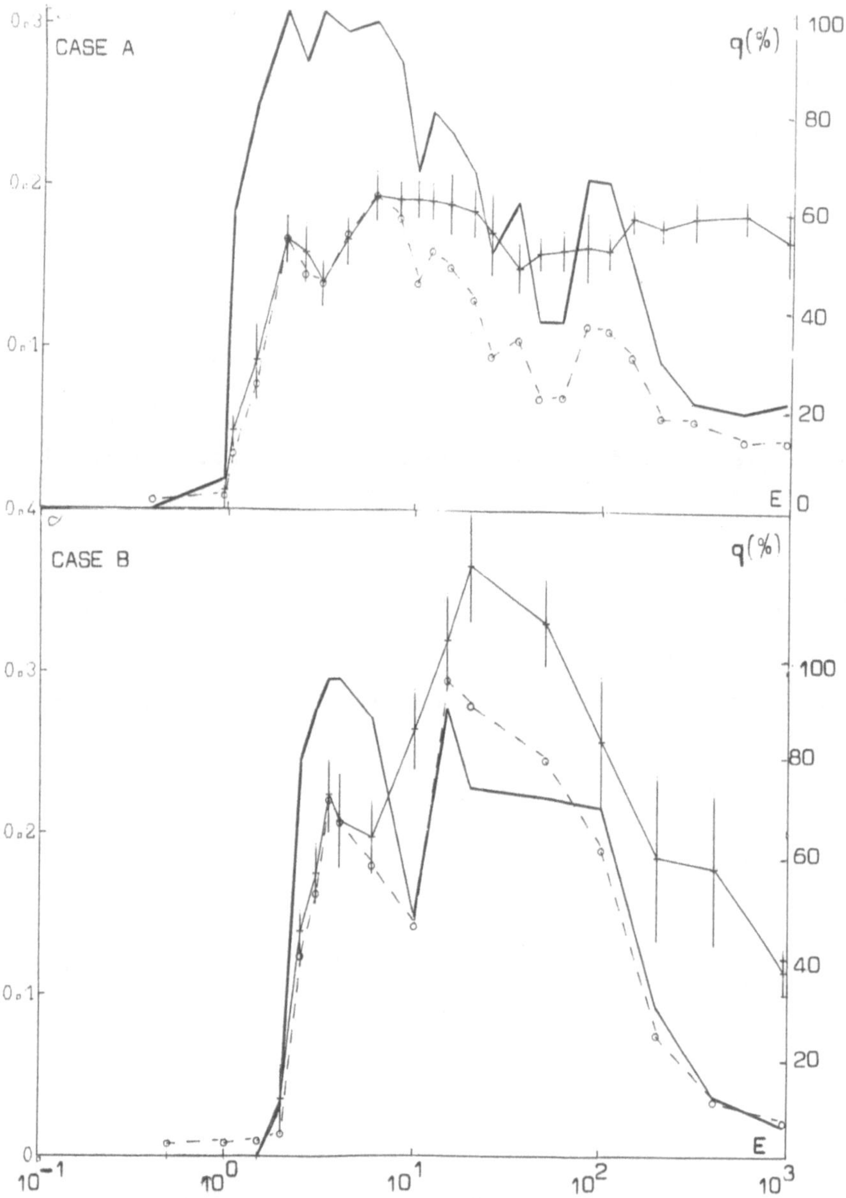

Fig. 2 : Variation, as a function of the energy E, of :
▬ : percentage q of irregular orbits
o --- o : estimation of the Kolmogorov entropy α of the full phase space
|—| : estimation of the Kolmogorov entropy α_{st} of the stochastic region
error bars on this estimation are mean square roots.

about the same values of E. However in case A the two curves are roughly parallel, but in case B they intersect, which means that the maximum of disorder does not necessarily appear when the volume of the stochastic zone is maximum.

To try to explain this result, we have calculated and plotted, the Kolmogorov entropy α_{st} of the stochastic zone only. This quantity α_{st} gives a measure of the disorder in the stochastic region. We also plotted error bars which are root mean squares and estimate the confidence interval of our statistical analysis. In case A, after a transition regime ($2. \lesssim E \lesssim 3.$) we notice that the entropy α_{st} remains roughly constant although the volume of the stochastic zone decreases. It is a consequence of the fact that for a test orbit of this zone, the value of the L.C.E. does not depend of the energy. Such a phenomenon is not observed in case B, where α_{st} is no more constant at high energy ($E \gtrsim 3.$) : we notice a bump of α_{st}, which reaches its maximum value at $E \cong 20$.

The different behaviours observed in the two cases may be due to the fact already pointed out by Ali an Somorjai, that the relative contribution of the interaction (coupling term) to the total energy of the system decreases more rapidly in case B with increasing energy.

CONCLUSION

The numerical study of these two non-integrable Hamiltonian systems allowed us to confirm the results obtained by Ali and Somorjai who calculated the surfaces of section at different energies. However their qualitative study gives only an approximate value of the volume of the stochastic region of the phase space. By the determination of the Kolmogorov entropy, we precise the value of the disorder inside this region. We find as the previous authors that from energies greater than $\cong 10.$ the number of irregular orbits decreases, i.e. the stochastic volume gets smaller. But the entropy of this volume, or degree of stochasticity α_{st}, is not necessarily a decreasing function: in case A we find that it does not change significantly from $E \cong 2$; this means that the volume of irregular orbits can be small but the degree of chaos in this region may be important. Case B clearly shows that for some energy ($E \cong 3.5$), although the full phase space looks stochastic ($q \cong 100\%$), the Kolmogorov entropy is less than for some other value ($E \cong 15.$) where only 80% of orbits are irregular.

Then the surface of section method give a visualisation and a measure of the stochastic region, while the calculation of the Kolmogorov entropy can estimate the degree of disorder of this zone.

REFERENCES

(1) Ali, M.K., Somorjai, R.L., Physica ID 383 (1980).
(2) Benettin, G., Galgani, L., Giorgili, A., Strelcyn, J.M., C.R. Acad. Sci. Ser. A, 286, 431 (1978).
(3) Burlish, R., Stoer, J., Numer. Math. 8, 1-13 (1966)

(4) Cours des Houches XXXVI, North Holland (1981)
(5) Froeschlé, C., Review paper in the same issue
(6) Hénon, M., Heiles, C., Astron. J., 69, 73 (1964)
(7) Pesin, Y.B., Sov. Math. Dokl., 17, 196 (1976)

GENERALIZATIONS OF THE JACOBIAN INTEGRAL

Victor Szebehely, Professor
D.D. Cockrell Chair

and

Arthur L. Whipple, Research Associate

The University of Texas at Austin
Austin, Texas, USA

ABSTRACT. The restricted problem of three bodies is generalized to the restricted problem of 2+n bodies. Instead of one body of small mass and two primaries, the system is modified so that there are several gravitationally interacting bodies with small masses. Their motions are influenced by the primaries but they do not influence the motions of the primaries. Several variations of the classical problem are discussed. The separate Jacobian integrals of the minor bodies are lost but a conservative (time-independent) Hamiltonian of the system is obtained. For the case of two minor bodies, the five Lagrangian points of the classical problem are generalized and fourteen equilibrium solutions are established. The four linearly stable equilibrium solutions which are the generalizations of the triangular Lagrangian points are once again stable but only for considerably smaller values of the mass parameter of the primaries than in the classical problem.

1. THE CLASSICAL RESTRICTED PROBLEM OF THREE BODIES

The classical restricted problem of three bodies may be described as the problem of determining the motion of a mass (m_1) acted upon by the Newtonian gravitational field of two much larger masses (M_1 and M_2) which are termed the primaries. The primaries are taken to be point masses which move about their center of mass on circular Keplerian orbits. The minor body (m_1) is influenced by, but does not influence the motion of the primaries. The geometry of the problem in a system rotating around the q_3 axis with the constant mean motion of the primaries is shown in Figure 1.

In the synodic system, the motion of the minor body is determined from the Hamiltonian

Celestial Mechanics **34** (1984) 125–133. 0008–8714/84.15

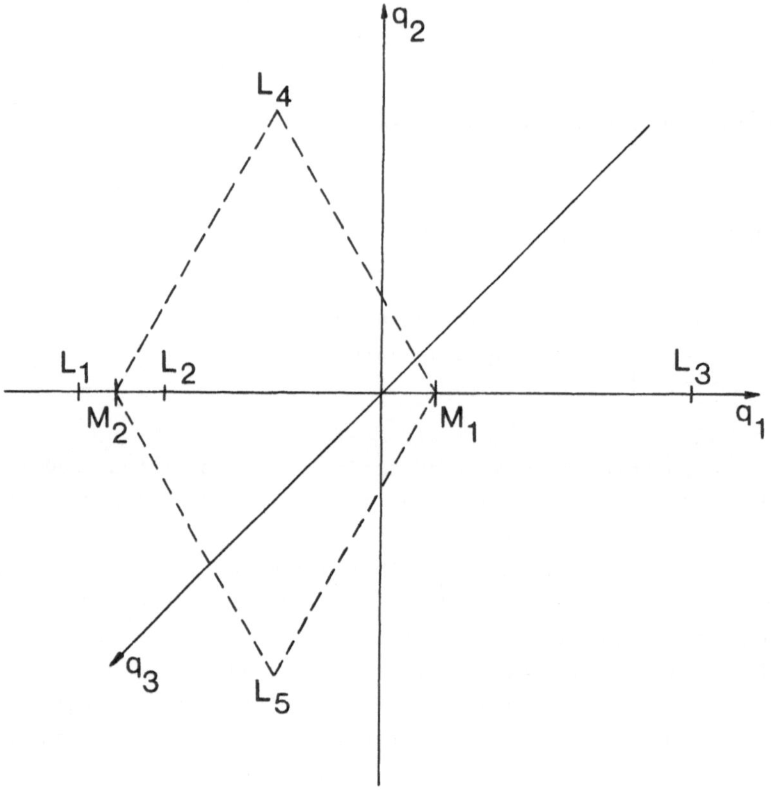

Figure 1. The restricted problem of three bodies in the synodic system.

$$H = \frac{1}{2}(p_1^2+p_2^2+p_3^2)+q_2p_1-q_1p_2-\Phi(q_1,q_2,q_3) \quad (1)$$

where q_1,q_2, and q_3 are the generalized dimensionless coordinates, p_1,p_2, and p_3 are the conjugate momenta,

$$\Phi = \frac{1-\mu}{[(q_1-\mu)^2+q_2^2+q_3^2]^{\frac{1}{2}}} + \frac{\mu}{[(q_1-\mu+1)^2+q_2^2+q_3^2]^{\frac{1}{2}}}$$

and

$$\mu = \frac{M_2}{M_1+M_2}$$

The Hamiltonian of the classical problem is time-independent since

$$\frac{\partial H}{\partial t} = \frac{dH}{dt} = 0$$

This constant Hamiltonian is closely related to the Jacobian integral of the classical problem (for example, see Szebehely, 1967). The Jacobian integral is

$$\dot{x}^2+\dot{y}^2+\dot{z}^2 = x^2+y^2+2\left[\frac{1-\mu}{[(x-\mu)^2+y^2+z^2]^{\frac{1}{2}}} + \frac{\mu}{[(x-\mu+1)^2+y^2+z^2]^{\frac{1}{2}}}\right]-C \quad (2)$$

where dots denote differentiation with respect to the dimensionless time. This integral has been applied to the study of the boundedness of certain solutions of the system through Hill's (1878) method. Hill's stability is based on the fact that for fixed values of the velocity of the minor body, the Jacobian integral divides the three-dimensional position space by two-dimensional surfaces.

Another well known property of the classical restricted problem of three bodies is the existence of five equilibrium solutions which are frequently called the Lagrangian points and are denoted L_i for i(1)5. The locations of the Lagrangian points are illustrated in Figure 1. Linear stability analysis reveals that the three collinear Lagrangian points (L_1,L_2, and L_3) are linearly unstable. The triangular Lagrangian points are found to be linearly stable provided the mass parameter of the primaries satisfies

$$\mu < \tfrac{1}{2}[1-(69)^{\frac{1}{2}}/9] \cong 0.03852$$

The nonlinear stability of these points has been verified for this range of μ by Leontovic (1962).

2. THE RESTRICTED PROBLEM OF 2+N BODIES

The main subject of this article is a generalization of the restricted problem of three bodies and the effects of this generalization on the classical results, discussed in the previous section.

The restricted problem of 2+n bodies (Whipple and Szebehely, 1984) assumes the known gravitational field of two massive point-mass primaries which move on circular orbits, as in the classical problem. In this gravitational field are moving n minor bodies ($m_i << M_2 \leq M_1$, $i(1)n$) which do not perturb the motion of the primaries. Finally, the gravitational interaction of the minor bodies is also included. A synodic system which rotates with the mean motion of the primaries is again employed. The geometry of the problem in the synodic system is shown in Figure 2.

Due to the coupling of the motion of the minor bodies through their mutual gravitational interaction, the equations of motion of each minor body show dependence on the other minor bodies. Consequently, the classical Jacobian integral for each body is lost. In place of this a new integral of the system is found. This integral is

$$\sum_{i=1}^{n} \mu_i(\dot{x}_i^2+\dot{y}_i^2+\dot{z}_i^2) = 2\sum_{i=1}^{n} \mu_i\left[\frac{1}{2}(x_i^2+y_i^2) + \frac{1-\mu}{R_{1i}} + \frac{\mu}{R_{2i}} + \frac{1}{2}\sum_{\substack{j=1\\ j\neq i}}^{n} \frac{\mu_j}{r_{ij}}\right] - C \qquad (3)$$

where

$$\mu_i = \frac{m_i}{M_1 + M_2}$$

$$R_{1i}^2 = (x_i-\mu)^2+y_i^2+z_i^2$$

$$R_{2i}^2 = (x_i-\mu+1)^2+y_i^2+z_i^2$$

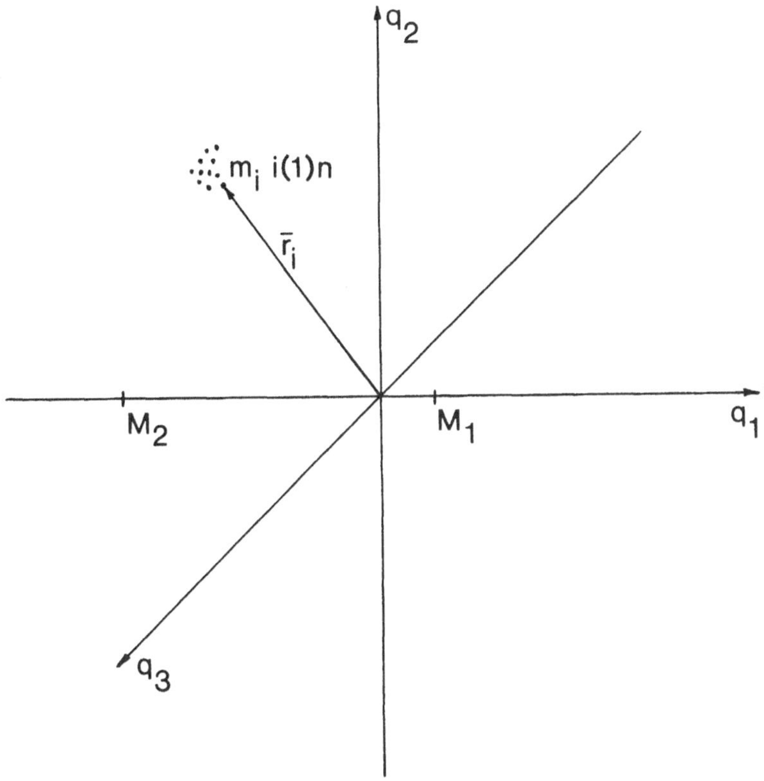

Figure 2. The restricted problem of 2+n bodies in the synodic system.

and

$$r_{ij}^2 = (x_i-x_j)^2+(y_i-y_j)^2+(z_i-z_j)^2$$

This integral is again closely related to the constant Hamiltonian of the system

$$H = \sum_{i=1}^{n} \left[\frac{1}{2\mu_i}\left(p_{i1}^2+p_{i2}^2+p_{i3}^2\right)+q_{i2}p_{i1}-q_{i1}p_{i2}-\mu_i\left(\frac{1-\mu}{R_{1i}} + \frac{\mu}{R_{2i}} + \frac{1}{2}\sum_{\substack{j=1\\ j\neq i}}^{n} \frac{\mu_j}{r_{ij}}\right)\right] \quad (4)$$

where q_{ik} and p_{ik} for $k(1)3$ are the generalized coordinates and momenta of the i-th minor body.

The integral given in Equation (3) may in principle be used in a similar fashion to the Jacobian integral for stability studies. However, note that if the interaction of the minor bodies is neglected then n integrals of the system exist (i.e. the n independent Jacobian integals). The inclusion of the gravitational interaction of the minor bodies has destroyed these independent integrals and given rise to one integral of the system of minor bodies. The change of the classical problem has therefore drastically reduced our knowledge of the system by leaving only one integral to partition a 3n-dimensional position space with a (3n-1)-dimensional hypersurface. This partitioning does not sufficiently bound the solutions of the system to permit conventional stability analysis along the lines of Hill's method.

3. EQUILIBRIUM SOLUTIONS

For the case of two minor bodies, the effect on the five Lagrangian points of the interaction between the minor bodies has been examined (Whipple, 1984). If the gravitational interaction of the minor bodies is neglected then both minor bodies can be placed in equilibrium with the primaries at any of the five Lagrangian points. The inclusion of the interaction between the minor bodies necessitates the balancing of forces due, not only to the primaries and centrifugal forces, but also due to the minor bodies themselves.

In the restricted problem of 2+2 bodies the existing fourteen equilibrium solutions are located in the plane of the revolving primaries. Each equilibrium solution involves a small displacement of both minor bodies from one of the five Lagrangian points. There are two equilibrium solutions around each of the three collinear Lagrangian points and four equilibrium solutions around both of the triangular Lagrangian points. For the case of $\mu=0.3$, $\mu_1=0.02$, and $\mu_2=0.01$, the equilibrium solutions are shown in Figure 3. From Figure 3 it may be seen that the center of mass of the minor bodies lies approximately at the Lagrangian point about which the particular equilibrium solution is arranged.

It is also of interest to consider the effect on the stability of these solutions of the gravitational interaction of the minor bodies. As expected, linear stability analysis reveals that all of the collinear equilibrium solutions are unstable for all values of μ, μ_1, and μ_2. In addition, the equilibrium solutions which are oriented approximately in-line with the center of mass of the primaries and the triangular Lagrangian points are topologically equivalent to the solutions about L_3 in the limiting case of $\mu=0$. Since stability of the triangular Lagrangian points in the classical problem is associated with small values of μ, this instability of the in-line triangular solutions at the smallest possible value of μ is indicative of the general instability of these solutions. In fact the in-line solutions are found to be unstable for all values of μ, μ_1, and μ_2.

Perhaps the most important variation of the properties of the classical restricted problem, that is due to this generalization, is the effect of the gravitational interaction of the minor bodies on the remaining four equilibrium solutions. The four solutions which are oriented approximately perpendicular to the line joining the center of mass of the primaries and the triangular Lagrangian points are found to be linearly stable for certain values of μ, μ_1, and μ_2. While this stability depends on the masses of all four bodies, it is found that for $\mu_1, \mu_2 < 10^{-12}$ the maximum value of μ for which the perpendicular solutions are linearly stable is independent of μ_1 and μ_2 to $O(10^{-5})$. This limiting value is $\mu=0.0119$. For any case where μ is greater than this value, one or both of the minor bodies will depart from the vicinity of the triangular Lagrangian point. Since this critical value of μ may be seen to be significantly smaller than the corresponding value in the classical problem, the incluson of the gravitational interaction of the minor bodies has a significant effect on the stability of the generalized equilibrium solutions.

4. CONCLUSIONS

If n minor bodies are taken to move independently in the gravitational field of two much more massive, circularly orbiting primary bodies, the motion of each minor body may be described by solutions of

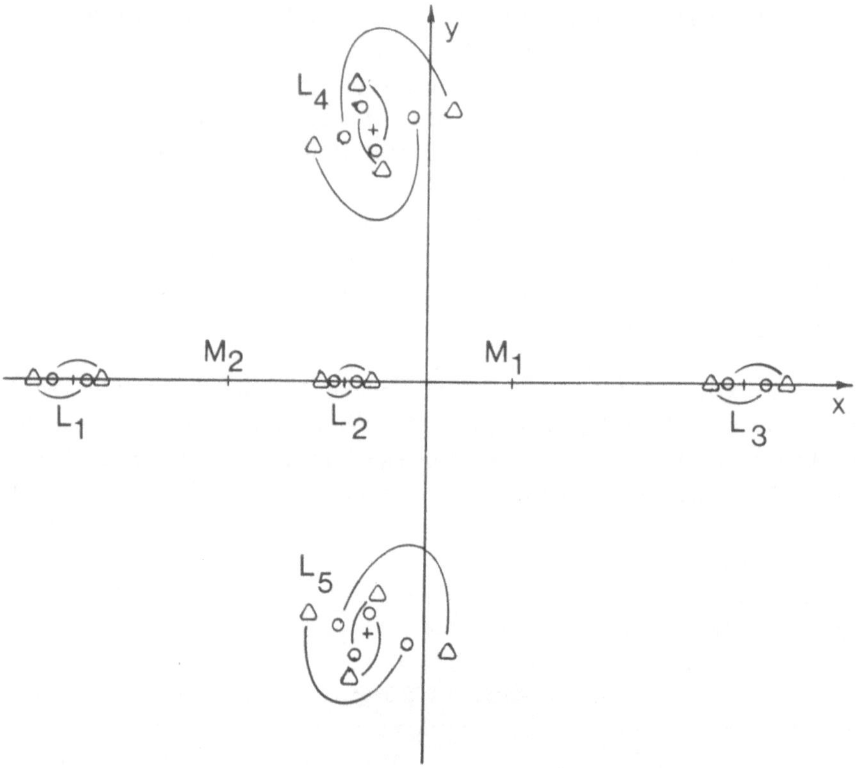

Figure 3. Equilibrium solutions of the restricted problem of 2+2 bodies for $\mu=0.3$, $\mu_1=0.02$, and $\mu_2=0.01$. Circles denote m_1 and triangles denote m_2. Pairing of the minor bodies in each equilibrium solution is indicated by a line joining the minor bodies.

the classical restricted problem of three bodies. This classical problem yields an integral of the motion of each body and five identical equilibrium solutions for each body. If the small perturbation of the gravitational interacton of the minor bodies is considered, then the n integrals of the motion of each minor body are lost and only one integral of the system of minor bodies is preserved. This small perturbation of the dynamics of the system therefore has a profound effect on the ultimate integrability of the system.

For the case of two minor bodies, there exist fourteen equilibrium solutions of the pair of minor bodies, which are generalizations of the five classical Lagrangian points. The solutions which generalize the collinear Lagrangian points and two solutions around each of the triangular Lagrangian points are unstable. There are also two solutions around the triangular Lagrangian points which are linearly stable for a range of values of the mass parameter of the primaries which has an upper limit which is significantly smaller than the critical value of μ in the classical problem. The inclusion of the gravitational interaction of the minor bodies, therefore, destabilizes the stable equilibrium solutions of the classical restricted problem of three bodies. This destabilization raises the question of the effects of gravitational interaction among minor bodies around Lagrangian points in the solar system such as the Trojan asteroids.

REFERENCES

Hill, G.W.: 1878, 'Researches in the Lunar Theory', Am. J. Math. 1, 5, 129-245.

Leontovic, A.M.: 1962, 'On the Stability of the Lagrangian Periodic Solutions for the Reduced Problem of Three Bodies', Dokl. Akad. Navk. USSR 143, 425-429.

Szebehely, V.: 1967, Theory of Orbits, Academic Press, New York.

Whipple, A.: 1984, 'Equilibrium Solutions of the Restricted Problem of 2+2 Bodies', Celes. Mech., 33, 271-294.

Whipple, A., and V. Szebehely, V.: 1984, 'The Restricted Problem of n+ν Bodies', Celes. Mech. 32, 137-144.

APPLICATION OF LIE-SERIES TO REGULARIZED PROBLEMS IN CELESTIAL MECHANICS

A. Hanslmeier
Inst. f. Astronomie
Universitätsplatz 5
A-8010 Graz
Austria

ABSTRACT. In the following paper we tried to apply the Lie-formalism to the regularized restricted three body problem. It will be shown that this algorithm leads to a very simple structure program which is also fast.

1. INTRODUCTION

The application of Lie-series to the n-body problem in celestial mechanics has been first investigated by W. Gröbner (1967), especially the acting of the "Kepler operator" to the coordinate and velocity vector for the two body problem. In two earlier works recursion formulae for the solution of the n-body problem have been derived by means of Lie-series (Hanslmeier 1983; Hanslmeier, Dvorak 1984). With these formulae we have a rapid numerical integration procedure. The structure of the computer program is very simple, so that any modification (e.g. automatic step adaption) can be done without great difficulties.

In this work we will try to show how it is possible and useful from the practical point of view to apply this algorithm to the regularized plane restricted three body problem. It will be seen that in this case too Lie-series provide a fast and very clear method for solving the regularized differential equations. Since regularization theory plays an important role when studying motions of natural and artificial bodies as well as for investigations concerning stability problems of dynamical systems, our investigations may also be of practical interest.

2. SHORT REMARKS ABOUT LIE-SERIES, REGULARIZED TWO BODY PROBLEM

The Lie-operator consists of partial derivatives and holomorphic functions $\theta_i(z)$, z being a complex variable. It is defined according to Gröbner (1967)

Celestial Mechanics **34** (1984) 135–143. 0008–8714/84.15

$$D = \theta_1(z)\frac{\partial}{\partial z_1} + \theta_2(z)\frac{\partial}{\partial z_2} + \ldots + \theta_n(z)\frac{\partial}{\partial z_n} \qquad (1)$$

We can act this linear differential operator on any function f(z) which has to be holomorphic within the same domain as are the functions $\theta_i(z)$. Thus we can write (1) as:

$$L(z,t) = \sum_{\nu=0}^{\infty} \frac{t^\nu}{\nu!} D^\nu f(z) \qquad (2)$$

where $D^\nu f(z) = D^{\nu-1}(Df(z))$ (2a)
Symbolically eq. (2) can be written in the form

$$e^{tD}f(z) = f(z) + tDf(z) + \frac{t^2D^2f(z)}{2!} + \ldots \qquad (3)$$

The proof of convergence of L(z,t) can be found in Gröbner (1967) where it is also shown that Lie-series are a generalization of Taylor-series. Let us demonstrate hown one can solve differential equations by means of Lie-series; as an example we solve the equations of the regularized two body problem (a detailed description of the regularized two body problem is given in Stiefel (1971)).

Let $\vec{x} = (x_1,x_2,x_3)$ be the position vector of mass m_2 with respect to the mass m_1, r being their mutual distance, t the real time, s the fictitious time. If we look at the ordinary equations of the two body problem, we see that they become singular at the point r=0 which corresponds to a collision. Therefore we regularize the equations of motion in two steps:

1) Transformation of time:

$$t = \int r ds$$

2) Transformation of coordinates and velocities according to Kustaanheimo and Stiefel:

$$\vec{x} = L(\vec{u})\vec{u} \qquad (4)$$

where $L(\vec{u})$ is the well known K-S matrix.
In the following sections ′ denotes differentiation with respect to s. The differential equations for the regularized two body problem are then:

$$u_j'' + \frac{h}{2}u_j = 0 \qquad j=1,2,..,4 \qquad (6)$$

h is the "Kepler energy":

$$h = \frac{K^2}{2} - \frac{1}{2}\dot{x}^2 \quad , \; K^2 = k^2(m_1 + m_2) \qquad (7)$$

The solution of eq. (6) by means of Lie-series is found by determining the functions $\theta_i(z)$:

$$\theta_j = u'_j$$
$$\theta'_j = u''_j = -\frac{h}{2} u_j \qquad j = 1,..,4 \tag{8}$$

Now we can write the Lie-operator (1) as:

$$D = \sum_{j=1}^{4} \left\{ u'_j \frac{\partial}{\partial u_j} - \frac{h}{2} u_j \frac{\partial}{\partial u'_j} \right\} \tag{9}$$

The solutions of eq. (6) with Lie-series are then:

$$u_j = (e^{tD} u_j)_{(0)}$$

$$D(u_j) = u'_j$$

$$D^2(u_j) = -\frac{h}{2} u_j$$

$$D^3(u_j) = -\frac{h}{2} D(u_j)$$

$$\vdots$$

$$D^n(u_j) = -\frac{h}{2} D^{n-2}(u_j) \qquad n = 4,5,...,k \quad j = 1,...,4 \tag{10}$$

The CPU-time in this simple case is for numerical integration about 1/3 than that needed for the ordinary two body problem and can be compared with the CPU-time needed for the integration with Stumpff- functions (Lichtenegger 1984).

Of course no automatic step adaption process is required when integrating very excentric orbits. In the next section we try to apply this formalism to the plane restricted three body problem.

3. REGULARIZED PLANE RESTRICTED THREE BODY PROBLEM

We consider three masses m_1, m_2, m_3, where m_3 does not have any gravitational influence on the two other masses. We further use the well known rotating coordinate system where the distance $\overline{m_1 m_2}$ remains fix. The units used are then:

unit of mass: m_1+m_2, $m_2=\mu$, $m_1=1-\mu$

unit of length: distance $\overline{m_1 m_2}$

unit of time: 1 period of revolution of m_2 about m_1 divided by

$$2\Pi \tag{11}$$

With these units chosen the value of the gravitational constant will be 1. Before applying the Lie-series let us briefly recall the Hamiltonian formalism.

Let H be the Hamiltonian function, x and y be the coordinates, p_x, p_y the momenta. The equations of motion are then:

$$\dot{x} = \frac{\partial H}{\partial p_x} \qquad \dot{p}_x = - \frac{\partial H}{\partial x}$$

$$\dot{y} = \frac{\partial H}{\partial p_y} \qquad \dot{p}_y = - \frac{\partial H}{\partial y} \tag{12}$$

In this case, the Lie-operator (1) can be written in the following form (Giacaglia, 1981):

$$D = \sum_i \left\{ \frac{\partial H}{\partial p_i} \frac{\partial}{\partial q_i} - \frac{\partial H}{\partial q_i} \frac{\partial}{\partial p_i} \right\} = [\ ,H] \tag{13}$$

q_i ... generalized coordinates

p_i ... generalized momenta

[ ]... Poisson brackets

If we act D on a function f(z), we get:

$$D(f) = [f,H]$$

$$D^2(f) = [[f,H],H]$$

$$D^3(f) = [[[f,H],H],H] \tag{14}$$

We follow the regularization with parabolic coordinates as has been given by A. Deprit and R. Broucke (1964).

Let x, y be the coordinates of the massless body m_3, ρ_1 be the distance to m_1, ρ_2 be the distance of m_3 to m_2.

$$\rho_1^2 = \left(x + \frac{1}{2} \right)^2 + y^2$$

$$\rho_2^2 = \left(x - \frac{1}{2} \right)^2 + y^2 \tag{15}$$

We finally derive that the Hamiltonian function H is equal to the expression:

$$H = \frac{1}{2} (p_x^2 + p_y^2) + (y p_x - x p_y) - (1 - \mu) / \rho_1 -$$

$$- (\frac{1}{2} - \mu) x - \mu / \rho_2 \tag{16}$$

From (13) we can deduce the equations of motion for the plane restricted three body problem.
It is obvious that we have two singular points now at m_1 and m_2 . The regularization is a conformal mapping of the complex plane $z = x + i y$ to the complex plane $\zeta = \xi + i\eta$.

If we consider regularization with respect to m_1, using parabolic coordinates, the conformal mapping has the form:

$$z = - \frac{1}{2} + \zeta_1^2 \tag{17}$$

whereas if we regularize with respect to m_2, we have to use the mapping

$$z = \frac{1}{2} + \zeta_2^2 \tag{18}$$

We can treat both cases if we introduce at the beginning of the program the following parameters:

Table 1: List of parameters

Parameter	at m_1	at m_2
ε	-1	1
ρ	ρ_1	ρ_2
σ	ρ_2	ρ_1
α	$1-\mu$	μ

We have the following relations:

$$x = \frac{1}{2}\varepsilon + \xi^2 - \eta^2 \qquad y = 2\xi\eta$$

$$2\rho p_x = \xi p_\xi - \eta p_\eta \qquad 2\rho p_y = \eta p_\xi + \xi p_\eta$$

$$\rho = \xi^2 + \eta^2$$

$$\sigma = (\rho^2 - 2 (\xi^2 - \eta^2) + 1)^{1/2} \tag{19}$$

The Hamiltonian function is then:

$$H = \frac{1}{2}\,(p_\xi^2 + p_\eta^2) + 2\,\{(\rho - \frac{1}{2}\varepsilon)\eta p_\xi - (\rho + \frac{1}{2}\varepsilon)\,\xi p_\eta\}$$
$$- 4\alpha - 4(1-\alpha)\,(\rho/\sigma) - 4\,(\frac{1}{2} - \mu)\,(\frac{1}{2}\varepsilon + \xi^2 - \eta^2)\rho$$
$$- 4h\rho \tag{20}$$

The Lie-operator is finally found according to (13):

$$D = [p_\xi + (2\rho - \varepsilon)\eta]\frac{\partial}{\partial\xi} - [4\xi\eta p_\xi - 4\xi^2 p_\eta - 2(\rho + \frac{1}{2}\varepsilon)\,p_\eta$$
$$- A\,(2\xi\sigma^{-1}(1 - \rho^2\,\sigma^{-2} + \rho\sigma^{-2})) - M(4\xi^3 + \varepsilon\xi) - 8h\xi]\,\frac{\partial}{\partial p_\xi}$$
$$+ [p_\eta - (2\rho + \varepsilon)\xi]\frac{\partial}{\partial\eta} - [4\eta^2 p_\xi - 4\eta\xi p_\eta + 2(\rho - \frac{1}{2}\varepsilon)\,p_\xi$$
$$- A(2\eta\sigma^{-1}\,(1 - \rho^2\sigma^{-2} - \sigma^{-2}\rho)) - M(\varepsilon\eta - 4\eta^3) - 8h\eta]\,\frac{\partial}{\partial p_\eta} \tag{21}$$

Here we introduce the abbreviations:

$$A = 4(1 - \alpha)$$
$$M = 4\,(\frac{1}{2} - \mu) \tag{22}$$

The solutions are of the form:

$$\xi = e^{tD}\,\xi_{(o)} \qquad \eta = e^{tD}\,\eta_{(o)} \tag{23}$$

The first term of the Lie-series can be found very easily.

$$D\xi = p_\xi + (2\rho - \varepsilon)\eta$$
$$D\eta = p_\eta - (2\rho + \varepsilon)\xi \tag{24}$$

Since the equations are similiar for the x and the y coordinates, we consider here only the Lie-series for x!

We also can evaluate the second Lie-term:

$$D^2\xi = Dp_\xi + D\eta(2\rho - \varepsilon) + \eta 2D\rho \tag{25}$$

We see that:

$$Dp_\xi = \theta_2 \tag{26}$$

At this point, we are able to give some kind of recursion relation for the $D^{n+2}\xi$, n=1,2,...:

$$D^{n+2}\xi = -\{4\sum_{\nu=o}^{n}\binom{n}{\nu}D^{\nu}\delta D^{n-\nu}p_{\xi} - 4\sum_{\nu=o}^{n}\binom{n}{\nu}D^{\nu}\xi^{2}D^{n-\nu}p_{\eta} -$$

$$-2\sum_{\nu=o}^{n}\binom{n}{\nu}D^{\nu}(\rho+\frac{1}{2}\varepsilon)D^{n-\nu}p_{\eta} - 2A\,[\sum_{\nu=o}^{n}\binom{n}{\nu}D^{\nu}\xi D^{n-\nu}\sigma^{-1} -$$

$$-\sum_{\nu=o}^{n}\binom{n}{\nu}D^{\nu}\rho^{2}D^{n-\nu}\sigma_{\xi} + \sum_{\nu=o}^{n}\binom{n}{\nu}D^{\nu}\rho D^{n-\nu}\sigma_{\xi}] - M\,(4D^{n}\xi^{3} +$$

$$+ D^{n}\xi) - 8hD^{n}\xi\} + \sum_{\nu=o}^{n}\binom{n}{\nu}D^{\nu}(2\rho-\varepsilon)D^{n-\nu}\eta \qquad (27)$$

For brevity we introduced:

$$S = \sum_{\nu=o}^{n}\binom{n}{\nu}$$

The expression (27) contains the following "derivatives" for which recursion relations can be found:

$$D^{n}\delta = D^{n}(\xi\eta) = S\,D^{\nu}\xi D^{n-\nu}\eta \qquad (28)$$

$$D^{n+1}\,\xi^{2} = 2\,S\,D^{\nu}\xi D^{n-\nu+1}\xi \qquad (29)$$

$$D^{n+1}p_{\xi} = -\{4\,S\,D^{\nu}\delta D^{n-\nu}p_{\xi} - 4\,S\,D^{\nu}\,\xi^{2}D^{n-\nu}p_{\eta} -$$

$$-2\,S\,D^{\nu}(\rho+\frac{1}{2}\varepsilon)\,D^{n-\nu}p_{\eta} - 2\,A\,[S\,D^{\nu}\xi D^{n-\nu}\sigma^{-1} -$$

$$-S\,D^{\nu}\,\rho^{2}D^{n-\nu}\sigma_{\xi} + S\,D^{\nu}\rho D^{n-\nu}\sigma_{\xi}] - M\,(4D^{n}\,\xi^{3} + \varepsilon D^{n}\xi) -$$

$$-\,8hD^{n}\xi\} \qquad (30)$$

$$D^{n+1}(\rho+\frac{1}{2}\varepsilon) = 2\,(S\,D^{\nu}\xi D^{n-\nu+1}\,\xi + S\,D^{\nu}\eta D^{n-\nu+1}\eta) \qquad (31)$$

$$D^{n+1}\,\rho^{2} = 2\,S\,D^{\nu}\rho D^{n-\nu+1}\rho \qquad (32)$$

The only remaining problem is to find a recursion relation for the $D^{n}\sigma$ which seems to be impossible. But it is not difficult to evaluate the terms:

$$\sigma = (\rho^2 - 2(\xi^2 + \eta^2) + 1)^{1/2}$$

$$D\sigma = \sigma^{-1}(\rho D\rho - 2\xi D\xi - 2\eta D\eta)$$

$$D^2\sigma = -\sigma^{-2}D\sigma(\rho D\rho - 2\xi D\xi - 2\eta D\eta) + \sigma^{-1}D(\rho D\rho - 2\xi D\xi - 2\eta D\eta)$$

$$= \sigma^{-1}(-(D\sigma)^2 + D(\rho D\rho - 2\xi D\xi - 2\eta D\eta))$$

$$D^3\sigma = \sigma^{-1}(-3D\sigma D^2\sigma + D^2(\rho D\rho - 2\xi D\xi - 2\eta D\eta)) \qquad (33)$$

It is evident, how the higher derivatives are evaluated. The terms $D^n\sigma^{-3}$ and $D^n\sigma^{-1}$ can be derived and the term appearing in (27) stands for:

$$\sigma_\xi = \sigma^{-3}\xi$$

$$D^n\sigma_\xi = S\, D^\nu\sigma^{-3}D^{n-\nu}\xi \qquad (34)$$

With these relations, we can in principle calculate all Lie-terms up to the n-th order. In the next section we show that for practical uses it will be sufficient to work with 5 terms.

4. DISCUSSION

If we use the appropriate unit system (see section 3.), than the numerical values of $\xi, \eta, p_\xi, p_\eta, \rho$ for a typical Trojan orbit are less or equal 1, σ being greater than 1. So the convergence of the Lie-series depends mainly on

$$\frac{s^n}{n!}$$

where $dt = 4\rho\, ds$.

It will be sufficient to work with 5 Lie-terms, where the computing-time needed to evaluate the terms will not be great. The advantage of this solution is that we avoided trigonometric series and that the solution itself is very simple in structure. The program works very fast and can be changed easily. A detailed description of the regularization of the restricted three body problem has been given by V.Szebehely (1967). In a future work we plan to give some numerical tests and we want to compare this method to other numerical integration techniques.

REFERENCES

Deprit,A., Broucke,R.: 1964, 'Regularization du Problème Restreint Plan des Trois Corps par Representations Conforms', Icarus 2, p. 207-219

Giacaglia,G.E.O.: 1981,'A note about Lie-Series and some Applications' Preprint, Sao Paulo

Gröbner,W.: 1967, Die Lie-Reihen und ihre Anwendungen, VEB Deutscher Verlag der Wissenschaften, Berlin

Gröbner,W., Knapp,H.: 1967a, Contribution to the Method of Lie-series Bibliograph.Inst., Mannheim

Hanslmeier,A.: 1983, Anwendungen der Lie-Reihen als numerische Integrationsmethode in der Himmelsmechanik, Dissertation, Karl-Franzens-Universität Graz

Hanslmeier,A., Dvorak,R.: 1984,'Numerical Integration with Lie-Series' Astron.Astrophys. 132, Nr.1

Lichtenegger,H.: 1983, Personal communication

Stiefel,E.L., Scheifele,G.: 1981, Linear and Regular Celestial Mechanics, Springer Verlag, Berlin-Heidelberg-New York

Szebehely,V.: 1967, Theory of Orbits, Academic Press, New York

INTEGRATION OF THE ELLIPTIC RESTRICTED THREE-BODY PROBLEM WITH LIE SERIES*

Magda Delva
Institut für Astronomie
Karl-Franzens-Universität Graz
Universitätsplatz 5
A-8010 Graz
Austria

ABSTRACT. The method of Lie series is used to construct a solution for the elliptic restricted three body problem. In a synodic pulsating coordinate system, the Lie operator for the motion of the third infinitesimal body is derived as function of coordinates, velocities and true anomaly of the primaries. The terms of the Lie series for the solution are then calculated with recurrence formulae which enable a rapid successsive calculation of any desired number of terms. This procedure gives a very useful analytical form for the series and allows a quick calculation of the orbit.

1. INTRODUCTION

In many cases in celestial mechanics, the series development of the disturbing function is a difficult and complicated problem. To facilitate this, the procedure can often be performed with an operator, which enables computer handling of the development. A special linear differential operator, the Lie operator, produces a Lie series; due to its algebraic properties, the method is of practical use in celestial mechanics (Gröbner,W.: 1967; Stumpff,K.: 1974). The convergence of the series is the same as for Taylor series, since the Lie series is only another analytical form of the Taylor series (Schneider,M.:1979).

The aim of our work is to find the motion of the infinitesimal body in the elliptic restricted three body problem with a Lie series. We construct the Lie operator and recurrence formulae for the terms of the series.

The algorithm can be used for any asteroidal or satellite motion in the planar restricted three body problem. An extension to the three dimensional treatment contains no difficulties and can be performed in the same way.

A change of the stepsize is easy; this is an important advantage for the treatment of problems which require a variable stepwith e.g.

*The project is supported by the Austrian Fonds zur Förderung der wissénschaftlichen Forschung under Project No. 4471.

Celestial Mechanics **34** (1984) 145–154. 0008–8714/84.15

masschanges of the primaries.

2. LIE OPERATOR FOR THE INFINITESIMAL BODY

An appropriate coordinate system for calculations of the Jacobian function was presented by Szebehely and Giacaglia (1964): The planar restricted three body problem is described in a barycentric, pulsating, non-uniformly rotating coordinate system (O;x,y). It is made dimensionless in taking the variable distance r between the primary bodies as the unit of length; the primaries then have fixed positions on the x-axis (Fig.1):

$$r = \frac{a(1-e^2)}{1+e\cos f} \tag{1}$$

(a,e,f being the semi-major axis, the eccentricity, and the true anomaly of the orbit of the primaries, respectively)

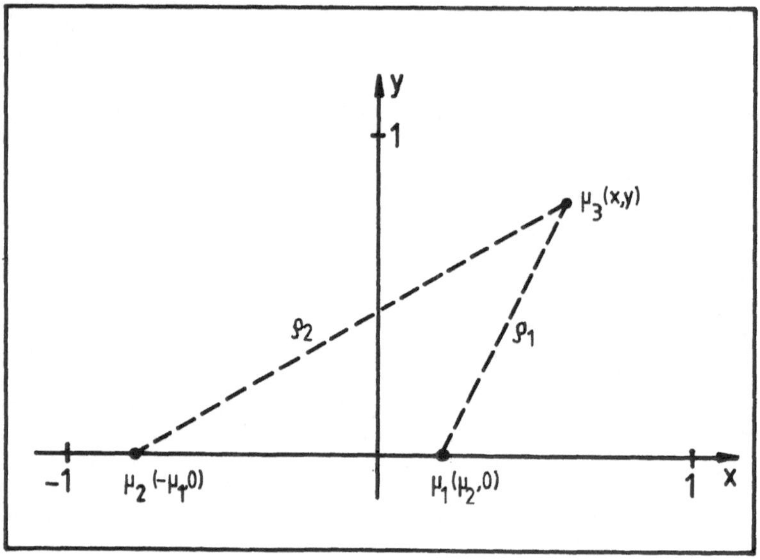

Fig.1: Barycentric pulsating coordinate system (O;x,y)

First body: mass μ_1; coordinates $(\mu_2,0)$
Second body: mass μ_2; coordinates $(-\mu_1,0)$
Third body: mass $\mu_3 = 0$; coordinates (x,y)

$(\mu_1 + \mu_2 = 1)$

If the true anomaly f is chosen to be the independent variable, the analytical form of the differential equations for the third body is the same in the elliptical and circular problem. This allows to treat the eccentricity e as a parameter ($0 \le e \le E \in R$).

The differential equations for the orbit of the third body then are:

$$\frac{d^2x}{df^2} = 2\frac{dy}{df} + \frac{\partial\omega}{\partial x} ,$$

$$\frac{d^2y}{df^2} = -2\frac{dx}{df} + \frac{\partial\omega}{\partial y} , \tag{2}$$

where

$$\omega = \frac{1}{1+e\cos f}\left\{\frac{\mu_1\rho_1^2 + \mu_2\rho_2^2}{2} + \frac{\mu_1}{\rho_1} + \frac{\mu_2}{\rho_2}\right\} \tag{3}$$

$$\rho_1^2 = (x-\mu_2)^2 + y^2$$

$$\rho_2^2 = (x+\mu_1)^2 + y^2 \tag{4}$$

The analytical solution now is approximated by a Lie series, a method which is described by Gröbner (1967), Stumpff (1974) and recently by Hanslmeier and Dvorak (1983). The terms of the series are generated by the Lie operator. The equations (2) are rewritten as a system of four first order differential equations in the coordinates (x,y) and velocities (x',y') with f as independent variable:

$$\frac{dx}{df} = x'$$

$$\frac{dy}{df} = y'$$

$$\frac{dx'}{df} = 2y' + \frac{1}{1+e\cos f}\{\mu_1(x-x_1) + \mu_2(x-x_2) - \\ - \frac{\mu_1(x-x_1)}{\rho_1^3} - \frac{\mu_2(x-x_2)}{\rho_2^3}\}$$

$$\frac{dy'}{df} = -2x' + \frac{1}{1+e\cos f}\{\mu_1 y + \mu_2 y - \frac{\mu_1 y}{\rho_1^3} - \frac{\mu_2 y}{\rho_2^3}\}. \tag{5}$$

The linear Lie operator D has the general form

$$D = \frac{d}{df} = \frac{\partial}{\partial x}\frac{dx}{df} + \frac{\partial}{\partial y}\frac{dy}{df} + \frac{\partial}{\partial x'}\frac{dx'}{df} + \frac{\partial}{\partial y'}\frac{dy'}{df} + \frac{\partial}{\partial f}\frac{df}{df} \tag{6}$$

or with eq. (5):

$$D = x' \frac{\partial}{\partial x} + y' \frac{\partial}{\partial y} + \{\frac{1}{1+e\cos f}\ [x - \frac{\mu_1(x-\mu_2)}{\rho_1^3} - \frac{\mu_2(x+\mu_1)}{\rho_2^3}] + 2y'\} \frac{\partial}{\partial x'} + \{ \frac{1}{1+e\cos f} [y - \frac{\mu_1 y}{\rho_1^3} - \frac{\mu_2 y}{\rho_2^3}] - 2x'\} \frac{\partial}{\partial y'} + \frac{\partial}{\partial f} \; . \tag{7}$$

The solution $\vec{X}(x,y,x',y',f)$ is then given by the Lie series:

$$\vec{X}(x,y,x',y',f) = [\{\exp[(f-f_o)D]\}\ X]_{\vec{X}=\vec{X}_o} = \sum_{j=o}^{\infty} [D^j\vec{X}]_{\vec{X}_o} \frac{(f-f_o)^j}{j!} \tag{8}$$

where $D^j\ \vec{X}$ is to be evaluated for the initial condition $\vec{X}_o\,(x_o,y_o,x'_o,y'_o,f_o)$. To find the terms of the series, it will be necessary to calculate the multiple action of D on the variables x,y,x',y',f. The single action on x,y produces x',y':

$$Dx = x' \;,$$

$$Dy = y' \;,$$

and hence, the multiple action gives

$$D^j x = D^{j-1}x' \qquad j \geq 1$$

$$D^j y = D^{j-1}y' \; .$$

The multiple action of D on x', y' is known to the order $(j-1)$ if that of D on x,y has been computed to the order j.

Due to the form (6) of D, its action on f is reduced to a partial differentiation with respect to f.

The operator D may also be used for the circular problem if one is aware of the following: for zero eccentricity, the explicit dependence of D on f through $(1+e\cos f)$ vanishes and the true anomaly will not occur in $D^j\vec{X}$ (except for $Df = 1$). Since the set of recurrence formulae (of section 2) was developed essentially for the elliptic problem, it will contain partial derivatives with respect to f. If the calculation of these derivatives is omitted in the circular case, the presented set of formulae will give a rapid method of orbit calculation also for this case.

3. TERMS OF THE LIE SERIES

To construct the terms of the Lie series (8), the Lie operator D must be applied repeatedly to the vector function $\vec{X}$ or to this components (x,y,x',y',f). If a general recurrent form for this can be found, a rapid iteration of any desired number of terms will be possible.

3.1. Lie operator acting on the x-component

The first two terms of the Lie series for x are calculated separately. Then a recurrence formula is found, which allows the calculation of the next term from all the previous in x and y. The first two terms are without difficulty:

$$Dx = x' \tag{9a}$$

$$D^2x = 2\ [Dy] + \frac{1}{1+e\cos f}\ \left[x - \frac{\mu_1(x-\mu_2)}{\rho_1^3} - \frac{\mu_2(x+\mu_1)}{\rho_2^3}\right] \tag{9b}$$

To be able to write the higher terms as a recurrence, new notations are introduced:

$$E = e \cos f$$

$$F = \frac{1}{1+E}$$

$$U_1 = \mu_1\ (x-\mu_2)$$

$$U_2 = \mu_2\ (x+\mu_1)$$

$$R_1 = -\frac{1}{\rho_1^3} \tag{10}$$

$$R_2 = -\frac{1}{\rho_2^3}$$

$$A = x + U_1R_1 + U_2R_2\ ;$$

Equation (9b) can then be written as

$$D^2x = 2\ [Dy] + F\ A\ . \tag{11}$$

Using the linearity of the Lie operator (Gröbner,W.: 1967) a general form for the powers of D acting on x is easily derived:

$$D^{p+2} x = 2 [D^{p+1} y] + \sum_{j=0}^{p} \binom{p}{j} [D^{p-j}F] [D^j A] , \quad p \geq 0 \qquad (12a)$$

where

$$\binom{p}{j} = \frac{p!}{j!(p-j)!} .$$

Equation (12a) allows an iterative calculation of all the terms of the Lie series, provided the factors $[D^{p-j}F]$ and $[D^jA]$ $(j = 0,\ldots,p)$ are known. The solution for x is then given by the series:

$$x(f) = [x]_{\vec{X}_o} + [Dx]_{\vec{X}_o} (f-f_o) + \sum_{p=o}^{\infty} \{2 [D^{p+1} y]_{\vec{X}_o} +$$

$$+ \sum_{j=o}^{p} \binom{p}{j} [D^{p-j}F]_{\vec{X}_o} [D^jA]_{\vec{X}_o} \} \{\frac{(f-f_o)^{p+2}}{(p+2)!}\} . \qquad (12b)$$

The still unknown factors in (12b) are calculated in the next sub-sections.

3.1.1. The $[D^{p-j}F]$ $(j=0,\ldots,p)$ are found without difficulty, if the factor $F = \frac{1}{1+E}$ is always separated from the rest:

$$DF = - F [DE] [D^oF]$$

$$D^2F = - F \{[D^2E] [D^oF] + 2 [DE] [DF]\}$$

$$\vdots$$

$$D^pF = - F \{\sum_{j=o}^{p-1} \binom{p}{j} [D^jF] [D^{p-j}E]\} , \quad p \geq 1 . \qquad (13)$$

3.1.2. For the $[D^{p-j}E]$ $(j=0,\ldots,p-1)$ the calculation of the first four is sufficient and

$$D^kE = D^{k-4}E, \quad k \geq 4 . \qquad (14)$$

3.1.3. Since A was defined as a sum by equation (10), the factors $[D^jA]$ $(j=0,\ldots,p)$ are found through $U_i, R_i (i=1,2)$:

$$D^pA = [D^px] + \sum_{i=1}^{2} \{\sum_{j=o}^{p} \binom{p}{j} [D^{p-j}U_i] [D^jR_i]\} , \quad p \geq 0 . \qquad (15)$$

3.1.4. No problems arise for $[D^{p-j}U_i]$ $(j=0,\ldots,p)$:

$$D^p U_i = \mu_i\, [D^p x] \,, \quad p \geq 1 \,, \; i = 1,2 \,. \tag{16}$$

3.1.5. The factors $[D^j R_i]$ $(j=0,\ldots,p)$ $(i=1,2)$ are more complicated to compute; for simplification a new $(p \times p)$ - matrix M of coefficients is introduced:

$$M=(m_{k\ell}) = \begin{pmatrix} m_{o,o} = -3 & 0 & \cdot\cdot & \cdot\cdot & 0 \\ -3 & -5 & 0 & \cdot\cdot & 0 \\ -3 & -8 & -7 & 0 & 0 \\ \cdot & & & & \\ \cdot & & & & \\ \cdot & & & & \\ m_{p-1,o} & \cdots\cdots & & & m_{p-1,p-1} \end{pmatrix}$$

The $(m_{k\ell})$ result from the use of the binomial formula and can be found iteratively:

$$\begin{aligned} & m_{oo} = -3 \\ k \geq 1 : \quad & m_{ko} = m_{k-1,o} \\ & m_{k\ell} = m_{k-1,\ell-1} + m_{k-1,\ell} \,, \quad 1 < \ell < k \\ & m_{kk} = m_{k-1,k-1} - 2 \\ & m_{k\ell} = 0 \,, \quad \ell > k \end{aligned} \tag{17}$$

Separating the factor $\frac{1}{\rho_i^{\,2}}$, and considering that

$$D\left(\frac{1}{\rho_i^{\,2}}\right) = -\frac{1}{\rho_i^{\,4}}\, D\,(\rho_i^{\,2})$$

one finds that

$$DR_i = \frac{1}{2\rho_i^{\,2}} \{m_{oo}\, [D^o R_i]\, [D(\rho_i^{\,2})]\}$$

$$D^2 R_i = \frac{1}{2\rho_i^{\,2}} \{\sum_{j=o}^{1} m_{1,j}\, [D^j R_i]\, [D^{2-j}(\rho_i^{\,2})]\}$$

or, in general, that

$$D^pR_i = \frac{1}{2\rho_i^2} \{\sum_{j=o}^{p-1} m_{p-1,j} [D^jR_i] [D^{p-j}(\rho_i^2)]\} , \; p\geq 1, \; i=1,2. \tag{18}$$

3.1.6. The factors $D^j(\rho_i^2)$ $(j=1,\ldots,p)$ are still unknown. It is easily proved that

$$D(\rho_1^2) = 2 \{x [Dx] + y [Dy] - \mu_2 [Dx]\}$$

$$D(\rho_2^2) = 2 \{x [Dx] + y [Dy] + \mu_1 [Dx]\} .$$

The linearity of the operator D now leads to the equations (19) and (20):

$$D^p(\rho_1^2) = 2 \{\sum_{j=o}^{p-1} \binom{p-1}{j} \{[D^{p-j}x] [D^jx] + [D^{p-j}y] [D^jy]\} - \mu_2 [D^px]\} , \qquad p \geq 1 \tag{19}$$

$$D^p(\rho_2^2) = 2 \{\sum_{j=o}^{p-1} \binom{p-1}{j} \{[D^{p-j}x] [D^jx] + [D^{p-j}y] [D^jy]\} + \mu_1 [D^px]\} , \qquad p \geq 1 . \tag{20}$$

All additional quantities for the calculation of the Lie series (12b) for x are known, except for the terms depending on $D^{p+1}y$. These will be calculated in the next sub-section.

3.2. Lie operator acting on the y-component

In analogy to Sect. 3.1., the recurrence formula for the multiple action of the operator D on y (D^3 up to any desired order D^{p+2}) is derived. According to equation (7), the first two terms are as follows:

$$Dy = y' \tag{21a}$$

$$D^2y = -2 [Dx] + \frac{1}{1+e \cos f} [y - \frac{\mu_1 y}{\rho_1^3} - \frac{\mu_2 y}{\rho_2^3}]. \tag{21b}$$

Using the similarity with the equations (9) and (10), new quantities are defined:

$$V_1 = \mu_1 y$$

$$V_2 = \mu_2 y$$

$$B = y + V_1 R_1 + V_2 R_2 \ . \qquad (22)$$

The higher powers of D applied to y are now given by an expression similar to equation (12a) for $D^{p+2}x$:

$$D^{p+2}y = -2\,[D^{p+1}x] + \sum_{j=0}^{p} \binom{p}{j} [D^{p-j}F]\,[D^jB] \ , \quad p \geq 0 \ . \qquad (23)$$

The factors $[D^jB]$ are calculated in analogy to the computation of $[D^jA]$ (equations (15) to (20)), where U_i (i=1,2) is to be replaced by V_i (i=1,2).

The solution for y is given by the Lie series:

$$y(f) = [y]_{\vec{X}_0} + [Dy]_{\vec{X}_0}(f-f_0) + \sum_{p=0}^{\infty} \{-2\,[D^{p+1}x]_{\vec{X}_0} + \sum_{j=0}^{p} \binom{p}{j} [D^{p-j}F]_{\vec{X}_0} [D^jB]_{\vec{X}_0}\} \{\frac{(f-f_0)^{p+2}}{(p+2)!}\} \ . \qquad (24)$$

3.3. Conclusion

The equations (12b) and (24) present the solution for the elliptic restricted three body problem in terms of Lie series. The formulae have an easy analytical structure and may be programmed without difficulty. Since any desired number of terms can be found by iteration, the series can be continued up to satisfactory convergence is reached.

As a test of the method, the orbits of long-period librating Trojans were calculated in the Sun-Jupiter system for 10 periods of Jupiter. It was found, that 15 terms and a stepsize of 5 steps per orbit of the primaries were enough to reach an accuracy up to the 15th digit in the dimensionless coordinates. A rigorous comparison of the Lie series method with another existing program of high accuracy is in preparation and we hope to present them in a following paper.

4. ORBIT CALCULATION WITH THE LIE SERIES FOR THE THIRD BODY

The orbit calculation uses the initial coordinates x_0, y_0 and velocities x_0', y_0' (with respect to the true anomaly f). The values of the terms in the Lie series (8) are calculated for the inital condition $\vec{X}_0(x_0, y_0, x'_0, y'_0, f_0)$. For each order in D, the terms in x and y are computed simultaneously: the first two explicitely by means of equations (9) and (21), the others are found through iteration (equations (12) and (24)). The velocities correspond to one order less in D than the coordinates.

The optimal number of terms needed will depend on the step size and on the desired accuracy. However, the next order term is easily computed, if necessary. A change of the step size $(f-f_0)$ during integration needs not more computer time, since no initial iterations are needed; only the powers $(f-f_o)^{p+2}$ must be calculated again. It is concluded, that the method of Lie series is a useful tool to calculate orbits in the restricted three body problem: only the infinitesimal body is integrated and the calculation is organized in a comfortable and rapid way. Moreover, any additional need for more accuracy can be fulfilled easily through a change of the step size or by taking more terms into the series.

REFERENCES

Gröbner,W.: 1967, Die Lie Reihen und ihre Anwendungen, VEB Berlin
Hanslmeier,A., Dvorak,R.: 1983, 'Numerical Integration with Lie series', Astron.Astrophys. 132, pp.203
Schneider,M.: 1979, Himmelsmechanik, BI Mannheim, p.408
Stumpff,K.: 1974, Himmelsmechanik, Vol.III, VEB Berlin
Szebehely,V., Giacaglia,G.E.O.: 1964, Astron.J.69, 270

THE STABILITY OF OUR SOLAR SYSTEM

P. J. Message
Dept. of Applied Mathematics & Theoretical Physics
Liverpool University
P.O. Box 147
Liverpool L69 3BX
England

ABSTRACT. Most investigations of the stability of the solar system have been concerned with the question as to whether the very long term effect of the gravitational attractions of the planets on each other will be to alter the nearly coplanar, nearly circular nature of the orbits in which they move. Analytical investigations in the traditions of Laplace, Lagrange, Poisson and Poincaré strongly indicate stability, though rely on asymptotic expansions with difficult analytical properties. The question is related to the existence of invariant tori, which have been proved to exist in certain motions. Numerical integration experiments have thrown considerable light on possible types of motions, especially in fictitious solar systems in which the planetary masses have been increased to enhance the perturbations, and in testing how critical are stability boundary estimates given by Hill surface type methods.

The real solar system in which we live is not, of course, simply a set of particles moving under their mutual gravitational attraction. That idealisation of the motions of the planets has however become recognised as so close an approximation to the real situation as far as planetary motions are concerned as to have become almost always identified with it, in the context of discussing the stability of the system. We should note, however, that however small any dissipative effects in the system may be, and one can scarcely doubt that there are some, they must ultimately disrupt the system in its present form, and in fact the most likely far and distant fate of the planets is to become part of the Sun itself. But non-gravitational forces are evidently of extremely small effect on motions in the system, as judged by the predictive success of planetary theories constructed on gravitational forces alone; only in the case of comets is it necessary to include non-gravitational forces in ephemeris construction. A further idealisation usually made in the context of the study of the stability of the system is to use Newton's inverse square law of gravitation and his equations of motion, and to suppose that the masses of the Sun and planets do not change, but we might notice that most of the conclusions which have been reached are unaltered by any change in the theory which essentially simply changes

Celestial Mechanics **34** (1984) 155–163. 0008–8714/84.15

the time scale, or which leaves the equations of motion in Hamiltonian form, as for example, in general relativity theory, using the next approximation beyond that given by Newton's law of gravitation.

In asking about the stability of the Solar System, it is usually understood that we are asking whether the planets may be expected to continue for ever to move in nearly circular, nearly coplanar orbits as they do at present, that is, whether the celestial clockwork is self-perpetuating, in the sense suggested by Nicholas of Oresme, or whether the gravitational attractions of the planets on each other could disrupt their motions, leading to the situation which Newton envisaged, in which it might be necessary for the Creator to put the planets back into their orbits from time to time. In the eighteenth century, however, Laplace and Lagrange found that the major semi-axes of the planetary orbits were, to first order in the perturbations, subject only to a set of superposed periodic changes, whose frequencies were linear combinations of the orbital mean motions of the planets. The investigations, also of Laplace and Lagrange, on the "secular variations" of the quantities ($e \cos \omega$, $e \sin \omega$, $\sin i \cos \Omega$, $\sin i \sin \Omega$) which describe the shape and orientation of the orbit of each planet showed that, to first order in these quantities, the long-period parts of the mutual planetary attractions give rise to superposed periodic changes of very long periods, up to millions of years, but with no tendency to disruption of the system. The Emperor Napoleon is said to have asked Laplace what part God played in his "Système du Monde", to which Laplace is reported to have answered that he had no need of that hypothesis. We could interpret that reply, in the context of Newton's anxieties, that the celestial clockwork has been properly made, and does not need periodic re-adjustment by its Creator.

Early in the following century, Poisson announced that the absence of a secular term (i.e. a term proportional to the time) in the perturbations of the major semi-axes of each planet extended also to the theory when extended to the second order in the masses of the planets, (which, expressed in units of the Sun's mass, are all less than 1/1000) though it has been since realised that this depends on the precise way in which the perturbation problem is set up. It is not difficult to see that Poisson's result is true in any formulation in which the equations of motion for the relative motions of the planets and the Sun can be described by a single Hamiltonian system, such as Jacobi's formulation, in which each planet's position is described by its position vector relative to the mass-centre of the Sun and all the planets in smaller orbits than its own. One has simply to consider the long-period Hamiltonian problem reached by a transformation of von Zeipel's or the Lie series type to remove those terms in the Hamiltonian which depend on the mean longitudes of the planets, and then in that long-period problem consider the solution in powers of the time, and relate it back to the original problem, to determine the perturbations in the major semi-axes, to first and second order in the planetary masses. Such considerations will also show that there will in general be secular terms in the major semi-axes of the third order, a result found with much more labour by

Haretu (1878) and Eginitis (1889). However, if each planet's position is described by its position vector relative to the Sun, then Poisson's result for the second order is not true if, for the constant μ in the expression μ/r^2 for the central force in the unperturbed Keplerian motion, one uses the constant of gravitation (G) times the sum of the masses of the Sun and the planet in question, in the usual way, but Poisson's result is true if one takes $\mu = G$ times the mass of the Sun. (See Duriez, 1978). The presence of secular perturbations in the major semi-axes, at any rate in the third order, appears to threaten the stability of the system. But this particular threat has in fact no basis, since these terms linear in the time, and in fact also the higher powers of the time which appear when perturbations of higher order are calculated, represent the beginning of Taylor series expansions of periodic expressions of longer period, as we shall see.

This procedure, of calculating the perturbations of successive orders in the planetary masses in turn, and which leads to secular terms, i.e. power series in the time, is sometimes known as "Poisson's method", and is the approach which has been most often used in the construction of analytical theories of the planets' motions, by Leverrier (1855), Newcomb (1895), Hill (1895), Clemence (1943, 1949, 1961), and others. The more recent work of Bretagnon, Chapront, and Simon (1975ff) in constructing "global" theories of all the eight major planets (apart from Pluto) makes use of iterative methods to determine the perturbations of higher orders for the four largest planets.

All of these types of theories give expressions for the orbital parameters or other quantities specifying the planets' positions, which contain short-period terms, that is periodic terms whose frequencies are linear combinations of the mean orbital angular motions of the planets, and also polynomials in powers of the time. But of course the secular variation theory of Laplace and Lagrange, which has been developed and improved successively by Stockwell (1874), Brouwer and van Woerkom (1950), Bretagnon (1974) and others, show that the parameters describing the orientation and shapes of the planetary orbits vary only by the superposition of periodic changes of very long periods, up to some millions of years. The frequencies of these periodic terms are all of the order of magnitude of the planets' mean orbital motions multiplied by the planets' masses (in units of the Sun's mass), and so any theory of planetary motions constructed by Poisson's method or a variant of it will represent such long-period terms by their Taylor series in powers of the time. This suggests that the secular terms in the major semi-axes given by Poisson's method may also be interpreted in this way, and Newcomb (1874) proposed expressions for the rectangular co-ordinates of the planets, comprised entirely of periodic terms, some of these of very long periods. In fact the question as to whether the major axes of the planets' orbits really possess secular terms was settled by Poincaré (1893a) who showed how expressions comprised entirely of periodic terms could be derived for the orbital parameters, by using a succession of canonical transformations. He includes two transformations of the type which have come to be called "von Zeipel's" (1916), and called by Poincaré "Lindstedt's method", though surely in reality due to Poincaré himself, in the form he uses. More recently, using the Lie series type of canonical transformation introduced by Hori (1966), explicit expressions for the orbital parameters have been able to be given, entirely in series of periodic terms, in which in principle any frequency may appear which is a linear combination of the mean orbital frequencies of the planets, and of the equal number of frequencies associated with the very long-period "secular variations" (Message 1976 and 1982). Geometrically, of course, such multiple Fourier series expressions correspond to an infinite series of epicycles for each planet, and it is clear that the threat to the stability of the system raised by the existence of secular terms in a Poisson method type of solution is only an apparent one, resulting from the use of that method. By beginning with the rectangular-type, or Poincaré, parameters for the shapes and inclinations of the orbits, as in the theory of the secular variations, it is shown that there are no singularities arising in the perturbation expressions when the eccentricities or inclinations of the orbits are zero, answering a concern of Poincaré (1893b). In reality, of course, circular motion, and coplanar motion, are especially simple types of motion, and only apparently singular if inappropriate types of parameter are used. Duriez (1977, 1979), following a method proposed by Brumberg and Chapront (1973), has constructed a numerical "general theory" of the perturbations of the four major outer planets, in which only periodic terms appear. To obtain a given precision in the prediction of planetary positions, many times more terms are needed than for a Poisson method type theory of the same precision, but the precision is retained for an interval of many hundreds of thousands of years.

However, all these types of theory use asymptotic series, to represent quasi-periodic functions by multiple Fourier series, and, while sets of initial conditions for which the series are convergent are everywhere dense in the space of possible initial configurations, so also are sets for which they are divergent, as was shown by Poincaré (1893c). So these series cannot be uniformly convergent, and while they indicate the stability of the system, they do not provide the mathematically rigorous proof of it. The difficulties arise in association with commensurabilities of the mean motions, where the divergencies occur. Error bounds for finite time intervals using strict and rigorous mathematical analysis prove to be many orders of magnitude too pessimistic, as shown by the agreement of the asymptotic series with numerical integration experiments, for very much longer time intervals than such analyses indicate (Message 1979).

Now in a separable Hamiltonian dynamical system (for example the "unperturbed" planetary problem, in which each planet is supposed attracted by the Sun, but not by the other planets), each possible motion may be shown to be in fact quasi periodic, with each co-ordinate expressible as a sum (usually infinite) of sines and cosines of linear combinations of a finite set of linear functions of the time, this being demonstrated by solving the Hamilton-Jacobi equation to provide action and angle variables, that is, angle co-ordinates (in each of which the positions are periodic with period 2π) each of which is a linear function of the time, and, as their conjugate momenta, "action" variables which are constants of the motion. Each dynamical motion in phase space is confined to a torus, corresponding to a particular set of values of the action momenta, and on which the angle variables are parameters. This solution is free of any infinite process, and so of any convergence difficulties such as beset the asymptotic series, just described, which result from the use of the von Zeipel or Lie series methods. The result of the use of the Lie series method, just described, indicates the motion to be of this type when the perturbations are considered also. It is of course the confinement of the solution to a torus, rather than the particular nature of the solution on the torus, which is relevant to the question of the stability of the system, and if two of the action momenta for each planet are G - L and H - G (in de Launay's notation) it is a region of the plane of these quantities near the origin which corresponds to nearly circular, nearly coplanar motion, and for such a region for each planet that we would like to be assured of stability. Now it has been proved, by a method of proof successively developed by Kolmogorov (1954), Arnol'd (1963), Moser (1962) and Rüssman (1970), that there do indeed exist invariant tori for sufficiently small values of the perturbation parameters (for example, the planetary masses in terms of that of the Sun), for sets of initial conditions whose complement set in a suitably chosen open set of phase space is of arbitrarily small measure. The formal proof has been given only for values of the perturbation parameters too small to be of relevance to the actual Solar System, but it is of course not excluded that it is true for large enough values. But of course we wish to be assured that the motions of our Solar System are confined in this way, and since we cannot know the parameters defining its motion with arbitrarily fine precision, we would like to be assured that all motions originating in an open region of phase space were confined to such invariant tori, that is, that the tori were stable, in the sense that a sufficiently small deviation from a point on an invariant torus still leaves one on an invariant torus. The tori do not enclose a region of phase space since for dynamical systems of more than two degrees of freedom they are of too small a dimension. So a motion originating between two invariant tori is not prevented from eventually moving to parts of phase space very distant from either torus, and corresponding to quite different types of motion. This is "Arnol'd diffusion", and the existence of the invariant tori does not exclude the possibility that it occurs, but it is clear that it can at most occur from a set of initial conditions of arbitrarily small measure, whithin suitably chosen regions of phase space. Considerations of this sort have led to the concept of "stability in measure", as introduced by Moser (1974), and of "dynamical stochastic metastability", introduced by Percival (1978): the property that the probability that a solution ever wanders more than an assigned distance from the torus can be made arbitrarily small by starting sufficiently close to the torus. We note that Jupp (1982) has found, in the study of the use of von Zeipel and Lie series methods in certain resonance problems, that, in the initial value problem, some sets of initial conditions in the original problem cannot be shown to correspond to solutions of the transformed problem.

We have seen that the difficulties of convergence in the von Zeipel and Lie series methods are associated with commensurabilities of the mean motions of the planets. We should note, however, that in the plane two planet problem (a case of the general gravitational problem of three bodies), there are, associated with each commensurability of orbital periods (that is each rational value of the ratio of the mean motions), one parameter families of periodic solutions (usually one family of stable solutions, and one family of unstable ones) of Poincaré's second sort (Message 1980). In a system with three or more planets, the counterpart of the periodic solutions of Poincaré's first sort are the solutions free of secular variations, that is in which the amplitudes of the secular variations are all zero, of the type found by Brumberg, Edvokimova, and Skripnichenko (1975).

Another instrument for the study of the motions in the Solar System is the direct integration of the equations of motion in rectangular co-ordinates, introduced by Cowell and Crommelin (1909) in their study of the motion of Halley's comet, and used by Brouwer, Clemence and Eckert (1950) to provide the ephemeris of the outer planets widely used since that time. The equations for the outer planets were integrated over a time span of one million years by Cohen, Hubbard, and Oesterwinter (1972) and analysis of their results shows the persistence over that time of the features indicated by the analytical theories. The more recent integration over five million years by Kinoshita and Nakai will enable analytical theories to be tested further. The power of numerical integration experiments in widening understanding of the phenomena presented by dynamical systems was shown by the work of Henon and Heiles (1964) in tracing solutions of a dynamical system which is related to motions in the galaxy. The results were displayed by the surface of section technique, that is, the successive intersections of a solution with a surface (usually a co-ordinate plane) in phase space were plotted. The existence of an invariant torus is shown in this method by the intersections lying on an invariant curve on the surface. What was generally found was that, as solutions corresponding to varying initial conditions were traced, there appeared to be a fairly sudden transition from solutions falling on invariant curves to those whose intersections appeared to scatter randomly over that part of the surface not occupied by the invariant curves. Such solutions have been designated "chaotic", or "quasi-ergodic" (they are not of course ergodic in the strict sense because of their exclusion from the region occupied by the invariant curves). Numerical experiments have been carried out by Hills (1970), Birn (1973), and others, with fictitious solar systems in which the planetary masses have been increased in order to "speed up" the mutual perturbations, to induce more significant effects to be shown within time spans accessible to practicable numerical integrations. Birn found indications of "bands" of stability and instability in the ratios of the mean motions, with the greatest instability being associated with the ratio of successive small integers. We should note however the results of Nacozy (1976) whose experimental integrations in a general three-body problem representing the Sun, Jupiter, and Saturn, indicated, as the planetary masses were increased, a fairly sudden transition from well-ordered stable type motions over the integration span of 100,000 years, and then, with the increase in the mass-augmentation factor from 29 to 29.25, to a motion in which the augmented Saturn was ejected in 10,000 years.

In the restricted problem of three bodies, the Jacobi integral provides, for values of the integral less than a certain critical value, regions of configurations space in which the third body is confined for all time. Confinement of this type has been called "Hill stability", since Hill (1878) used it to argue that, within the limitations of the approximation involved, the Moon could not escape from the vicinity of the Earth. More recently a stability criterion of like kind has been developed for the general problem of three bodies by Smale (1970), Easton (1971, 1975), Marchal and Saari (1975), Szebehely and Zare (1977), and Milani and Nobili (1983). The critical parameter in this criterion is the square of the angular momentum times the energy, in the relative motion of the three bodies (c^2h). For values of this parameter less than a critical value which depends on the masses of the bodies, the relative paths can be shown never to cross. This leads to the concept of "hierarchical stability", that is the situation in which the planetary orbits remain for all time non-overlapping. The investigations of Szebehely and McKenzie (1977), together with the numerical experiments of Nacozy already mentioned, show that the transition from stable-type motion occurs for mass-augmentation ratios rather larger than that for which this criterion ceases to guarantee hierarchical stability. There is of course no analytical reason why the transition should occur for that value (or at all). Numerical integration experiments by Walker and Roy (1983) for a variety of systems indicate that, for given mass values, as the ratio of major axes is increased, the transition takes place for a value a little beyond that for which the criterion ceases to guarantee hierarchical stability. Motions with small-integer commensurability of mean motion appear to display greater stability than those around them, and this appears to be related to the existence of periodic solutions in which the relative motion of the three bodies does not take them into any close approaches.

More recently Milani and Nobili (1983b), finding bounds for the rates of change of the "c^2h" parameters of the three-body subsystems in a four body system, have found lower bounds for the time-span for which hierarchical stability can be guaranteed for the four-body system. For the system comprised of the Sun, Mercury, Venus, and Jupiter, they find a value of 1.1×10^8 years. This is of course a lower bound, arrived at by making the most pessimistic assumption at each stage, and we should note that this result is quite consistent with belief in the stability of the system for all time.

REFERENCES

Arnol'd, V.I., 1963, Russian Mathematical Surveys, 18, 86.
Birn, J., 1973, Astronomy and Astrophysics, 24, 283.
Bretagnon, P., Astronomy and Astrophysics, 30, 141.
Bretagnon, P. with Chapront, J., and Simon, J.L., 1975, Astronomy and Astrophysics, 42, 259, & A. & A. Suppl., 22, 107, also Astron. & Astrophys., 69, 369, 1978, l.c., 84, 329, 1980, l.c., 101, 342, 1981, l.c., 103, 223, 1981, and l.c., 114, 278, 1982.
Brouwer, D., Clemence, G.M., and Eckert, W., 1950 Astron. Pap. Amer. Ephem., 12
Brouwer, D., and van Woerkom, A.J.J., 1950, Astron. Papers American Ephemeris 13, part 2, 81.
Brumberg, V.A., and Chapront, J., 1973, Celest. Mech., 8, 335.
Brumberg, V.A., Edvokimova, L.S., and Skripnichenko, V.I., 1975, Astronomiyeski Zhurnal, 52, 420.
Clemence, G.M., 1943, Astron. Papers American Ephemeris, 11, and 16, pt.2, 1961.
Cohen, C.J., Hubbard, E.C., and Oesterwinter, C., 1972, Astron. Pap. Amer. Ephemeris, 22, pt.1,9.
Duriez, L., 1977, Astronomy and Astrophysics, 54, 93.
Duriez, L., 1978, l.c., 68, 199.
Duriez, L., 1979, Thesis, Univ. des Sciences Techniques de Lille.
Easton, R., 1971, J. Diff. Equns., 10, 371, and 1975, l.c., 19, 258.
Eginitis, M.D., 1889, Annales de l' Obs., Paris, Memoires, 19.
Henon, M., and Heiles, C., 1964, Astron. Journ., 69, 73.
Haretu, S.C., 1878, Thesis, Fac. de Sciences, Paris.
Hill, G.W., 1878, Amer. Journ. Math., 1, 5.
Hill, G.W., 1895, Astron. Pap. Amer. Ephemeris, 7.
Hills, J.G., 1970, Nature, 225, 840.
Hori, G.I., 1966, Publ. Astr. Soc. Japan, 18, 287.
Jupp, A.H., 1982, Celest., Mech., 26, 413.
Kolmogorov, A.N., 1954, Proc. Int. Math. Congress, Amsterdam, 1, 315, also Dokl. Akad. Nauk. SSSR, 98, 527.
Leverrier, U.J.J., 1855 ff., Annales de l' Obs Paris, Memoires, 1 ff.
Lagrange, J.L., 1781, Nouveaux Mem. de l' Acad. Roy. de Sci. et Belles Ll. de Berlin.
Marchal, C. and Saari, D., 1975, Celest. Mech., 12, 115.
Message, P.J., 1976, in Long Time Predictions in Dynamics, ed. V. Szebehely and B. Tapley, 279.
Message,P.J., 1979, in Instabilities in Dynamical Systems, ed. V. Szebehely, 165.
Message, P.J., 1980, Celest. Mech., 21, 55.
Message, P.J., 1982, Celest. Mech., 26, 25.
Moser, J., 1962, Nachr. Akad. Wiss. Gottingen Math.Phys. Klasse II.1.
Moser, J., 1974, in The Stability of the Solar System and of Small Stellar Systems, ed. Y. Kozai, 1.
Milani, A., and Nobili, A., 1983a, Celest. Mech., 31, 213.
Milani, A., and Nobili, A., 1983b, Celest. Mech., 31, 241.
Nacozy, P., 1976, Astron. Journ., 81, 787.

Newcomb, S., 1874, Smithsonian Contribns., No. 281.
Newcomb, S., 1895, Astron. Pap. Amer. Ephem. 6.
Percival, I.V., 1978, Report of Dept. of Appl. Math., Queen Mary Coll., London.
Poincaré, H., 1893a, Methodes Nouvelles de la Mecanique Celeste, (Gauthiar Villars, Paris), vol. II, Chs. IX, X, and XI.
Poincaré, H., 1893b, l.c. Ch. XII.
Poincaré, H., 1893c, l.c. Ch. XIII.
Poisson, D., 1809, Journ. de l' Ecole Polytechnique, Paris, 15, 1.
Russmann, H., 1970, Nachr. Akad. Wiss. Gottingen Math. Phys. Kl.II 67.
Stockwell, J.N., 1874, Smithsonian Contribns., No. 282.
Smale, S., 1970, Invent. Math., 10, 305 and 11, 45.
Szebehely, V., and McKenzie, R., 1977, Astron. Journ., 82, 79.
Szebehely, V., and Zare, K., 1977, Astron. and Astrophys., 58, 145.
Walker, I., and Roy, A.E., 1983, Celest. Mech., 29, 117.
von Zeipel, H., 1916, Ark, fur Mat., Ast. and Fysik, 11, No. 1.
Zare, K., 1976, Celest. Mech., 14, 73.

APPROXIMATION METHODS IN CELESTIAL MECHANICS. APPLICATION TO PLUTO'S MOTION.

J. CHAPRONT
Bureau des Longitudes. Equipe de Recherche Associée au CNRS.
77, avenue Denfert Rochereau, 75014 Paris. France.

ABSTRACT. Various methods of approximation for the computation of planetary perturbations are investigated: Fourier, Fourier-Chebyschev and Legendre. Application is made to Pluto's motion. Based upon JPL's ephemeris DE200, Pluto's mean elements are provided and also a compact ephemeris covering 2 centuries.

1. INTRODUCTION

The elaboration of an analytical or a semi-analytical theory of Pluto's motion is a difficult problem which never received a complete solution, leading to an ephemeris. It mainly exists three types of difficulties. On one side, large values of the eccentricity (0.25) and inclination (17°) slow down considerably the numerical convergence of the semi-analytical developments, either in the case of perturbations of Pluto by Neptune or those produced by the other planets. On the other side, in the case of the pair of planets Neptune and Pluto alone, orbits intersect in projection; the usual developments of the inverse distance with Laplace coefficients, in the ratio of the semi-major axes, are not any longer feasible. Petrovskaia (1970) proposes a solution to overcome this difficulty which leads to cumbersome developments. At last, the resonance induced by the small divisor $2\bar{\lambda}_N - 3\bar{\lambda}_P$ ($\bar{\lambda}_N$ and $\bar{\lambda}_P$ are the mean longitudes of Neptune and Pluto respectively) whose period is about 25 000 years, does not allow us to build a solution with classical methods used in theories with secular variations as those of Le Verrier, or modern ones (Bretagnon, 1982) for the whole set of planets of the solar system.

Here, we are not concerned by the evolution and stability on the long range for Pluto's motion. Those studies are based upon results given by numerical integrations as the one which has been carried out over one million years by (Cohen et al., 1972), or by averaging methods (Giacaglia and Nacozy, 1970). Our goal is to build a precise solution for Pluto's motion over a few periods of revolution ($\sim$ 1000 years) which complements modern theories with secular variations for planets. Two approaches are sought:

Celestial Mechanics 34 (1984) 165–184. 0008–8714/84.15

1.1 Approximation for the pair of planets Neptune and Pluto by Fourier-Chebyschev series.

In order to get rid of the resonance $\rho = 2\bar{\lambda}_N - 3\bar{\lambda}_P$, we choose to represent the motion over a time interval which is small as compared to the period of ρ. Then, we look for a solution in the form of periodic functions of the time whose coefficients are slowly variable, and can be approached by developments in Chebyschev series.

We have already discussed this method for approximating semi-analytical solutions for the motion of the Moon and planets (Chapront, 1982). Here, the method is used to solve the equations of motion, in elliptical variables, for the couple Neptune-Pluto. The long periods being all developed with respect to the time, we do not distinguish between resonance and secular variations for nodes and perihelia. The Fourier terms are obtained by Fourier approximation. The large values of Pluto's eccentricity and inclination do not bring peculiar difficulties; if the large value of the ratio of the semi-major axes slow down the numerical convergence, we will see later that it is possible nevertheless to reach a very high accuracy with a small number of terms. This proceeding generalizes the method to solve a differential system of equations with series of Chebyschev polynomials, when performing a direct computation of planetary ephemerides (Carpenter, 1966), (Chapront, 1977). The developments in Fourier-Chebyschev series remain valid as long as the interval of approximation is small as compared to long periods (secular variations and/or resonances). This type of approximation is intermediate between a tabulated ephemeris as it results from a numerical integration, and the developments in quasi-periodic functions of the time.

Several years ago, such a method has already been investigated by Goodrich and Carpenter (1966) in the case of the computation of general planetary perturbations for resonance cases, and applied to the same problem of the perturbations of Neptune due to Pluto, and also for the perturbations of Hilda due to Jupiter. Thanks to P.K. Seidelmann (1984), this reference has been discovered after performing the present work.

1.2 Approximation of a tabulated ephemeris.

If our ultimate goal is to produce an ephemeris for Pluto, valid over a few centuries, the N-bodies numerical integration remains the simplest and less costly tool. Then we may look for a function of approximation resulting from this numerical integration, which has to be compact and which preserves the handiness of a semi-analytical formulation. In that respect, we have built a solution for the complete problem of Pluto's motion based upon JPL's numerical integration DE200-LE200 (Standish and Williams, 1982), which is valid over one period of revolution. Over a finite time interval, one approaches DE200 by an array of Poisson's terms (mixed functions of the time in polynomial and sinusoidal form), the norm of the approximation being quadratic mean (Chapront and Vu, 1984). Because of the simplicity of the result we gave up the other forms, from the point of view of the ephemerides.

Before discussing this last approximation, we first present the attempts which have been made to solve the problem of the couple

Neptune-Pluto alone by means of Fourier approximation, in terms of quasi-periodic functions of the time.

2. FOURIER'S APPROXIMATION.

At Bureau des Longitudes, theories have been built to represent planetary motions with secular variations : VSOP82 (Bretagnon, 1982) for the eight planets, and TOP82 (Simon, 1983) for the 4 major planets. At the beginning, we have chosen the same approach to construct at the first order of planetary masses, Pluto's perturbations by the whole set of planets.

We denote the variables by :

λ : mean longitude,

a : semi-major axis,

$k + \sqrt{-1}\, h = e \exp \sqrt{-1}\, \tilde{\omega}$, (e : eccentricity; $\tilde{\omega}$: longitude of perihelion),

$q + \sqrt{-1}\, p = \sin \frac{i}{2} \exp \sqrt{-1}\, \Omega$, (i : inclination; Ω : longitude of the node).

σ standing for any one of the elliptical variables, one writes the system of Lagrange's equations :

$$\frac{d\sigma_P}{dt} = \sum_{i=1}^{8} \mu_i F_i(\sigma_P , \sigma_i) \quad . \tag{1}$$

σ_P are the elements belonging to Pluto :

$$\sigma_P = \{ a_P , \lambda_P , k_P , h_P , q_P , p_P \} \quad .$$

σ_i are the elements corresponding to the ith disturbing planet and μ_i the masses with respect to the Sun (i = 1 to 8). The reader will find in (Chapront et al., 1975) an explicit formulation of equations (1). We use for σ_i mean values for the epoch J2000.0, as they appear in VSOP82, this last solution being compared and fitted to DE200.

At the first order of planetary masses, it is easy to elaborate a solution to system (1), for $i \leq 7$. As an illustration, we give in table 1 , Pluto's perturbations larger than 20 arcseconds. We denote them by $\tilde{\sigma}_P$. Pluto's elements are gathered in table 2. We denote them by $\sigma_P^{(0)}$. They are derived from a comparison with DE200 (see below).

In the case of the pair Neptune and Pluto (i = 8), the construction of the right hand members in (1) with series whose arguments are linear combination of the mean longitudes involves the computation of the inverse distance $1/\Delta$ between the two planets. For its development, we use Newton's approximation :

$$(1/\Delta)_k = \frac{1}{2} (1/\Delta)_{k-1} \times \{3 - (1/\Delta)_{k-1}^2 \Delta^2\} \quad .$$

k is the iteration number. Δ^2 is a series which is built once for all, easy to obtain and rapidly convergent when it is developed in multiples of the mean longitudes λ_N (Neptune) and λ_P (Pluto).

Table 1. First order perturbations with respect to planetary masses of Pluto's elements $\tilde{\sigma}_P$ by outer planets. J, S, U and P are mean longitudes of Jupiter, Saturn, Uranus and Pluto reckoned from J2000.0. Coefficients are expressed in arcseconds.

$\tilde{a}_P/a_P^{(0)}$	=	+ 1083 cos (J - P)	- 1 sin (J - P)
		- 199 cos (J - 2P)	+ 193 sin (J - 2P)
		+ 255 cos (S - P)	
		- 99 cos P	- 97 sin P
		- 50 cos (S - 2P)	+ 48 sin (S - 2P)
		+ 2 cos (J - 3P)	- 80 sin (J - 3P)
		+ 50 cos (2J - P)	+ 13 sin (2J - P)
			+ 35 sin 2P
		+ 16 cos (J - 4P)	+ 18 sin (J - 4P)
		+ 32 cos (U - P)	
		- 21 cos (U - 3P)	- 5 sin (U - 3P)
		- 16 cos (2U - 6P)	+ 8 sin (2U - 6P)
		+ 1 cos (S - 3P)	- 21 sin (S - 3P)
$\tilde{\lambda}_P$	=		+ 1037 sin (J - P)
		+ 34 cos (U - 3P)	- 628 sin (U - 3P)
		- 212 cos J	+ 217 sin J
		- 114 cos (2U - 6P)	- 237 sin (2U - 6P)
			+ 222 sin (S - P)
		- 64 cos (J - 2P)	- 66 sin (J - 2P)
		- 47 cos S	+ 48 sin S
		- 13 cos (2J - P)	+ 49 sin (2J - P)
		- 24 cos (3U - 9P)	+ 26 sin (3U - 9P)
		- 17 cos P	+ 19 sin P
		- 13 cos (S - 2P)	- 13 sin (S - 2P)
			+ 23 sin (U - P)
		- 13 cos 2J	+ 8 sin 2J
$\tilde{q}_P$	=	+ 14 cos (J - P)	- 37 sin (J - P)
		+ 13 cos (J + P)	- 33 sin (J + P)
		+ 8 cos J	- 19 sin J
$\tilde{p}_P$	=	- 36 cos (J - P)	- 14 sin (J - P)
		+ 33 cos (J + P)	+ 12 sin (J + P)
		+ 7 cos J	- 19 sin J

Table 1. (continued)

$\tilde{k}_P$	=	+ 785 cos J	+ 11 sin J
		+ 255 cos P	− 3 sin P
		+ 248 cos (J − 2P)	− 5 sin (J − 2P)
		+ 173 cos S	+ 2 sin S
		+ 48 cos (J − P)	+ 96 sin (J − P)
		− 70 cos (J − 3P)	+ 69 sin (J − 3P)
		− 47 cos 2P	− 45 sin 2P
		+ 62 cos (S − 2P)	− 1 sin (S − 2P)
		+ 37 cos 2J	+ 10 sin 2J
		+ 23 cos (J + P)	+ 22 sin (J + P)
		− 18 cos (S − 3P)	+ 18 sin (S − 3P)
		+ 1 cos (J − 4P)	− 34 sin (J − 4P)
		+ 11 cos (S − P)	+ 23 sin (S − P)
		+ 21 cos (2U − 6P)	+ 12 sin (2U − 6P)
		− 10 cos (U − 3P)	+ 19 sin (U − 3P)
$\tilde{h}_P$	=	+ 11 cos J	+ 762 sin J
		− 3 cos P	+ 256 sin P
		+ 5 cos (J − 2P)	− 248 sin (J − 2P)
		+ 2 cos S	+ 169 sin S
		+ 46 cos (J − P)	− 100 sin (J − P)
		+ 67 cos (J − 3P)	+ 71 sin (J − 3P)
		+ 46 cos 2P	− 47 sin 2P
		+ 1 cos (S − 2P)	− 62 sin (S − 2P)
		− 9 cos 2J	+ 36 sin 2J
		− 22 cos (J + P)	+ 22 sin (J + P)
		+ 17 cos (S − 3P)	+ 18 sin (S − 3P)
		− 33 cos (J − 4P)	− 1 sin (J − 4P)
		+ 11 cos (S − P)	− 24 sin (S − P)
		+ 12 cos (2U − 6P)	− 21 sin (2U − 6P)
		+ 23 cos (U − 3P)	+ 9 sin (U − 3P)

When the starting approximation $(1/\Delta)_0$ is a development with Laplace coefficients where the ratio of the semi-major axes is : $a_N/a_P = 0.76$, the process is not convergent. Petrovskaia (1977) demonstrates the convergence of Newton's process when the first approximation is : $(1/\Delta)_0 = 1/\sqrt{\max \Delta^2}$.

Ivanova (1978) has undertaken such an approximation in the case of Neptune-Pluto pair. On our side, we have reproduced such a method. After 12 iterations we are led to an absolute convergence in evaluating $1/\Delta$ of $2.10^{-5} AU^{-1}$, with series of 3500 terms (!). If now, we raise this series to the third power, $(1/\Delta)^3$, as it arises in the disturbing force, the precision remains $2.10^{-3} AU^{-3}$, with 3200 terms.

Table 2. Pluto's constants (mean elements related to Δt = 1000 years) determined by fitting to DE200 over one century the variables $\sigma_P^{(0)}$. The mean longitude is : $\bar{\lambda}_P = \nu_P t + \varepsilon_P^{(0)}$. t is expressed in years from J2000.0

$a_P^{(0)}$	= 39.544 674 UA	$\varepsilon_P^{(0)}$	= 4.166 868 radians
$k_P^{(0)}$	= - 0.178 744	$h_P^{(0)}$	= - 0.173 426
$q_P^{(0)}$	= - 0.051 699	$p_P^{(0)}$	= 0.139 782
ν_P	= 0.025 333 30 radian per year.		

Apart from the resonance, it is not in the slow convergence that the difficulty lies. In spite of the bulkiness of the series, it is nevertheless possible to integrate system (1), and to obtain all short periodic inequalities with an amplitude of 0".2. Obviously such an accuracy is not preserved when summing up the series. The main difficulty lies in the resonance $\rho = 2\bar{\lambda}_N - 3\bar{\lambda}_P$ and its harmonics $k\rho$. For the longitude in the integration of the right-hand members of (1) the amplitudes depend on values of the mean motions assigned to Neptune (ν_N) and Pluto (ν_P).

Neptune's mean motion for J2000.0, according to VSOP82, is :

$$\nu_N = 0.038\,133\,036 \text{ radian per year.}$$

If one chooses for ν_P a value derived from the numerical integration of (Brouwer et al., 1951) as it appears in *Connaissance des Temps* (1979)

$$\nu_P = 0.\,025\,368\,11 \text{ radian per year.} \qquad (2)_a$$

We then obtain,

$$\nu_\rho = 2\nu_N - 3\nu_P = 0.000\,161\,74 \text{ (period} \sim 39\,000 \text{ years).}$$

If one chooses for ν_P the value that we got (Table 2.) after comparison with DE200,

$$\nu_P = 0.025\,333\,30 \text{ radian per year,} \qquad (2)_b$$

we find,

$$\nu_\rho = 0.000\,266\,17 \text{ (period} \sim 24\,000 \text{ years).}$$

With the double integration in the longitude, we are led to perturbations whose amplitudes reach several radians. The corresponding amplitudes in the harmonics $k\rho$ decrease slowly. For k = 16, the coefficients of the inequality $32\bar{\lambda}_N - 48\bar{\lambda}_P$ (period: 1500 years) is as large as 10^{-3} radian. The very small value of ν_ρ for J2000.0, inducing finite values for the perturbations with respect to planetary masses, forbids us all approaches with the classical methods used in theories with secular variations.

Besides, to get rid of the resonance, all developments with respect to time of ρ and its multiples cannot be obtained with the preceding series because of the lack of convergence. Fourier's approximation in multiples of planetary mean longitudes is bound to fail for a semi-analytical treatment, even on a short time interval.

3. FOURIER-CHEBYSCHEV'S APPROXIMATION.

In an earlier work (Chapront, 1982) we introduced the concept of time series approximation in Celestial Mechanics by Fourier-Chebyschev (F-T) developments. Given a series f(t), let us suppose that we know its explicit form, as it results from a semi-analytical theory, in terms of quasi-periodic functions of the time.

$$f(t) = \sum_{l=1}^{L} C_l(t).\exp\sqrt{-1}\omega_l t . \qquad (3)$$

$C_l(t)$ is a slowly variable time function. More generally it is the development of secular inequalities. Here ω_l are frequencies : $0 \leqq \omega_l \leqq \omega_{MAX}$. If we set a basic frequency μ, which verifies, when p is an integer,

$$(p - 1)\mu \leqq \omega_l < p\mu \qquad ; \qquad p = 1, 2, \ldots, P$$

$$P\mu \geqq \omega_{MAX} ,$$

it is shown that over a finite interval of time Δt, (3) can be written,

$$f(t) = \sum_{p=1}^{P} \sum_{n=0}^{N(p)} s_n^{(p)} T_n(t) \exp\sqrt{-1}(p\mu - \tfrac{\mu}{2})t . \qquad (4)$$

$$t \in \Delta t \qquad ; \qquad s_n^{(p)} \in \mathbb{C}$$

T_n are Chebyschev polynomials of degree $n \leqq N$.

Here, we shall introduce the Chebyschev polynomials with the concept of least squares polynomial approximation (Hildebrand, 1956). They are the set of orthogonal polynomials which possess the orthogonality property:

$$\int_{-1}^{+1} \frac{T_r(x).T_s(x)}{\sqrt{1 - x^2}} dx = \delta_s^r \ ; \quad \text{(weighting function: } 1/\sqrt{1-x^2})$$

They may be determined from the recurrence formula,

$$T_{n+1}(x) = 2xT_n(x) - T_{n-1}(x) \quad ; \quad T_0 = 1,\ T_1 = x.$$

In (4), the partial sums

$$\sum_{n=0}^{N(p)} s_n^{(p)} T_n$$

represent the slowly variable functions of time over the interval Δt;

their periods T satisfy,

$$T \geqq \frac{4\pi}{\mu} \quad .$$

Given a precision ε, if Δt is sufficiently small, the integers N(p) remain small :

$$N(p) \leqq N(\varepsilon) \qquad ; \qquad p = 1, 2, \ldots, P.$$

This formulation has been used to approximate lunar and planetary coordinates given by analytical theories built at Bureau des Longitudes. To generalize such a method, we have been brought to set the following problem : Find directly the solution of differential equations like (1), with F-T approximation.

3.1 The couple Neptune-Pluto.

In the case of Neptune-Pluto, we have been guided by the physical nature of the problem. We have assigned to the basic frequency μ the value, once for all :

$$\mu = \frac{\nu_N}{3} = 0.012\,711\,012 \text{ radian per year,}$$

corresponding to ν_N , Neptune's mean motion from VSOP82. Hence,

$$\nu_P = 2\mu + \eta_P \quad .$$

We have initiated our solution with the value $(2)_a$ for ν_P. η_P gets the numerical value,

$$\eta_P = -5.39\ 10^{-5} \text{ radian per year.}$$

We have chosen the time interval Δt = 1000 years, centered on the epoch J2000.0. The difference between the mean longitude $\bar{\lambda}_P$ and its first approximation :

$$(2\mu + \eta_P)t + \varepsilon_P \ ,$$

remains small on the interval. The other elements σ_P for Pluto have been derived from the numerical integration of Oesterwinter and Cohen (1972). It is useless to reproduce them here.

In Eqs. (1), σ_N and σ_P are constants for the set : {a, k, h, q, p}. The mean longitudes are,

$$\bar{\lambda}_N = 3\mu t + \varepsilon_N \ , \qquad\qquad (5)$$

$$\bar{\lambda}_P = 2\mu t + \varepsilon_P + \eta_P t \quad .$$

Except for the term $\eta_P t$, the right-hand members of Lagrange's equations (1) are periodic functions in

$$\theta = \mu t \quad .$$

Provisionally, denote by τ the time variable entering (5) through $n_P t$.

$$\bar{\lambda}_N = 3\theta + \varepsilon_N \quad ,$$

$$\bar{\lambda}_P = 2\theta + \varepsilon_P + n_P\tau \quad .$$

For a fixed value of τ, we perform a Fourier analysis of the right-hand members (1), i.e., we develop the functions,

$$\frac{d\sigma}{dt} = F_\sigma(\theta , \tau)$$

under the form of Fourier series,

$$F_\sigma(\theta , \tau) = \mathrm{Re}\ \{ \sum_{k=0}^{K} (C_k(\tau) - \sqrt{-1}\ S_k(\tau)) \exp \sqrt{-1}k\theta \} . \quad (6)$$

C_k and S_k are Fourier coefficients obtained by classical integrals, approximated by sums,

$$C_k + \sqrt{-1}\ S_k = \frac{1}{K(1+\delta_k^0 + \delta_k^K)} \sum_{j=0}^{2K-1} F_\sigma(j\frac{\pi}{K} , \tau) \exp \sqrt{-1}\, jk\frac{\pi}{K}$$

$$k = 0, 1, \ldots, K \quad .$$

Practically, in heliocentric coordinates, F_σ for Pluto is written,

$$F_{\sigma_P} = \frac{1}{n_P a_P^2} D_{\sigma_P} \vec{V}_P \cdot \vec{\mathrm{grad}} R_P \quad , \quad (7)$$

and,

$$\frac{d\lambda_P}{dt} = n_P + F_{\lambda_P} \quad .$$

D_{σ_P} is an elementary operator in terms of σ_P , acting on the position vector $\vec{V}_P$; R_P is the disturbing function with,

$$\frac{1}{n_P a_P^2} \vec{\mathrm{grad}}\ R_P = \frac{n_P a_P m_N}{1 + m_P} \left(\frac{\vec{V}_N - \vec{V}_P}{||\vec{V}_N - \vec{V}_P||^3} - \frac{\vec{V}_N}{||\vec{V}_N||^3} \right)$$

n stands for the osculatory meanmotions and m for the masses. The Fourier analysis is directly performed on the close form (7).

Over the time interval $\tau \in \{-\frac{\Delta t}{2} , +\frac{\Delta t}{2}\}$, we consider the linear transformation :

$$\tau = \frac{\Delta t}{2} x \quad .$$

We denote by x_m , $(m = 0, 1, \ldots, M-1)$ the Chebyschev abscissa over the set {-1, +1} , i.e.,

$$x_m = \cos \frac{2m + 1}{2M} \pi \ .$$

The corresponding times are denoted by τ_m . In formula (6) the coefficients,

$$U_k(\tau) = C_k(\tau) - \sqrt{-1}\ S_k(\tau)$$

are approximated by the developments in Chebyschev series :

$$U_k(\tau) = \sum_{m=0}^{M-1} u_m^{(k)} T_m(x) \quad ; \quad x = \frac{2\tau}{\Delta t} \ .$$

Coefficients $u_m^{(k)}$ are obtained by the classical sums,

$$u_m^{(k)} = \frac{2 - \delta_m^0}{M} \sum_{j=0}^{M-1} U_k\left(\cos \frac{2j+1}{2M}\pi \right) \times \cos m \frac{2j+1}{2M}\pi \ ,$$

$$m = 0, 1, \ldots, M-1 \ .$$

Finally, over the time interval Δt, the second members $F_\sigma(\theta, \tau)$ are approximated by the double sums,

$$F_\sigma(\theta, t) = \mathrm{Re}\left\{ \sum_{k=0}^{K} \sum_{m=0}^{M-1} u_m^{(k)} T_m(\tau) \exp \sqrt{-1}\ k\theta \right\} . \tag{8}$$

$$\tau = t \in \Delta t$$

It remains to integrate formula (8) to come back to variables σ_P :

$$\sigma_P = \int F_{\sigma_P}(\mu t, t)\, dt + \text{constant} \ . \tag{9}$$

We have established the following formula for the integration :
Let us consider the integral,

$$I_m = \int T_m(t) \exp \sqrt{-1}\ gt\, dt \tag{10}$$

which enters (8) and (9) with $gt = k\theta = k\mu t$.
We know the following property for the integrals of Chebyschev polynomials,

$$\int T_m dt = \frac{1}{2}\left(\frac{T_{m+1}}{m+1} - \frac{T_{m-1}}{m-1} \right) \quad ; \quad m > 1 \ .$$

An integration by parts of (10) leads us to the recurrence relation,

$$I_m = \sqrt{-1}\ \frac{2m}{g} I_{m-1} + \frac{m}{m-2} I_{m-2}$$
$$- \frac{\sqrt{-1}}{g}\left(T_m - \frac{m}{m-2} T_{m-2}\right) \exp \sqrt{-1}\ gt \ . \tag{11}$$

Let us put,

$$I_m = K_m \exp \sqrt{-1}\ gt .$$

We finally obtain the following recursive formulae,

$$K_0 = -\frac{\sqrt{-1}}{g} \quad , \quad K_1 = \frac{\sqrt{-1}}{g} K_0 - \frac{\sqrt{-1}}{g} T_1 \quad ,$$

$$K_2 = 4\frac{\sqrt{-1}}{g} K_1 - \frac{\sqrt{-1}}{g} T_2 \quad ,$$

and for m > 2,

$$K_m = 2m \frac{\sqrt{-1}}{g} K_{m-1} + \frac{m}{m-2} K_{m-2} - \frac{\sqrt{-1}}{g} (T_m - \frac{m}{m-2} T_{m-2}) \quad .$$

The integration (9) of the second members $F_\sigma(\theta, \tau)$ leads to a formulation of σ_P of the same nature as (8), i.e.,

$$\sigma_P(\theta, t) = \sigma_P^{(0)} + \mathrm{Re}\ \{ \sum_{k=0}^{K} \sum_{m=0}^{M-1} s_m^{(k)} T_m(\tau) \exp \sqrt{-1}\ k\theta \} \quad (12)$$

$$\tau = t \in \Delta t \quad .$$

For λ_P , in (12), $\sigma_P^{(0)}$ should be replaced by : $2\theta + \varepsilon_P^{(0)} + \eta_P t$.

The integration constants $\sigma_P^{(0)}$ and η_P will be determined later, by comparison of our solution with an external model : the numerical integration DE200 fitted to observations.

In (12), we evaluate $\sigma_N(t)$ like $\sigma_P(t)$. Starting with this first approximation for the elements of Neptune and Pluto, we come back to Eqs. (1) to undertake a new approximation. When τ is fixed in (12), a new Fourier analysis of the right-hand members is performed, next a development in Chebyschev polynomials, and lastly a new integration. Let us notice that the algorithm is very simple. Whether it concerns Fourier analysis with an argument θ , or the numerical substitution of a Chebyschev abscissa τ_m in $U_k(\tau)$, we are always led to sum up series of the type :

$$G_N(z) = \sum_{n=0}^{N} \alpha_n z^n \quad ; \quad z \in \mathbb{C} \quad ,$$

where α_n are numerical coefficients. Horner's scheme is well fitted and particularly efficient :

$$G_n = \alpha_{N-n} + zG_{n-1} \quad ; \quad 0 \leqq n \leqq N \quad ; \quad G_{-1} = 0 \ .$$

Practically in the developments (8) and (12) we fixed the upper values K = 50 and M = 10. We performed 6 iterations on the system of Lagrange's equations. The neglected moduli of $s_m^{(k)}$ are always less than

10^{-11} radians, which ensures a very high precision on the coordinates. If it is superabundant for Pluto's motion, it has been kept in order to prove the efficiency of the method.

Table 3. The pair of planets Neptune and Pluto alone.
Convergence of the coefficients $s_m^{(0)}$ in Pluto's mean longitude, λ_P^*. Units : 10^{-5} radians.

Iteration number	$s_1^{(0)}$	$s_2^{(0)}$	$s_3^{(0)}$
1	- 229.4142	+ 131.7719	- 229.2160
2	- 228.4474	+ 130.9045	- 228.5833
3	- 228.4507	+ 130.9059	- 228.5871
4	- 228.4507	+ 130.9059	- 228.5871
	$s_3^{(0)}$	$s_4^{(0)}$	$s_5^{(0)}$
1	+ 6.255521	- 0.099006	+ 0.002478
2	+ 6.046955	+ 0.067595	- 0.006280
3	+ 6.047105	+ 0.067950	- 0.006263
4	+ 6.047105	+ 0.067955	- 0.006264

Table 4. The pair of planets Neptune and Pluto alone.
Secular variations over Δt = 1000 years for Pluto's elliptical elements : σ_P^*. t is expressed in julian centuries from J2000. $t \in \{-5, +5\}$. Coefficients are in arcsecond.

$$< a_P^* / a_P^{(0)} > = +19''770\, t - 0''314\, t^2 - 0''0019\, t^3 + 0''00009\, t^4$$

$$< \lambda_P^* > = +46''506\, t - 37''764\, t^2 + 0''4012\, t^3 + 0''00177\, t^4 - 0''00007\, t^5$$

$$< k_P^* > = -16''256\, t + 0''120\, t^2 + 0''0008\, t^3 - 0''00003\, t^4$$

$$< h_P^* > = +1''973\, t + 0''140\, t^2 + 0''0005\, t^3 - 0''00004\, t^4$$

$$< q_P^* > = +1''045\, t - 0''006\, t^2 + 0''0001\, t^3$$

$$< p_P^* > = +0''440\, t + 0''006\, t^2$$

Table 5. The pair of planets Neptune and Pluto alone.
Periodic perturbations in Pluto's mean longitude λ_P^* .

t is expressed in julian centuries from J2000.0 . $t \in \{-5, +5\}$.
Coefficients are in arcseconds.
$\mu = 1.271\,101\,2$ radian per century.

$$
\begin{array}{lllllll}
\lambda_P^* & = & (- 15''.57 & - 3''.153\,t & - 0''.017\,t^2 & + 0''.0025\,t^3 &)\cos \mu t \\
 & + & (+ 25''.83 & - 0''.870\,t & + 0''.005\,t^2 & + 0''.0003\,t^3 &)\sin \mu t \\
 & + & (+ 15''.82 & - 4''.864\,t & - 0''.003\,t^2 & + 0''.0003\,t^3 &)\cos 2\mu t \\
 & + & (+ 24''.18 & - 0''.251\,t & & + 0''.0002\,t^3 &)\sin 2\mu t \\
 & + & (- 14''.74 & - 0''.082\,t & & + 0''.0001\,t^3 &)\cos 3\mu t \\
 & + & (+ 6''.08 & - 0''.149\,t & & + 0''.0001\,t^3 &)\sin 3\mu t \\
 & + & (- 4''.08 & + 0''.080\,t & + 0''.001\,t^2 & &)\cos 4\mu t \\
 & + & (- 4''.50 & - 0''.026\,t & & &)\sin 4\mu t \\
 & + & (+ 1''.95 & + 0''.021\,t & & &)\cos 5\mu t \\
 & + & (- 3''.87 & + 0''.073\,t & + 0''.001\,t^2 & &)\sin 5\mu t \\
 & + & (+ 2''.47 & - 0''.046\,t & - 0''.001\,t^2 & &)\cos 6\mu t \\
 & + & (+ 0''.53 & + 0''.044\,t & & &)\sin 6\mu t \\
 & + & (+ 0''.07 & - 0''.043\,t & & &)\cos 7\mu t \\
 & + & (+ 1''.48 & - 0''.020\,t & - 0''.001\,t^2 & &)\sin 7\mu t \\
 & + & (- 0''.98 & + 0''.006\,t & & &)\cos 8\mu t \\
 & + & (+ 0''.27 & - 0''.033\,t & & &)\sin 8\mu t \\
 & + & (- 0''.34 & + 0''.025\,t & & &)\cos 9\mu t \\
 & + & (- 0''.61 & - 0''.002\,t & & &)\sin 9\mu t \\
 & + & (+ 0''.34 & + 0''.006\,t & & &)\cos 10\mu t \\
 & + & (- 0''.32 & + 0''.018\,t & & &)\sin 10\mu t \;.
\end{array}
$$

Below the elements related to the pair Neptune and Pluto alone are denoted by σ_P^* . We give in table 3 an illustration of the process of convergence on the coefficients $s_m^{(0)}$ ($m \leqq 6$). These coefficients represent the difference between the longitude λ_P^* and its linear secular variation $\bar{\lambda}_P$. The rapidity of numerical convergence is obvious. The computing time on a NAS-9080 for the whole process is 35 seconds.

Table 4 provides the "secular variations" of the variables : $a_P^*/a_P^{(0)}$, k_P^* , h_P^* , q_P^* , p_P^* , in the sense of an approximation which is valid over a finite interval $\Delta t = 1000$ years.

Finally, table 5 is an illustration of F-T series (12) belonging to the longitude λ_P^* . A comparison of our solution for the pair of planets Neptune-Pluto has been done with an internal numerical integration, over 1000 years. Table 6 shows the maximum errors for the 6 elliptical elements. This internal comparison makes obvious the high accuracy of our approximation method.

Table 6. The pair of planets Neptune and Pluto alone.
Internal precision of the solution resulting from a comparison over 1000 years with a numerical integration. Maximum differences $\Delta\sigma^*$ between Fourier-Chebyschev's approximation (F-T) and numerical integration.
6 iterations, K = 50, M = 10.

Δa^* (UA)	$\Delta\lambda^*$ (rad.)	Δk^*	Δh^*	Δq^*	Δp^*
5.10^{-10}	4.10^{-11}	3.10^{-12}	1.10^{-11}	2.10^{-12}	3.10^{-12}

3.2 Pluto's perturbations by other planets.

As we mentioned earlier, Fourier's method allows us to build easily Pluto's perturbations by planets other than Neptune. We have added this new solution $\tilde{\sigma}_P$ to F-T approximation σ_P^* , for the pair Neptune-Pluto. For the epoch J2000.0, a fit of the integration constants has been performed by comparison with DE200 over one century. Table 2 summarizes the results. One has to take care of the meaning of $\sigma_P^{(0)}$. These numbers are not mean elements in a strict sense, since they are obtained as mean values for a solution,

$$\sigma_P = \sigma_P^{(0)} + \tilde{\sigma}_P + \sigma_P^* .$$

which is valid over 1000 years only, in the case of the pair of planets Neptune and Pluto : σ_P^* . Furthermore, our comparison is carried out over one century. In table 7, the maximum discrepancies between our solution and DE200 show that the theory should be extended beyond the first order of planetary masses in the case of the perturbations due to major planets, in order to reduce the differences between theory and "observations". The construction of a second and third order is a long and difficult work. Such a research does not seem to be appropriate since the elaboration of a Pluto's ephemeris can be undertaken using a very much less cumbersome method, as we will see below.

Table 7. Differences with DE200 of the complete solution : $\Delta\sigma_P$, where $\sigma_P = \sigma_P^{(0)} + \tilde{\sigma}_P + \sigma_P^{*}$.

σ_P^{*} : The pair of planets Neptune and Pluto alone.

$\tilde{\sigma}_P$: Perturbations of Pluto by planets from Mercury to Uranus, at the first order of planetary masses.

Maximum differences and residual trends :

$\Delta\sigma_P$	Δa (UA)	$\Delta\lambda$ (rad.)	Δk	Δh	Δq	Δp
Differences σ_P - DE200	$2.5\ 10^{-3}$	$4.5\ 10^{-5}$	$3.2\ 10^{-5}$	$3.4\ 10^{-5}$	$0.3\ 10^{-5}$	$0.3\ 10^{-5}$
Trends (per year)	$-9.\ 10^{-6}$	-----	$-1.5\ 10^{-7}$	$2.0\ 10^{-7}$	$-0.7\ 10^{-8}$	$0.5\ 10^{-8}$

4. AN EPHEMERIS FOR PLUTO OVER 2 CENTURIES.

The goal of the method developed in the preceding chapter was to show that it is possible to solve completely the problem of Neptune-Pluto provided that we limit ourselves to a few thousands of years. The method is general, and could probably be applied to other problems when, in a Fourier analysis with one or several arguments, we are able to isolate slowly variable functions of the time, which can be approximated by Chebyschev series. If now, we simply are looking for an ephemeris for Pluto, covering few centuries, there exists a much simpler way to proceed.

Numerical integration is the most easily handable tool; we may consider the result of such an integration as obtained, and search for an approximation of Pluto's coordinates under a semi-analytical form, over a finite given time interval. A new approximation method has been conceived by Bacchus (1981), and developed by (Chapront and Vu, 1984), in case of planetary and natural satellites motions. We briefly describe here its most essential features.

The polynomial approximation like Chebyschev's is unaware of the "natural frequencies" of the motion. The representation of rapidly varying quasi-periodic functions of the time requires a choice of small intervals of approximation, with polynomials in terms of the time. Hence, it is natural to search for a basis of approximating functions which takes into account the short periods of the motion. For short writing, let us consider even functions of a variable,

$$x = \frac{2}{\Delta t}\, t .$$

Instead of the basis over $\{-1, +1\}$,

$$\{1, x^2, x^4, \dots, x^{2n}\},$$

let us choose,

$$\{\phi_i\} = \{1, x^2, \dots, x^{2n}, \dots, \cos\omega x, \cos 2\omega x, \dots$$
$$\dots, x\sin\omega x, x\sin 2\omega x, \dots$$
$$\dots, x^2\cos\omega x, x^2\cos 2\omega x, \dots\}.$$

ω is the main "natural frequency" of the body. Other frequencies ω', ω'' etc... can also be added. Polynomials in x, as well as mixed terms like $x^k \cos \ell\omega x$ or $x^{k+1} \sin \ell\omega x$ are used to enlarge the domain of frequency variations, or to represent long periods over the interval Δt. $f(x)$ being the time function to be approximated, in the basis $\{\phi_i\}$, one writes,

$$f(x) \cong \sum_{i=0}^{N} a_i \phi_i(x) .$$

The evaluation of the coefficients a_i is obtained by solving the matrix equation,

$$M(a_i) = (b_i)$$

where,

$$b_i = \int_{-1}^{+1} f(x)\phi_i(x)\,dx$$

and,

$$((m_{i,j})) = M \quad ; \quad m_{i,j} = \int_{-1}^{+1} \phi_i(x)\phi_j(x)\,dx .$$

This is Legendre's approximation with the norm of quadratic mean.

In the case of Pluto's coordinates, it is sufficient to take as a basis,

$$\{\phi_i\} = \{\underbrace{1, \cos\omega x, \dots, \cos N\omega x}_{\dots\text{ even }\dots} ; \underbrace{x, \sin\omega x, \dots, \sin N\omega x}_{\dots\text{ odd }\dots}\},$$

$$N \leqq 24.$$

The chosen variables are V, β, and r, which are heliocentric ecliptic mean coordinates of the date. The numerical integration that served as a source is DE200.

Table 8 represents Pluto's ephemeris from 1805 to 2030. θ is the time in years reckoned from J2000.0. ν is Pluto's frequency :

$$\nu = 0.025\,333\,30 \text{ radian per year.}$$

The limiting julian dates are :

2 380 000.0 (02/13/1804, 12^h ET) to 2 463 000.0 (05/13/2031, 12^h)

The maximum differences between the source DE200 and its approximation are,

$$\Delta V = 0''.9 \quad ; \quad \Delta\beta = 0''.2 \quad ; \quad \Delta r = 13.\,10^{-5} \text{ AU.}$$

Table 8. Pluto's coordinates (from DE200).
V : heliocentric longitude (Coefficients in arcseconds).
β : heliocentric latitude (Coefficients in arcseconds).
r : heliocentric distance (Coefficients in astronomical units)
Mean equinox and ecliptic of date. θ in julian years from J2000.0. ν = 0.025 333 30 radian per year. Phases in radians.

$$
\begin{aligned}
V = \; & 860\,205.103 + 5\,275.674\,22\,\theta \\
& + 100\,356.396 \sin(\nu\theta + 0.243\,784\,810) \\
& + 18\,307.800 \sin(2\nu\theta + 0.667\,151\,398) \\
& + 4\,809.091 \sin(3\nu\theta + 1.053\,563\,71) \\
& + 1\,409.088 \sin(4\nu\theta + 1.431\,098\,78) \\
& + 447.997 \sin(5\nu\theta + 1.877\,613\,4) \\
& + 179.308 \sin(6\nu\theta + 2.301\,947\,9) \\
& + 79.692 \sin(7\nu\theta + 2.403\,065) \\
& + 14.606 \sin(8\nu\theta + 2.597\,485) \\
& + 3.937 \sin(9\nu\theta + 5.296\,34) \\
& + 9.329 \sin(10\nu\theta + 4.640\,37) \\
& + 7.759 \sin(11\nu\theta + 3.806\,41) \\
& + 5.630 \sin(12\nu\theta + 2.701\,47) \\
& + 4.631 \sin(13\nu\theta + 1.502\,29) \\
& + 3.967 \sin(14\nu\theta + 0.377\,01) \\
& + 3.040 \sin(15\nu\theta + 5.558\,86) \\
& + 2.165 \sin(16\nu\theta + 4.028\,24) \\
& + 3.169 \sin(17\nu\theta + 2.574\,60) \\
& + 5.979 \sin(18\nu\theta + 2.000\,75) \\
& + 8.684 \sin(19\nu\theta + 2.220\,68) \\
& + 23.676 \sin(20\nu\theta + 2.886\,576) \\
& + 3.522 \sin(21\nu\theta + 0.395\,60) \\
& + 0.717 \sin(22\nu\theta + 0.045\,3) \\
& + 0.211 \sin(23\nu\theta + 5.347\,5) \\
& + 0.060 \sin(24\nu\theta + 4.250) \; .
\end{aligned}
$$

Table 8. (continued).

$$
\begin{aligned}
\beta = & - 14\,129.118 - 1.239\,03\,\theta \\
& + 57\,494.716 \sin(\nu\theta + 2.250\,256\,128) \\
& + 13\,980.356 \sin(2\nu\theta + 2.498\,628\,906) \\
& + 3\,955.667 \sin(3\nu\theta + 2.789\,709\,34) \\
& + 1\,209.070 \sin(4\nu\theta + 3.100\,094\,58) \\
& + 389.942 \sin(5\nu\theta + 3.408\,800\,9) \\
& + 127.010 \sin(6\nu\theta + 3.684\,916\,1) \\
& + 38.173 \sin(7\nu\theta + 3.929\,326) \\
& + 10.261 \sin(8\nu\theta + 4.616\,340) \\
& + 6.410 \sin(9\nu\theta + 5.029\,80) \\
& + 3.877 \sin(10\nu\theta + 4.858\,68) \\
& + 1.944 \sin(11\nu\theta + 4.030\,50) \\
& + 1.357 \sin(12\nu\theta + 2.726\,15) \\
& + 1.246 \sin(13\nu\theta + 1.598\,73) \\
& + 1.122 \sin(14\nu\theta + 0.548\,45) \\
& + 1.112 \sin(15\nu\theta + 5.891\,09) \\
& + 1.039 \sin(16\nu\theta + 5.251\,34) \\
& + 0.658 \sin(17\nu\theta + 5.579\,53) \\
& + 1.532 \sin(18\nu\theta + 0.481\,21) \\
& + 3.381 \sin(19\nu\theta + 0.757\,38) \\
& + 0.764 \sin(20\nu\theta + 4.518\,32) \\
& + 3.008 \sin(21\nu\theta + 1.960\,07) \\
& + 0.436 \sin(22\nu\theta + 1.957\,6) \\
& + 0.079 \sin(23\nu\theta + 1.657) \\
& + 0.018 \sin(24\nu\theta + 1.181).
\end{aligned}
$$

$$
\begin{aligned}
r = 10^{-6}\times\{ & 40\,668\,607 - 431.433\,\theta \\
& + 9\,561\,707 \sin(\nu\theta + 4.969\,369\,006) \\
& + 1\,160\,913 \sin(2\nu\theta + 5.248\,149\,55) \\
& + 226\,897 \sin(3\nu\theta + 5.539\,118\,1) \\
& + 57\,319 \sin(4\nu\theta + 5.649\,361\,9) \\
& + 12\,595 \sin(5\nu\theta + 5.247\,527) \\
& + 4\,460 \sin(6\nu\theta + 3.268\,626) \\
& + 5\,142 \sin(7\nu\theta + 2.246\,966) \\
& + 5\,311 \sin(8\nu\theta + 0.700\,032) \\
& + 3\,330 \sin(9\nu\theta + 6.121\,29) \\
& + 2\,861 \sin(10\nu\theta + 5.065\,35) \\
& + 2\,392 \sin(11\nu\theta + 4.024\,34) \\
& + 1\,976 \sin(12\nu\theta + 2.974\,36) \\
& + 1\,632 \sin(13\nu\theta + 1.922\,06) \\
& + 1\,342 \sin(14\nu\theta + 0.876\,74) \\
& + 1\,080 \sin(15\nu\theta + 6.096\,65) \\
& + 960 \sin(16\nu\theta + 4.948\,10) \\
& + 870 \sin(17\nu\theta + 3.906\,76) \\
& + 976 \sin(18\nu\theta + 3.174\,43) \\
& + 975 \sin(19\nu\theta + 3.391\,72) \\
& + 4\,282 \sin(20\nu\theta + 4.519\,66) \\
& + 1\,072 \sin(21\nu\theta + 1.721\,27) \\
& + 137 \sin(22\nu\theta + 2.144\,3) \\
& + 30 \sin(23\nu\theta + 2.298) \\
& + 12 \sin(24\nu\theta + 2.485)\}.
\end{aligned}
$$

Such a precision is sufficient for Pluto's ephemeris whose physical motion is very much less known. In barycentric coordinates (the center of coordinates being the barycenter of the solar system), the solution series are more rapidly convergent on a numerical point of view. With the same number of basic functions (N = 24) a factor by 50 is gained in precision with respect to heliocentric coordinates, as well as in errors on ΔV, $\Delta\beta$ and Δr. Such expressions are given in (Chapront and Vu, 1984). The proposed results are simpler to use for a direct computation of the ephemeris.

5. CONCLUSION.

We have reviewed three approximation methods : Fourier, Fourier-Chebyschev and Legendre. All of them present complementary advantages. It is worth noticing that if Fourier's approximation has been widely used, the two other methods are less usual. They allow us to represent solutions with a great economy of means, and formal compactness, on time intervals of a few periods of revolution. Fourier-Chebyschev (F-T) has been used to solve directly a system of differential equations. Legendre's approximation could be used in the same way.

We consider that it is worthwhile developing such methods when the natural difficulties of a problem lead to bulky solutions uneasy to handle for practical uses. It is of particular importance when constructing solutions intended to be run on small computers or desk calculators.

REFERENCES.

Bacchus, P.: 1981, Pub. IRMA Lille, 3, 7.
Bretagnon, P.: 1982, Astron. Astrophys., 114, 278.
Brouwer, D., Clemence, G.M., and Eckert, W.J.: 1951, Astronomical Papers of the American Ephemeris, Vol. 12. US Government Printing Office. Washington.
Carpenter, L.: 1966, NASA TN. D-3168.
Chapront, J.: 1977, Astron. Astrophys., 61, 7.
Chapront, J. : 1982, Celes. Mech., 28, 415.
Chapront, J., Bretagnon, P. and Mehl, M.: 1975, Celes. Mech., 11, 379.
Chapront, J. and Vu, D.T.: 1984, Astron. Astrophys., to be published.
Cohen, C.J., Hubbard, E.C. and Oesterwinter, C.: 1972, Astronomical Papers of the American Ephemeris, Vol. 22, Part.1, US Government Printing Office. Washington.
Connaissance des Temps: 1979, Gauthier-Villars, Paris.
Giacaglia, G.E.O. and Nacozy, P.E.: 1970, in Periodic Orbits, Stability and Resonances, Ed. G. Giacaglia, D. Reidel Publ. Co., Dordrecht, Holland.
Goodrich, E.F. and Carpenter, L.: 1966, NASA TN. X643-66-133.
Hildebrand, F.B.: 1956, Introduction to Numerical Analysis, Mc Graw-Hill Book Company, Inc. New York.

Ivanova, T.V.: 1978, in Dynamics of Planets and Satellites, Ed. V. Szebehely, D. Reidel Publ. Co., Dordrecht, Holland.
Nacozy, P.E.: 1980, Celes. Mech., 22, 19.
Oesterwinter, C. and Cohen, C.J.: 1972, Celes. Mech., 5, 317.
Petrovskaia, M.S.: 1970, Celes. Mech., 3, 121.
Petrovskaia, M.S.: 1977, Celes. Mech., 15, 125.
Seidelmann, P.K.: 1984, Private communication.
Simon, J.L.: 1982, Astron. Astrophys., 120, 197.
Standish, E.M. and Williams, J.G., 1982, Development Ephemeris DE200-LE200, Magnetic tape.

TERMES SECULAIRES DANS LA THEORIE DES PERTURBATIONS

Takeshi INOUE
Kyoto Sangyo University, Kamigamo, Kyoto, 603, JAPAN

ABSTRACT. The convergence property is examined for the series which appear in the theory of the planetary motions. With another point of view, we obtain a result different from that already given by Gyldén : The almost divergency in the meaning of category is concluded when the ratio of the mean motions is irrational using as Gyldén the continuous fractions. The author explains the reason why the series are always useful in practical applications : Because the long evolution of the system of interest under the influence of the small irreversible effects has led to many resonances or quasi-resonances, i.e. conditions favorable to a good convergence of suitable series.

INTRODUCTION

Depuis longtemps, on a essayé de supprimer les inégalités séculaires dans la théorie du mouvement des corps célestes. Malgré tous les efforts, la plupart des séries habituellement utilisées ne sont pas uniformément convergentes à cause de l'existence de termes séculaires ainsi que d'inégalités à longues périodes accompagnées de petits diviseurs (Poincaré, 1884). Le problème de la convergence des séries est étroitement lié à celui de l'apparition de ces termes et de ces inégalités.

D'autre part, on sait depuis Gyldén que les séries de la théorie du mouvement des planetès sont presque toutes convergentes dans le cas où le rapport des moyens mouvements est irrationnel. Nous concluons pourtant que les séries sont convergentes seulement dans un ensemble de première catégorie (c'est-à-dire un ensemble "rare" au sens des catégories) sous la même condition que celle de Gyldén en profitant des propriétés des fractions continues. Cela signifie que les séries devraient être inutiles pour prévoir les positions des planètes. Pour quelle raison peut-on le faire? H.Poincaré a expliqué cette circonstance par le fait que certaines séries même divergentes deviennent bien valables si l'on cesse le calcul à un nombre de termes convenables.

Nous la comprenons différemment : L'utilité de ces séries se produit du fait que nous avons à considérer des systèmes physiquement évolués, comme le système solaire, où les petites irréversibilités ont

Celestial Mechanics **34** (1984) 185–191. 0008–8714/84.15

conduit à un grand nombre de résonnances ou quasi-résonnances. Cette situation est beaucoup plus favorable à la convergence de séries appropriées que celle que l'on obtiendrait à partir de conditions initiales absolument arbitraires. On est alors obligé d'admettre l'apparition de commensurabilités exactes ainsi que de termes séculaires de la forme $\kappa^s(t-t_\circ)$. Ici, la puissance s est déterminée selon la précision des données et l'on peut lui donner en pratique une valeur de l'ordre de 10^5. Si nous substituons respectivement 0,9995 et 10^{13} (jours) aux quantités κ et $(t-t_\circ)$, le terme séculaire donnera d'une grandeur de 5×10^{-4} secondes de degré qui est complètement négligeable. Le reste des séries est uniformément convergent et ainsi la théorie du mouvement réussit toujours à prédire les positions des planètes.

SERIES DU PROBLEME

Si nous aborderons dans les premières lignes ce que l'on trouve ailleurs sans peine, c'est à cause du fait qu'il nous est nécessaire de définir des quantités à notre manière pour mieux énoncer notre résultat qui est diamétralement opposé à celui déjà connu (Charlier, 1907).

Dans la théorie du mouvement des planètes, on rencontre une équation différentielle de la forme suivante :

$$\frac{dE}{dt} = \sum_{j=1}^{\infty} B_j \cos(\lambda_j t + \beta_j) \ , \tag{1}$$

E exprimant un des éléments de la planète considérée, B_j et β_j étant des quantités déterminées par les six éléments des planètes. Supposons ici que la série $\Sigma_j B_j$ soit absolument convergente. La quantité λ_j s'exprime, si le nombre des planètes est m, comme il suit :

$$\lambda_j = n_1 q_{1,j} + n_2 q_{2,j} + \cdots + n_m q_{m,j} \ , \tag{2}$$

où $n_1, n_2, \cdots, n_m$ sont les moyens mouvements et $q_{1,j}, q_{2,j}, \cdots, q_{m,j}$ peuvent prendre toutes les valeurs des nombres entiers.

Habituellement, en supposant qu'il n'y ait pas de relation commensurable entre les moyens mouvements, on obtient la solution approximative du premier ordre de l'équation (1) :

$$E = E_\circ + B(t-t_\circ) + \sum_{j=1}^{\infty} \frac{B_j}{\lambda_j} \{ \sin(\lambda_j t + \beta_j) - \sin(\lambda_j t_\circ + \beta_j) \} \tag{3}$$

où B montre la valeur du coefficient B_j pour lequel les nombres $q_{1,j}, \cdots q_{m,j}$ sont à la fois nuls. On sait que la série (3) est absolument convergente pour tout intervalle de temps défini $(t-t_\circ)$.

Pour examiner si la convergence de la série susdite est uniforme ou non, nous considérerons en réduisant le problème au cas de deux planètes seulement :

$$\sum_{j=1}^{\infty} \frac{|B_j|}{|\lambda_j|} = \sum_{j=1}^{\infty} \frac{A_j \ \kappa^{*|q_{1,j} + q_{2,j}|}}{|n_1 q_{1,j} + n_2 q_{2,j}|} \ , \ (0<A_j, \ 0<\kappa^*<1). \tag{4}$$

Poson maintenant pour simplifier :

$$\rho \equiv \frac{n_1}{n_2}\ , \ (0<n_1<n_2)\ ;\ r_j \equiv \frac{-q_{2,j}}{q_{1,j}}\ , \tag{5}$$

et introduisons l'expression de la fraction continue régulière :

$$\begin{aligned}\alpha &= k_0 + \frac{1}{k_1} + \frac{1}{k_2} + \cdots + \frac{1}{k_{\nu-1}} + \frac{1}{\xi_\nu} \\ &\equiv (k_0;k_1,k_2,\cdots,k_{\nu-1},\xi_\nu)\ ,\end{aligned} \tag{6}$$

où α et k_0 sont respectivement un nombre réel quelconque et un nombre entier, et que k_1, $k_2,\cdots$, $k_{\nu-1}$ sont des entiers positifs et ξ_ν est une quantité supérieure à 1, en général un nombre réel. Dans le cas où le nombre α est irrationnel, on peut croître arbitrairement le nombre ν des termes et spécialement comme il suit :

$$\alpha = (k_0;k_1,k_2,\cdots,k_{\nu-1},k_\nu,\xi_{\nu+1})\ ,$$

avec

$$\begin{aligned}k_\nu &\equiv [\xi_\nu]\ ,\quad (\ [\ \ ]\ :\ \text{symbole de Gauß})\ , \\ \xi_{\nu+1} &\equiv \frac{1}{\xi_\nu - k_\nu}\ ,\quad (1<\xi_{\nu+1})\ .\end{aligned} \tag{7}$$

Si le nombre α est rationnel, la valeur du dernier terme ξ_ν deviendra un entier positif lorsque le nombre ν arrive à un nombre fini, disons ν^*.

Revenons à notre sujet. Puisque l'on a obtenu l'intégrale de l'équation (1) sous la forme (3), la valeur du rapport des deux moyens mouvements ρ doit sans exception être irrationnel. Si nous tenons compte des réduites du nombre ρ parmi les nombres rationnels r, cela suffira pour étudier la convergence de la série (4). Ecrivons le rapport ρ et la réduite $r(\nu)$ sous la forme des fractions continues :

$$\begin{aligned}\rho &= (0;k_1,k_2,\cdots,k_{\nu-1},\xi_\nu)\ ,\quad \xi_\nu = (k_\nu;\xi_{\nu+1})\ ; \\ r(\nu) &= (0;k_1,k_2,\cdots,k_{\nu-1},k_\nu)\ .\end{aligned} \tag{8}$$

On peut facilement calculer la différence des deux nombres :

$$\rho - r(\nu) = \frac{(-1)^\nu}{s^*(\nu;\nu)\ s(\nu;\nu)}\ \frac{1}{\xi_{\nu+1}}\ , \tag{9}$$

à l'aide de suites auxiliaires définies par :

$$\begin{aligned}&s^*(\nu;0) = 1,\ s^*(\nu;1) = \xi_\nu\ ; \\ &s^*(\nu;\mu+1) = k_{\nu-\mu}s^*(\nu;\mu) + s^*(\nu;\mu-1)\ ;\end{aligned}$$

(10)

$$s(\nu;0) = 1,\ s(\nu;1) = k_\nu\ ;$$
$$s(\nu;\mu+1) = k_{\nu-\mu}s(\nu;\mu) + s(\nu;\mu-1)\ ;$$
$$(\mu = 1,\ 2,\cdots,\ \nu-1)\ .$$

THEOREME DE BRUNS

Comme nous le pouvons, nous allons égaler respectivement les deux nombres $q_{1,j}$ et $q_{2,j}$ aux quantités $s(\nu;\nu)$ et $-s(\nu;\nu-1)$ et nous aurons ainsi :

$$\sum_{j=1}^{\infty}\frac{|B_j|}{|\lambda_j|} > \sum_{\nu}\frac{A_\nu}{n_2}\xi_{\nu+1}\ s^*(\nu;\nu)\ \kappa^{*s(\nu;\nu)\ -\ s(\nu,\nu-1)}$$
$$> \frac{1}{n_2}\sum_{\nu}A_\nu\ \xi_{\nu+1}\ s(\nu;\nu)\ \kappa^{s(\nu;\nu)}\ ,\quad (\kappa \equiv \kappa^{*1-\rho} < 1)\ , \tag{11}$$

puisque nous savons qu'il existe une inégalité $\xi_\nu > k_\nu$ ainsi que $s^*(\nu;\nu) > s(\nu;\nu)$ et que la différence des deux nombres ρ et $r(\nu)$ diminuera indéfiniment quand croîtra le nombre ν des termes. Les coefficients A_ν ne jouent pas de grand rôle dans les études de la convergence et nous les remplaçons tous par l'unité. D'ailleurs, les deux séries :

$$\sum_{\nu}\xi_{\nu+1}\ s(\nu;\nu)\ \kappa^{s(\nu;\nu)} \quad \text{et} \quad \sum_{\nu}\xi_{\nu+1}\ \kappa^{s(\nu;\nu)}$$

ont de mêmes caractères de convergence et nous étudions notre problème par l'entremise de la dernière série, à savoir :

$$\sum_{\nu}\xi_{\nu+1}\ \kappa^{s(\nu;\nu)}\ . \tag{12}$$

Sous cette forme, on est aisément capable de vérifier ce que signifie le théorème de Bruns : Supposons que la série (12) converge pour un nombre irrationnel ρ^*, on peut toujours trouver un autre nombre irrationnel ρ^{**} tel qu'il soit aussi voisin que l'on veut du premier ρ^*, pour lequel la série (12) diverge. On peut le dire également d'une manière inverse.

Etant le rapport ρ^* irrationnel, on peut augmenter sans limite le nombre ν des termes et déterminer chaque fois le dernier terme $\xi^*_{\nu+1}$. Dans le cas même où toutes les valeurs de ces derniers termes sont inférieures à une quantité positive finie, soit K, le nombre ainsi déterminé est définitivement irrationnel. Pour ce nombre irrationnel, la série (12) converge évidemment. L'ensemble de tels nombres irrationnels ne peut pas être fini. Par contre, il existe une infinité de nombres irrationnels ρ^{**} dont les derniers termes ont toujours des valeurs supérieures à la même quantité positive finie K. Si l'on connait les valeurs des termes $k_1,\ k_2,\ \ldots,\ k_\nu$, on peut calculer la valeur de la quantité $s(\nu;\nu)$ ainsi que celle de la quantité $\zeta^{s(\nu;\nu)}$ où l'on a fait $\zeta \equiv \kappa^{-1}$, $(1<\zeta)$. Chaque fois que l'on augmente le nombre des termes ν, on peut donner au dernier terme $\xi^{**}_{\nu+1}$ une valeur supérieure à $\zeta^{s(\nu;\nu)}$. Cela signifie que la quantité $\xi^{**}_{\nu+1}\,\kappa^{s(\nu;\nu)}$ est supérieure à 1 et que série (12) diverge dans ce cas. En même temps, on aura la relation suivante en choisissant convenablement un nombre positif N qui dépend d'un nombre positif arbitrairement donné δ :

$$| \rho^* - \rho^{**} | = \frac{\xi^{**}_{\nu+1} - \xi^{*}_{\nu+1}}{s^*(\nu+1;\nu+1)\; s(\nu+1;\nu+1)} < \delta ,$$

$$1 < \xi^{*}_{\nu+1} < K < \xi^{**}_{\nu+1} , \quad (N < \nu) . \tag{13}$$

Tout ce qui précède a démontré l'énoncé du théorème de Bruns.

DIVERGENCE DES SERIES

L'ensemble des nombres ρ^* et celui des nombres ρ^{**} sont tous les deux infinis. Il est cependant clair par la relation (13) que l'ensemble des nombres $\xi^{*}_{\nu+1}$ est bien fini en tant que l'on fixe la valeur ν. L'ensemble des nombres $\xi^{**}_{\nu+1}$ est nettement infini pour la même valeur fixe ν. C'est ainsi que nous pouvons en conclure comme il suit : La série (12) est partout divergente à l'exception d'un ensemble de première catégorie. C'est très différent du résultat obtenu par Gyldén (Charlier, 1931 ; Hagihara, 1961). Il a calculé la prababilité de la convergence de la série en se basant sur la valeur du rapport de deux longueurs des intervalles définis par $(\frac{1}{\infty}, \kappa^{s(\nu;\nu)})$ et $(\frac{1}{\infty}, \frac{1}{1})$. Selon la définition des fractions continues (6), il vaut mieux étudier la topologie en comparant la répartition des nombres entiers contenus dans les deux intervalles $(1, \zeta^{s(\nu;\nu)})$ et $(\zeta^{s(\nu;\nu)}, \infty)$.

De toute façon, nous avons appris que la série (3) converge absolument mais non pas presque uniformément. Cela veut dire que l'on ne peut pas espérer d'obtenir une haute précision même si l'on calcule de nombreux termes dans la série (3). On est pourtant capable de prédire les positions des planètes d'une manière suffisante. Comment doit-on comprendre ce fait-ci? Poincaré a expliqué cette utilité de notre série en rattachant à la théorie des parties finies de séries divergentes (1957). Au point de vue théorique, on peut donner n'importe quelle valeur d'un nombre rationnel ou irrationnel, dans notre cas par exemple, pour la valeur du rapport des moyens mouvements ρ. Or, on ne sait juger l'utilité ou l'inutilité d'une théorie approximative qu'après avoir calculé des positions des astres et comparé avec des observations.

Si l'on applique ce point de vue à notre problème, on aura d'abord :

$$\rho = \rho(\nu) = (0;k_1,k_2,\cdots,k_{\nu-1},k_\nu) , \tag{14}$$

ensuite :

$$r(\nu) = \frac{-q_{2,j}}{q_{1,j}} = \frac{p\times s(\nu;\nu-1)}{p\times s(\nu;\nu)} = (0;k_1,k_2,\cdots,k_{\nu-1},k_\nu) \; , \qquad (15)$$

pour toutes les combinaisons de deux nombres naturels $(q_{1,j}, -q_{2,j})$:

$$(q_{1,j}, -q_{2,j}) = (p\times s(\nu;\nu),\ p\times s(\nu;\nu-1)),\ (p = 1,2,\cdots). \qquad (16)$$

On voit alors des commensurabilités exactes dans l'équation (1) ainsi que des inégalités séculaires dans la solution (3). On doit en effet ajouter la quantité $\Sigma\, B_p \cos\beta_p$ à la constante B. Nous obtiendrons aisément la limite de l'influence de ces termes commensurables :

$$\sigma(1+\sigma+\sigma^2+\cdots)(t-t_\circ) = \frac{\sigma}{1-\sigma}(t-t_\circ) \quad , \ (\sigma \equiv \kappa^{s(\nu;\nu)}). \qquad (17)$$

Ce n'est pour la plupart qu'une quantité négligeable si l'on a des données numériques d'une précision considérable.

Quant aux inégalités à longues périodes sous cette hypothèse, on aurait une amplitude grande d'une des expressions de la forme suivante :

$$\frac{B_j}{\lambda_j}\{ \sin(\lambda_j t + \beta_j) - \sin(\lambda_j t_\circ + \beta_j)\} \quad , \qquad (18)$$

avec

$$\lambda_j = n_2 q_{1,j} \ \{\rho(\nu) - r(\nu-1)\} = \frac{(-1)^{\nu-1} n_2 p}{k_\nu \ s(\nu-1;\nu-1)} \quad , \qquad (19)$$

$$r(\nu-1) = (0;k_1,k_2,\cdots,k_{\nu-1}) = \frac{p\times s(\nu-1;\nu-2)}{p\times s(\nu-1;\nu-1)} \ , \ (p=1,2,\cdots). \quad (20)$$

CONCLUSION

L'étude topologique des séries de la théorie des mouvements planétaires conclut donc en général à une mauvaise convergence de ces séries. La bonne convergence rencontrée dans le cas de systèmes concrets comme le système solaire doit sans doute tenir à sa longue évolution et à l'apparition d'un certain nombre de résonnances ou de quasi-résonnances.

Il serait intéressant de vérifier à contrario ce résultat et de voir si un système choisi arbitrairement conduit à une mauvaise convergence des séries et à de gros écarts avec les résultats d'une intégration numérique.

Nous sommes très reconnaissant à Monsieur C. Marchal qui a bien voulu nous donner les conseils précieux grâce auxquels nous avons pu perfectionner notre recherche.

BIBLIOGRAPHIE

Charlier, C.L. : 1907, *Die Mechanik des Himmels*, Band II, Verlag von Veit & Comp., 321

Charlier, C.L. : 1931, *Application de la théorie des probabilité à l'astronomie*, Gauthier-Villars, 34

Hagihara, Y. : 1961, dans *Planets and Satellites* ed. par Kuiper, G.P. et Middlehurst, B.M., The University of Chicago Press, 126

Poincaré, H. : 1884, *Bulletin astronomique*, 1, 319

Poincaré, H. : 1957, *Les méthodes nouvelles de la mécanique céleste*, Tome II, Dover, 1

AMELIORATION DES THEORIES PLANETAIRES ANALYTIQUES

P. Bretagnon
Bureau des Longitudes
77 avenue Denfert-Rochereau
75014 Paris
France

ABSTRACT. The VSOP82 and TOP82 theories intend to represent the motion of planets, with a satisfactory accuracy, over an interval of 1000 years from and after J2000.0. The precision of the newtonian part of the solutions for the system of the sun and eight point masses is given in table 1. We present the construction of complements in order to keep this accuracy over one thousand years for the real motion : the relativistic perturbations, the perturbations by the minor planets, the perturbations by the Moon. Besides, we have undertaken the improvement of the solutions through lengthening the interval of validity up to six thousand years from and after J2000.0.

1. INTRODUCTION

Les théories VSOP82 (Bretagnon, 1982) et TOP82 (Simon, 1983) développées au Bureau des Longitudes cherchent à représenter le mouvement de l'ensemble des planètes, à une bonne précision, sur un intervalle de 1000 ans de part et d'autre de l'origine J2000.0.

Elles sont construites jusqu'à l'ordre 3 par rapport aux masses pour les planètes inférieures Mercure, Vénus, la Terre et Mars. Les théories des grosses planètes Jupiter, Saturne, Uranus et Neptune sont construites jusqu'à l'ordre 6 soit par une méthode itérative soit par analyse harmonique.

Des comparaisons de ces solutions à des intégrations numériques internes ou aux intégrations numériques du JPL, DE102 (Newhall *et al*, 1983) et DE200 (Standish, 1982), nous permettent de donner une estimation de la précision des longitudes sur un siècle et sur 1000 ans. Nous donnons dans le tableau I cette précision pour la partie newtonienne des solutions du système constitué du soleil et de huit masses ponctuelles. Pour conserver ces précisions sur 1000 ans pour le mouvement réel, nous avons construit des compléments : les perturbations relativistes, les perturbations par les petites planètes, les perturbations par la Lune. Par ailleurs, nous avons entrepris l'amélioration des solutions par le prolongement de leur intervalle de validité à 6000 ans de part et d'autre de l'origine J2000.0.

Celestial Mechanics **34** (1984) 193–201. 0008–8714/84.15

Tableau I. Précision des longitudes en secondes de degré sur un siècle et sur 1000 ans.

Planète	Mercure	Vénus	Terre	Mars
100 ans	0,0013	0,0064	0,0050	0,0100
1000 ans	0,0600	0,0240	0,1150	0,3500

Planète	Jupiter	Saturne	Uranus	Neptune
100 ans	0,0200	0,0600	0,0100	0,0100
1000 ans	0,0970	0,1900	0,0770	0,0640

2. PERTURBATIONS RELATIVISTES

Elles ont été déterminées (Lestrade, Bretagnon, 1982) au premier et au deuxième ordre, c'est-à-dire qu'elles contiennent les termes proportionnels à μ et à $\mu \times m$ et μ^2 où μ est le rayon gravitationnel du soleil et m une masse planétaire. Ces perturbations sont contenues dans VSOP82 et dans TOP82 et sont suffisantes sur 1000 ans. Sur un intervalle plus grand, il est nécessaire de faire intervenir les termes d'ordre 3 qui produisent, essentiellement, des termes en T^3 dans les variables en excentricité et inclinaison.

Rappelons l'ordre de grandeur des perturbations relativistes. La perturbation la plus importante est évidemment l'avance des périhélies que nous donnons dans le tableau II en secondes par milliers d'années.

Tableau II. Avance séculaire des périhélies en secondes par milliers d'années.

Planète	Mercure	Vénus	Terre	Mars
$\Delta\tilde{\omega}$	429,81	86,25	38,39	13,51

Les perturbations relativistes périodiques sont également très importantes. Nous donnons dans le tableau III la plus grosse perturbation périodique, en coordonnées isotropiques, de la longitude, en secondes et la valeur correspondante en mètres.

Tableau III. Perturbations relativistes périodiques.

Planète	Mercure	Vénus	Terre	Mars
"	0,0218	0,0004	0,0007	0,0025
m	6109	205	506	2816

Au deuxième ordre, on trouve des termes en T^2 importants correspondant à des variations de l'excentricité, par exemple dans le cas de Mercure :

$$\Delta e_{Me} = 0{,}000\,011\,472\,2\,T^2$$

Les termes en T^3 du troisième ordre de la relativité sont négligeables sur l'intervalle de 1000 ans mais doivent être pris en considération si on veut augmenter l'intervalle de validité de la théorie.

3. PERTURBATIONS PAR LES PETITES PLANETES

Nous avons calculé les perturbations au premier et au deuxième ordre par rapport aux masses de l'ensemble des planètes principales par les petites planètes Cérès, Pallas, Vesta en utilisant les masses $5{,}9 \times 10^{-10}$, $1{,}1 \times 10^{-10}$ et $1{,}2 \times 10^{-10}$ respectivement.

Ce calcul présente de nombreuses difficultés : Vesta a une orbite proche de celle de Mars et l'inverse de la distance Mars-Vesta est très lentement convergent ce qui limite la précision ; les planètes Pallas et Jupiter sont en résonance pratiquement stricte : l'argument $7\lambda_{Pallas} - 18\lambda_J$ est de période environ 70 000 ans.

Nous avons rassemblé dans le tableau IV les perturbations les plus importantes au premier ordre par rapport aux masses dans la longitude de Mars en donnant l'amplitude en secondes et en mètres.

Tableau IV. Perturbations de la longitude moyenne de Mars. Premier ordre par rapport aux masses.

Argument	Amplitude en "	Amplitude en m	Période en années
$\lambda_M - 2\lambda_{Vesta}$	0,003893	4302	52
$2\lambda_M - 5\lambda_{Pallas}$	0,002416	2670	49
$2\lambda_M - 5\lambda_{Cérès}$	0,002180	2409	44
$\lambda_M - 2\lambda_{Cérès}$	0,000714	789	10
$2\lambda_M - 4\lambda_{Vesta}$	0,000601	664	26
$4\lambda_M - 10\lambda_{Pallas}$	0,000267	295	24
$9\lambda_M - 22\lambda_{Pallas}$	0,000174	192	61
$3\lambda_M - 7\lambda_{Pallas}$	0,000160	177	13
$2\lambda_M - 4\lambda_{Cérès}$	0,000157	174	5,1
$2\lambda_M - 3\lambda_{Cérès}$	0,000135	149	2,4

Les amplitudes sont assez voisines de celles déterminées par

Williams (1984). Certaines différences peuvent s'expliquer par les valeurs des masses qu'il utilise, en particulier $1{,}38 \times 10^{-10}$ pour Vesta.

Le tableau V contient, toujours pour la longitude de Mars, les perturbations les plus importantes au deuxième ordre par rapport aux masses. Ce tableau est essentiellement composé de termes à longue ou très longue période mais les amplitudes sont encore sensibles à la précision des mesures de distance Terre-Mars.

Tableau V. Perturbations de la longitude moyenne de Mars. Second ordre par rapport aux masses.

Argument	Amplitude en "	Amplitude en m	Période en années
$\lambda_M - 4\lambda_{Cérès} + 4\lambda_J$	1,743887	1927135	18821
$2\lambda_M - 6\lambda_{Vesta} + 7\lambda_J$	0,001774	1960	3038
$7\lambda_M - 16\lambda_{Pallas} - 3\lambda_J$	0,001774	1960	1298
$2\lambda_M - 3\lambda_{Pallas} - 5\lambda_J$	0,006680	752	119
$9\lambda_M - 24\lambda_{Pallas} + 5\lambda_J$	0,000413	456	221
$\lambda_M - \lambda_{Vesta} - 3\lambda_J$	0,000309	342	308
$2\lambda_M - 6\lambda_{Cérès} + 3\lambda_J$	0,000206	228	77
$4\lambda_M - 11\lambda_{Pallas} + 3\lambda_J$	0,000206	228	214
$5\lambda_M - 13\lambda_{Cérès} + 2\lambda_J$	0,000186	205	306
$4\lambda_M - 9\lambda_{Pallas} - 2\lambda_J$	0,000186	205	137
$3\lambda_M - 7\lambda_{Vesta} + 4\lambda_J$	0,000125	138	280

Dans la comparaison que nous avions faite de la solution VSOP82 à DE200, les perturbations par les petites planètes ne figuraient pas (Bretagnon, 1982). Les résidus que nous avions pour la longitude de Mars s'expliquent en grande partie par le terme de période 52 ans dû à Vesta.

Les plus grosses perturbations des petites planètes sur les planètes principales autres que Mars sont de l'ordre de quelques 10^{-4}".

4. PERTURBATIONS DUES A LA LUNE

4.1. Perturbations du premier ordre (proportionnelles à la masse de la Lune)

Nous avons déjà déterminé une solution des perturbations de la Lune sur le barycentre Terre-Lune (BTL) (Bretagnon, 1980). Cette solution avait été construite à l'aide d'une théorie du BTL contenant uniquement les perturbations planétaires du premier ordre des masses et d'une théorie du mouvement de la Lune limitée au problème principal (Chapront-Touzé,

1980). Nous avons repris ce travail avec une solution complète du mouvement de la Lune ELP-2000/82 (Chapront-Touzé, Chapront, 1983) et la théorie du BTL au troisième ordre des masses : VSOP82. La différence essentielle provient des termes planétaires contenus dans la théorie de la Lune et de quelques termes à très longue période de la théorie du BTL.

Nous avons retrouvé les termes à courte période de la première solution. Par exemple, dans la longitude moyenne :

$$\begin{aligned}\Delta\lambda_T = & \; 0\overset{''}{,}00304 \sin 2D \\ & - 0\overset{''}{,}00118 \sin (2D - l) \\ & - 0\overset{''}{,}00014 \sin \lambda_3 - 0\overset{''}{,}00059 \cos \lambda_3 \\ & - 0\overset{''}{,}00039 \sin l \\ & + 0\overset{''}{,}00005 \sin (\lambda_3 - 2D) + 0\overset{''}{,}00021 \cos (\lambda_3 - 2D) \\ & - 0\overset{''}{,}00018 \sin (2D - 2l) \\ & + 0\overset{''}{,}00011 \sin (2D + 2l)\end{aligned}$$

où les angles D, F, l sont les arguments de Delaunay ; λ_3 est la longitude moyenne de la Terre.

Nous avons, de plus, trouvé des termes à longue période provenant de la théorie du BTL et de la théorie de la Lune. Nous donnons les plus importants dans le tableau VI, concernant la longitude moyenne.

Tableau VI. Perturbations à longue période dans la longitude du barycentre Terre-Lune dues à la Lune.

Argument	Amplitude en "	Période en années
$10\lambda_3 - 19\lambda_4 + 3\lambda_6$	0,08850	6409
$8\lambda_3 - 16\lambda_4 + 4\lambda_5 + 5\lambda_6$	0,01645	93462
$10\lambda_2 - 3\lambda_3 - l$	0,00223	1889
$4\lambda_3 - 8\lambda_4 + \lambda_5 + 5\lambda_6$	0,00185	1750
$4\lambda_3 - 8\lambda_4 + 3\lambda_5$	0,00074	1783
$5\lambda_2 - 6\lambda_3 - 4\lambda_4$	0,00017	1138
$2\lambda_5 - 5\lambda_6$	0,00016	883

Les termes les plus importants sont toutefois les polynômes du temps que nous avons déterminés jusqu'à un degré élevé :

$$\begin{aligned}\Delta^1\lambda_T \times 10^{10} &= + 102{,}05\,T^2 - 260{,}92\,T^3 + 0{,}29\,T^4 \\ \Delta^1 k_T \times 10^{10} &= - 51\,898{,}49\,T + 981{,}66\,T^2 + 36{,}02\,T^3 + 2{,}94\,T^4 \\ \Delta^1 h_T \times 10^{10} &= - 11\,963{,}00\,T - 1302{,}08\,T^2 + 27{,}75\,T^3 - 4{,}27\,T^4\end{aligned}$$

où $k = e\cos\tilde{\omega}$; $h = e\sin\tilde{\omega}$; T en milliers d'années juliennes. Les termes en T^2 des variables k et h donnent, dans la longitude vraie au bout de 1000 ans, une perturbation de $3261 \times 10^{-10} = 0''\!,0673$.

Enfin, la théorie de la Lune contient les perturbations dues au potentiel de la Terre. Elles sont donc proportionnelles à J_2 et dépendent de la précession. Ces perturbations de la Lune produisent dans les variables en inclinaison du BTL un terme de la période de la précession :

$$\Delta^1 q_T \times 10^{10} = 33{,}14 \sin(\lambda_3 + D - \zeta) - 308{,}28 \cos(\lambda_3 + D - \zeta)$$

$$\Delta^1 p_T \times 10^{10} = -308{,}34 \sin(\lambda_3 + D - \zeta) - 33{,}12 \cos(\lambda_3 + D - \zeta)$$

de période 25 770 ans. On a $q = \sin\frac{i}{2}\cos\Omega$ et $p = \sin\frac{i}{2}\sin\Omega$.

L'angle ζ est défini par :

$$\zeta = w_1 + PT$$

où w_1 est la longitude moyenne de la Lune et P la constante de la précession.

Ceci entraîne dans l'inclinaison de l'orbite terrestre un terme de période 21 970 ans et d'amplitude $0''\!,0128$.

4.2. Perturbations du deuxième ordre (proportionnelles au produit de la masse de la Lune et d'une masse planétaire)

Nous avons calculé les perturbations sur toutes les planètes, y compris le BTL, dues aux variations des éléments du BTL que nous venons de calculer. On trouve à nouveau dans les variables q et p le terme de précession :

$$\Delta^2 q_T \times 10^{10} = +7{,}74 \sin(\lambda_3 + D - \zeta) - 72{,}04 \cos(\lambda_3 + D - \zeta)$$

$$\Delta^2 p_T \times 10^{10} = -73{,}79 \sin(\lambda_3 + D - \zeta) - 7{,}93 \cos(\lambda_3 + D - \zeta)$$

Au total nous trouvons, pour l'inclinaison, la perturbation de période 21 970 ans et d'amplitude $0''\!,0158$.

Dans le cas des quatre planètes Mercure, Vénus, le barycentre Terre-Lune et Mars, nous trouvons des perturbations périodiques qui atteignent quelques $10^{-3}''$. Ces perturbations du deuxième ordre sont négligeables pour les autres planètes.

Ici encore, les perturbations les plus importantes sont les variations séculaires. Nous trouvons :

Mercure : $\Delta\lambda_{Me} \times 10^{10} = -37{,}87\,T^2$

$$\Delta k_{Me} \times 10^{10} = -17{,}12\,T^2$$

$$\Delta h_{Me} \times 10^{10} = +56{,}95\,T^2$$

Vénus : $\Delta\lambda_V \times 10^{10} = +\ 33{,}39\,T^2 \ +\ 2{,}13\,T^3$

$\Delta k_V \times 10^{10} = -\ 184{,}80\,T^2 \ -\ 12{,}64\,T^3$

$\Delta h_V \times 10^{10} = +\ 713{,}66\,T^2 \ -\ 8{,}78\,T^3$

BTL : $\Delta\lambda_T \times 10^{10} = \ -\ 14{,}28\,T^2$

$\Delta k_T \times 10^{10} = \ +\ 370{,}93\,T^2$

$\Delta h_T \times 10^{10} = -\ 1572{,}19\,T^2$

Mars : $\Delta\lambda_M \times 10^{10} = +\ 224{,}14\,T^2 \ -\ 5{,}98\,T^3$

$\Delta k_M \times 10^{10} = \ -\ 35{,}99\,T^2 \ -\ 3{,}54\,T^3$

$\Delta h_M \times 10^{10} = +\ 225{,}05\,T^2 \ -\ 3{,}11\,T^3$

où T est en milliers d'années juliennes.

Ces perturbations sont très importantes en particulier pour la Terre. L'ensemble des variations séculaires dues à la Lune pour les variables k et h de la Terre (BTL) correspond à :

$$\Delta e_T = -\ 40{,}01 \times 10^{-10}\,T \ -\ 3090{,}33 \times 10^{-10}\,T^2 \ + \ \ldots$$

$$\Delta\tilde{\omega}_T = +\ 65''{,}74776\,T \ -\ 0''{,}82883\,T^2 \ + \ \ldots$$

Le terme en T^2 de l'excentricité modifie la longitude de la Terre de 0",13 au bout de 1000 ans.

On peut comparer le terme séculaire du périhélie aux valeurs trouvées par Le Verrier et par Newcomb :

Le Verrier : $\Delta\tilde{\omega}_T = +\ 70''{,}02\,T$

Newcomb : $\Delta\tilde{\omega}_T = +\ 76''{,}80\,T$

La forme analytique de ce terme est :

$$\Delta\tilde{\omega}_T = \frac{3}{4}\ n_T \left(\frac{a_L}{a_T}\right)^2 \mu\,(1 - \mu) \times (1 + \ldots)$$

où : $\mu = \dfrac{m_L}{m_T + m_L} = 0{,}012\,155\,057$; n_T est le moyen mouvement de la Terre ; a_L et a_T sont les demi-grands axes de la Lune et de la Terre respectivement.

Le terme principal donne un résultat voisin de celui de Newcomb mais il est modifié de manière importante par les autres paramètres, en particulier l'excentricité de la Lune.

L'application de cette expression aux quatre satellites galilées de

Jupiter donne une avance du périhélie de Jupiter :

$$\Delta\tilde{\omega}_J = + 0''\!,0366\,T$$

L'action de Titan donne une avance du périhélie de Saturne :

$$\Delta\tilde{\omega}_S = + 0''\!,0051\,T$$

5. PROLONGEMENT A 6000 ANS DE L'INTERVALLE DE VALIDITE DES THEORIES PLANETAIRES.

A partir des solutions TOP82 (Simon, 1983), VSOP82 (Bretagnon, 1982), nous avons construit une solution JASON84 (Simon, Bretagnon, 1984) de Jupiter et de Saturne valable 6000 ans de part et d'autre de J2000.0. Cette solution est construite essentiellement par analyse harmonique suivant les méthodes développées par Simon et Francou (1982) pour les perturbations mutuelles de Jupiter et Saturne et utilise des compléments pour les perturbations par les autres planètes venant de TOP82 et VSOP82.

Pour obtenir tous les termes du couple Jupiter-Saturne plus grands que $0''\!,01$, il a fallu développer les solutions jusqu'aux termes de Poisson de degré 20.

La précision de la solution complète est meilleure que 8" pour Jupiter et 23" pour Saturne sur l'intervalle - 4000, + 8000.

Pour les planètes inférieures Mercure, Vénus, la Terre et Mars, l'intervalle de validité peut être augmenté par l'amélioration des variations séculaires des éléments. Les éléments issus de la théorie générale de Laskar (1984a), développés jusqu'en T^6 (Laskar, 1984b) permettent de conserver une précision de quelques secondes sur un intervalle de 6000 ans de part et d'autre de J2000.0.

6. CONCLUSION

Les perturbations que nous venons de décrire : perturbations relativistes, perturbations par les petites planètes, par la Lune, ont des parties périodiques de faible amplitude, de l'ordre de $0''\!,01$ et sont certainement bien déterminées. L'amélioration des théories analytiques, en particulier dans le cas de Mars, passe donc par le calcul des perturbations du quatrième ordre et des ordres suivants.

BIBLIOGRAPHIE

Bretagnon, P.: 1980, *Astron. Astrophys.* **84**, 329.
Bretagnon, P.: 1982, *Astron. Astrophys.* **114**, 278.
Chapront-Touzé, M.: 1980, *Astron. Astrophys.* **83**, 86.
Chapront-Touzé, M., Chapront, J.: 1983, *Astron. Astrophys.* **124**, 50.
Laskar, J.: 1984a, *Astron. Astrophys.* à paraître.
Laskar, J.: 1984b, communication privée.
Lestrade, J.-F., Bretagnon, P.: 1982, *Astron. Astrophys.* **105**, 42.

Newhall, X. X., Standish, E. M., Williams, J. G.: 1983, *Astron. Astrophys.* **125**, 150.
Simon, J.-L.: 1983, *Astron. Astrophys.* **120**, 197.
Simon, J.-L., Bretagnon, P.: 1984, *Astron. Astrophys.* à paraître.
Simon, J.-L., Francou, G.: 1982, *Astron. Astrophys.* **114**, 125.
Standish, E. M.: 1982, 'DE200', bande magnétique.
Williams, J. G.: 1984, *Icarus* **57**, 1.

MOTIONS OF THE PERIHELIONS OF NEPTUNE AND PLUTO

HIROSHI KINOSHITA AND HIROSHI NAKAI
Tokyo Astronomical Observatory
1-21-1 Osawa, Mitaka, Tokyo, Japan

Abstract Five outer planets are numerically integrated over five million years in the Newtonian frame. The argument of Pluto's perihelion librates about 90 degrees with an amplitude of about 23 degrees. The period of the libration depends on the mass of Pluto: 4.0×10^6 years for $M_{pluto}=2.78\times10^{-6}M_{sun}$ and 3.8×10^6 years for $M_{pluto}=7.69\times10^{-9}M_{sun}$, which is the newly determined mass. The motion of Neptune's perihelion is more sensitive to the mass of Pluto. For $M_{pluto}=7.69\times10^{-9}M_{sun}$, the perihelion of Neptune does circulate counter-clockwise and for $M_{pluto}=2.78\times10^{-6}M_{sun}$, it does not circulate and the Neptune's eccentricity does not have a minimum. With the initial conditions which do not lie in the resonance region between Neptune and Pluto, a close approach between them takes place frequently and the orbit of Pluto becomes unstable and irregular.

1. Introduction

Cohen and Hubbard (1965) integrated outer planets over 120,000 years and found that the critical argument of Pluto with Neptune librates about 180 degrees with an amplitude of 76 degrees and a period of 20,000 years. Due to this libration, the close approach of Pluto with Neptune takes place near the Pluto's aphelion and the minimum distance between them is about 18 a.u. If the argument of Pluto's perihelion librates about 90 degrees, the aphelion of Pluto does not come in the orbital plane of Neptune and the configuration of Pluto and Neptune is more stable than the case that the argument of Pluto's perihelion circulates. Cohen, Hubbard, and Oesterwinter (1973) numerically integrated five outer planets over one million years. During this period, the argument of Pluto's perihelion changes from increase to decrease. One million years is not enough to see whether the argument of Pluto's perihelion librates or circulates.

Hori and Gicaglia (1968) studied the motion of Pluto with the perturbation method based on a canonical transformation that eliminates periodic terms. They treated this problem as a restricted three body problem, in which Neptune is an only disturbing planet, and eliminated periodic terms with numerical method and found that the argument of

Celestial Mechanics **34** (1984) 203–217. 0008–8714/84.15

Pluto's perihelion circulates with a period 30,000,000 years. William and Benson(1971) integrated numerically Pluto's motion as a restricted problem in which orbits of other four outer planets are given and concluded that the argument of pluto's perihelion librates about 90 degrees with an amplitude of 24 degrees and a period of 3.955×10^6 years. Nacozy and Diehl(1978a,b) applied Hori and Gicaglia's method(1968) to the same problem with including perturbations due to Jupiter,Saturn,and Uranus other than Neptune and found that the argument of Pluto's perihelion librates about 90 degrees with an amplitude of 23 degrees and a period of 3.775×10^6 years. Nacozy and Diehl's theory (1978) is a first-order secular perturbation. William and Benson(1971) integrated averaged equations of Pluto's motion,which means that their treatment is equivalent to the first-order secular perturbation. In this paper,in order to find how mutual perturbations among five outer planets contribute the long-periodic behaivior of orbital motion of planets and especially motions of perihelions of Pluto and Neptune,we extended one million years integration by Cohen *et al.*(1973) to five million years integration. Also we compare our results with the secular perturbation theory by Bretagnon(1974).

2. Method of Numerical Integration and Its Accuracy

The coordinate reference frame, masses of the planets(see the column A in Table1),and initial values of planets used in this study are the same as those adopted by Cohen *et al.* (1973). The x-y plane is the invariable plane of the solar system,which is perpendicular to the total angular momentum of the solar system. The masses of the inner planets are added to the Sun. The mass of Pluto is 1/360000 of the Sun's mass,which is larger 360 times than the newly determined mass of Pluto.

Table 1

Reciprocal Masses of Outer Planets

	A	B	C
Jupiter	1,047.355	1,047.355	1,047.35
Saturn	3,501.6	3,498	3,498.0
Uranus	22,869	22,869	22,960
Neptune	19,314	19,314	19,314
Pluto	360,000	——	130,000,000

A: Cohen,Hubbard,and Oesterwinter(1973)
B: Bretagnon(1974)
C: DE200(Jet Propulsion Laboratory)

The integration formula was the Störmer's multistep integrator,which is one of predictor type integrators and was used by Cohen *et al.*(1973). We did not apply a correction by a corrector type integrator for saving computer time,since we found that the integration by only use of a predictor is enough to get 1 degree accuracy of positions of the planets over five million years. The numerical calculation was carried out by the FACOM380R computer with a double precision(a 56bit mantissa) and a 40-day step size. The total computer time was about 4 hours.

At first we integrated outer planets over 10,000 years by the extrapolation method developped by Gragg (1965). The orbits thus obtained were chosen as standard orbits for checking the accuracy of orbits integrated by the Störmer's integrator. The accuracy of the standard orbits themselves were examined with the aid of the time reversal invariance : we integrated the motions over 10,000 years in backward with the positions and velocities at the last date and examined how the positions at the initial epoch are close to the initial positions. The integration by the extrapolation method were carried out with a stepsize of 250 days and quatre precision arithmetic that is about 33 decimal in floating point arithemtic calculation. The differences in the longitudes between forward and backward integrations increase as the square of time and the maximum difference in the longidude of Jupiter over 10,000 years is only 3×10^{-8} arcsecond. The standard orbits,therefore,are accurate enough to test the accuracy of the orbits integrated by the multistep method. The computer time was 27 minitues,which means that the computer time over 5 million years by the extrapolation method will be 224 hours.

Table 2

Errors in longitudes over 10,000 years

	quatre precision		double precision	
step size	20 days	40 days	20 days	40days
	$''$	$''$	$''$	$''$
Jupiter	6.2×10^{-7}	4.6×10^{-3}	1.5×10^{-2}	1.4×10^{-2}
Saturn	1.5×10^{-6}	1.0×10^{-2}	5.4×10^{-3}	8.3×10^{-3}
Uranus	3.3×10^{-9}	2.3×10^{-5}	8.0×10^{-3}	3.8×10^{-3}
Neptune	2.7×10^{-10}	1.8×10^{-6}	5.0×10^{-3}	1.8×10^{-3}
Pluto	2.7×10^{-10}	1.8×10^{-6}	6.4×10^{-3}	1.9×10^{-3}
Computer time	630sec	317sec	70sec	29sec

We integrated outer planets over 10,000 years with both double and quatre precision and with stepsizes of 20 days and 40 days, and compared them with the standard orbits. The errors in the longitudes over 10,000 years are listed in Table 2 with the computer time. We also integrated outer planets with a corrector and found that the orbits obtained with quatre precision are more accurate than those obtained without a corrector and ,however,in double precision there is no improvement,since the round-off errors are comparable with the discretization errors in double precision. The similar situation appears in the orbits with stepsizes of 20 days and 40 days in double precision as shown in Table 2. Assuming that the errors in the mean longitudes increase as the square of time and other element's errors increase proportionally with time, we estimate errors of orbital elements over five million years from the comparison of orbits integrated with only use of a predictor and 40 days and double precision with the standard orbits. The estimated errors are listed in Table 3. From Table 3, we see that orbits thus obtained have the accuracy of 1 degree in position over five million years. In addition, the last column in Table 3 show the advances of perihelions due to the general relativity effect, which is the same order as the accumulation error of the mean longitude in the Newtonian frame. The secular effect on the mean longitude due to the general relativity effect can be absorbed in a little change of the semi-major axis. The general relativity effect, therefore, can be neglected in the discussion of orbital motions of outer planets over a time scale of several million years or more.

Table 3

Estimated errors in orbital elements over 5,000,000 years

(double precision, step size=40days, Störmer method)

	$\Delta\lambda$ (degree)	$\Delta a/a$	Δe	ΔI (degree)	$\Delta\omega$ (degree)	$\Delta\Omega$ (degree)	$\Delta\tilde{\omega}_{rel.}$ (degree)
Jupiter	0.83	2.5×10^{-9}	1.3×10^{-8}	2.5×10^{-8}	1.5×10^{-5}	2.5×10^{-6}	0.86
Saturn	0.52	2.5×10^{-9}	1.5×10^{-8}	2.0×10^{-8}	2.5×10^{-5}	6.0×10^{-6}	0.19
Uranus	0.26	9.0×10^{-9}	2.5×10^{-9}	2.5×10^{-9}	4.0×10^{-5}	1.2×10^{-5}	0.04
Neptune	0.13	1.0×10^{-8}	2.5×10^{-9}	2.5×10^{-9}	2.5×10^{-5}	1.3×10^{-5}	1.1×10^{-2}
Pluto	0.09	5.0×10^{-9}	2.5×10^{-8}	1.5×10^{-6}	2.2×10^{-5}	3.8×10^{-5}	5.8×10^{-3}

3. Orbital Elements over Five Million Years

Averaged orbital elements, the semi-major axes, the eccentricities, the inclinations, the arguments of perihelions, the longitudes of nodes, and the longitudes of perihelions of outer planets are plotted in Figures 1,2,3,4, and 6. Averaged elements represent the mean of osculating elements over 500 years in order to eliminate short periodic variations.

Fig.1 Jupiter's Orbital Elements (facing page)
500 years average elements referred to the barycenter of the solar system are plotted.
A: a semi-major axis (relative variation), a_{mean}=5.18$a.u.$
B: e eccentricity,
C: I inclination,
D: ω argument of perihelion,
E: Ω longitude of node,
F: $\tilde{\omega}$ longitude of perihelion,
G-1: $K= e \cos \tilde{\omega}, H= e \sin \tilde{\omega}$; Numerical Integration,
G-2: $K= e \cos \tilde{\omega}, H= e \sin \tilde{\omega}$; Bretagnon's Theory,
H-1: $Q= \sin I/2 \cos \Omega, H= \sin I/2 \sin \Omega$; Numerical Integration,
H-2: $Q= \sin I/2 \cos \Omega, H= \sin I/2 \sin \Omega$; Bretagnon's Theory.

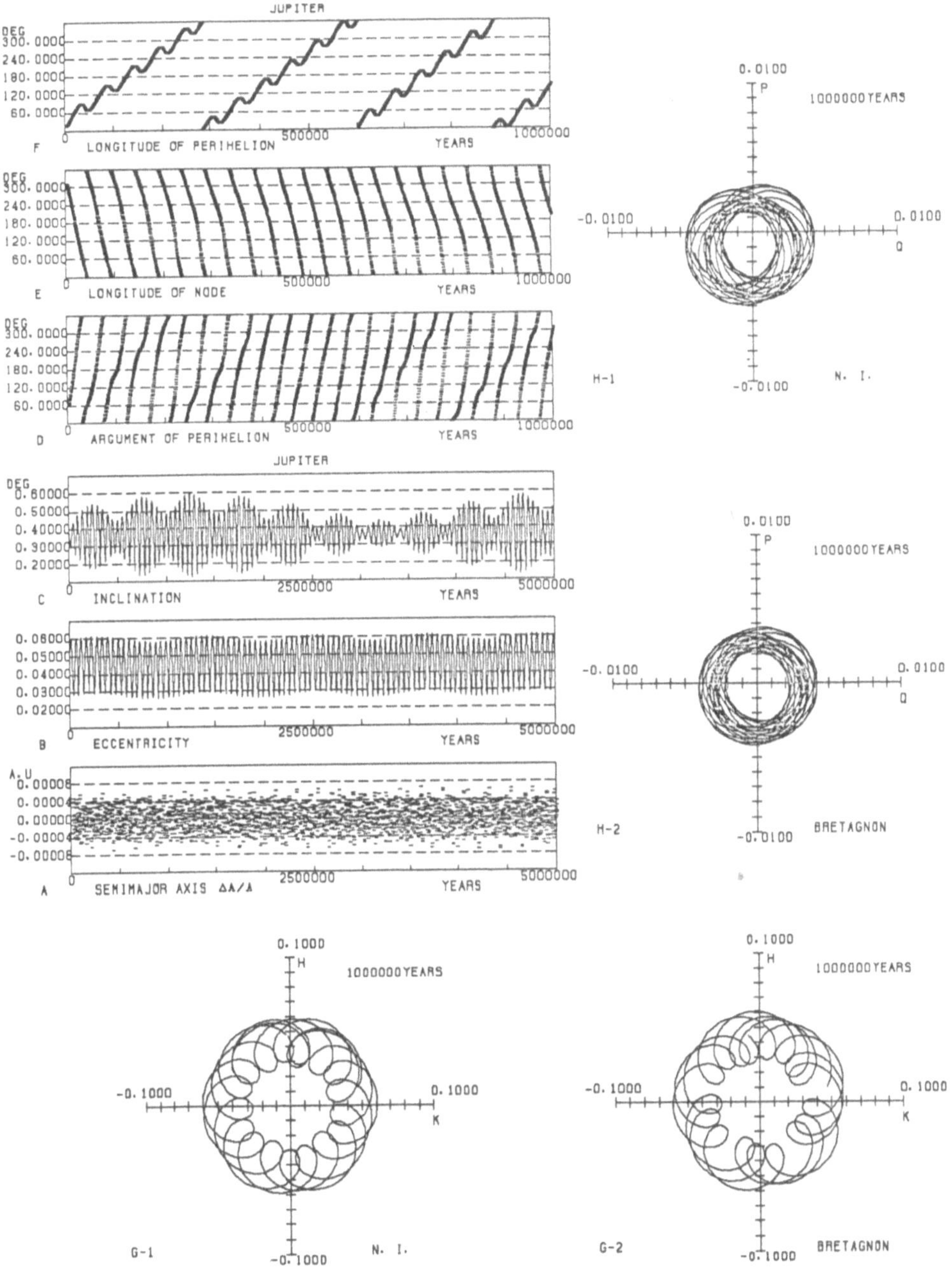
JUPITER
DEG
300.0000
240.0000
180.0000
120.0000
60.0000
0
500000
1000000
F
LONGITUDE OF PERIHELION
YEARS
E
LONGITUDE OF NODE
D
ARGUMENT OF PERIHELION
0.60000
0.50000
0.40000
0.30000
0.20000
2500000
5000000
C
INCLINATION
0.06000
0.05000
0.04000
0.03000
0.02000
B
ECCENTRICITY
A.U
0.00008
0.00004
0.00000
-0.00004
-0.00008
A
SEMIMAJOR AXIS ΔA/A
0.0100
P
1000000YEARS
-0.0100
Q
H-1
N. I.
H-2
BRETAGNON
0.1000
H
-0.1000
K
G-1
G-2

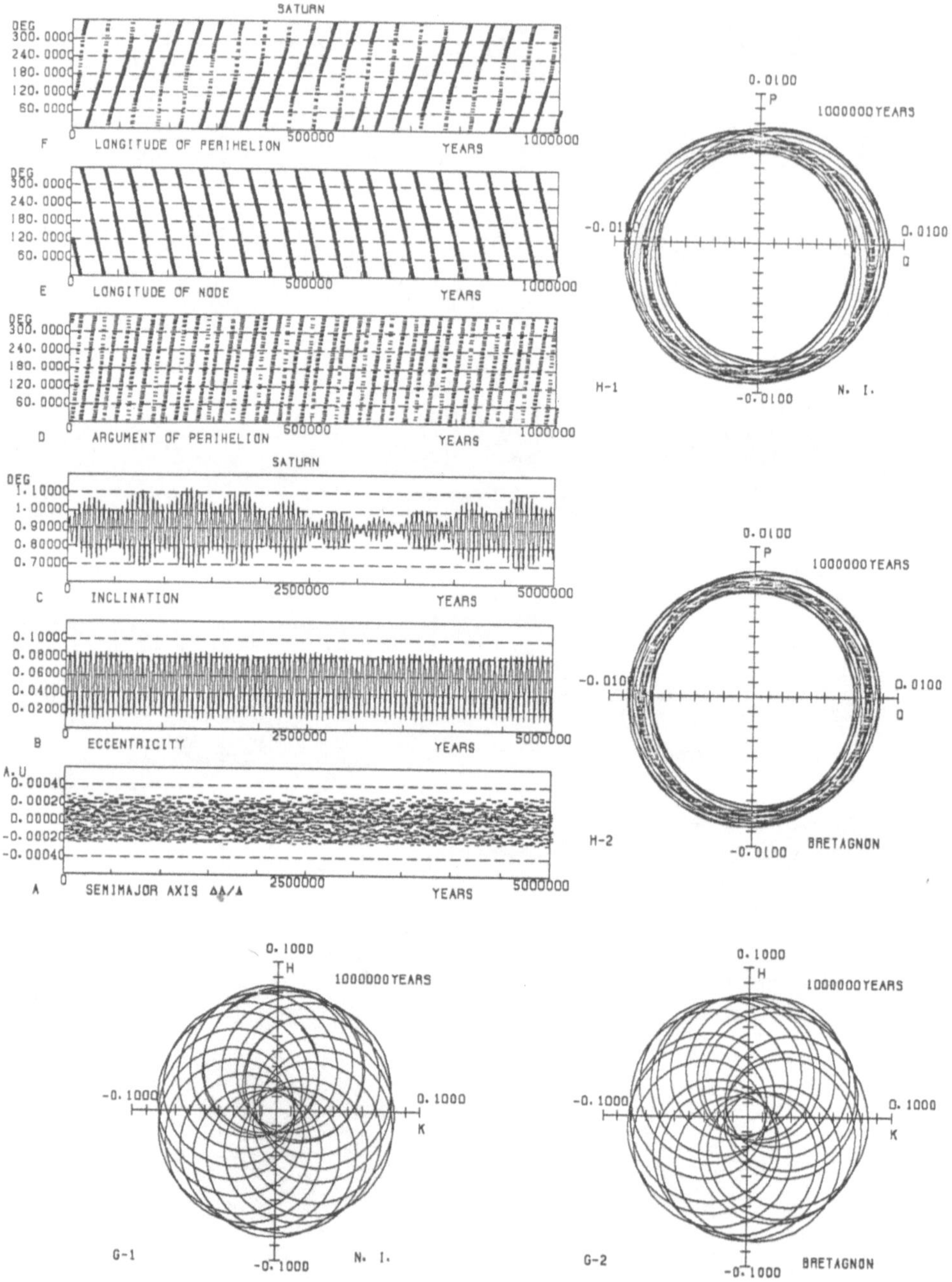

Fig.2 Saturn's Orbital Elements
(see the legend in Fig.1)
a_{mean}=9.54$a.u.$

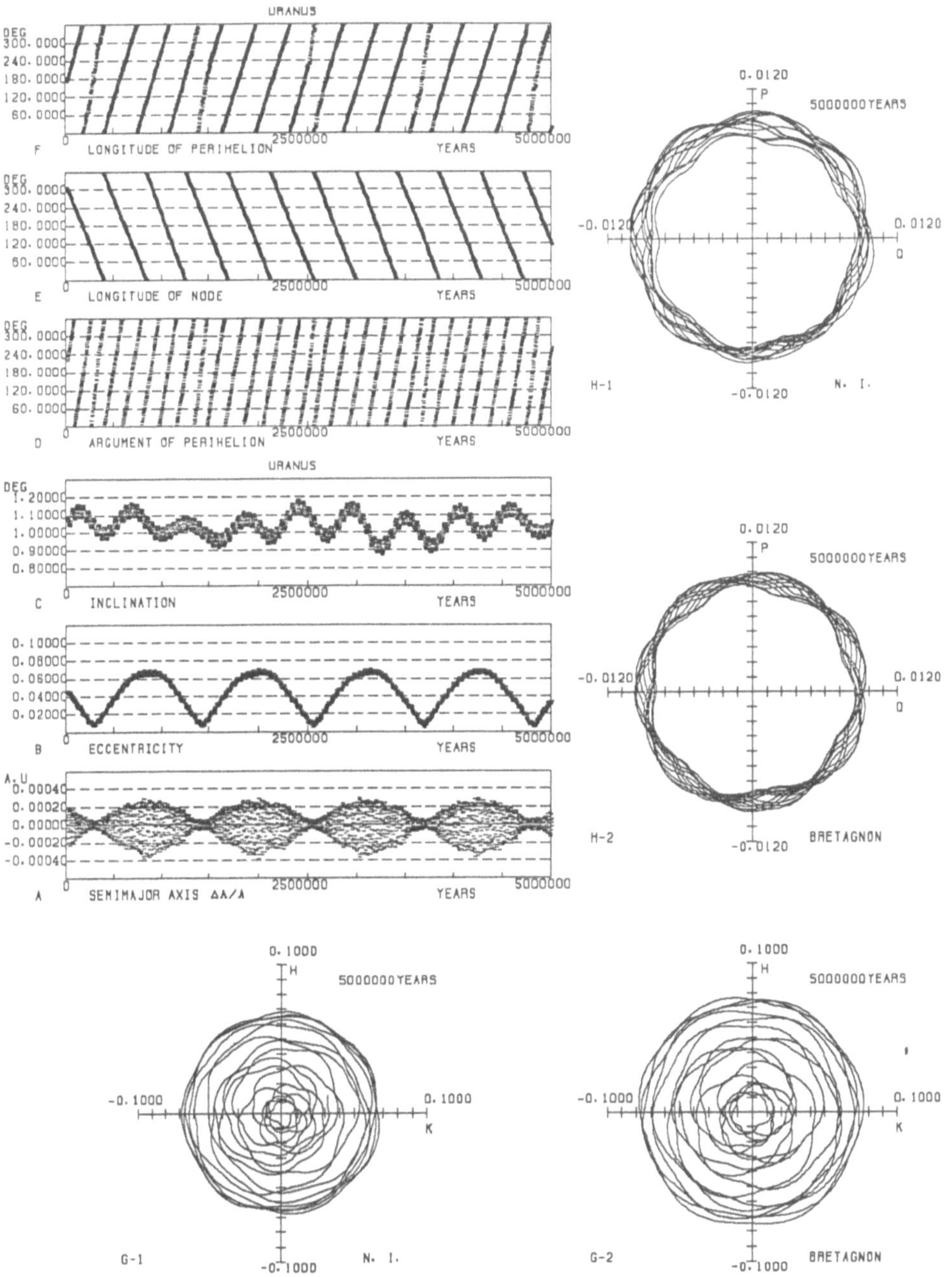

Fig.3 Uranus's Orbital Elements
(see the legend in Fig.1)
a_{mean}=19.21a.u.

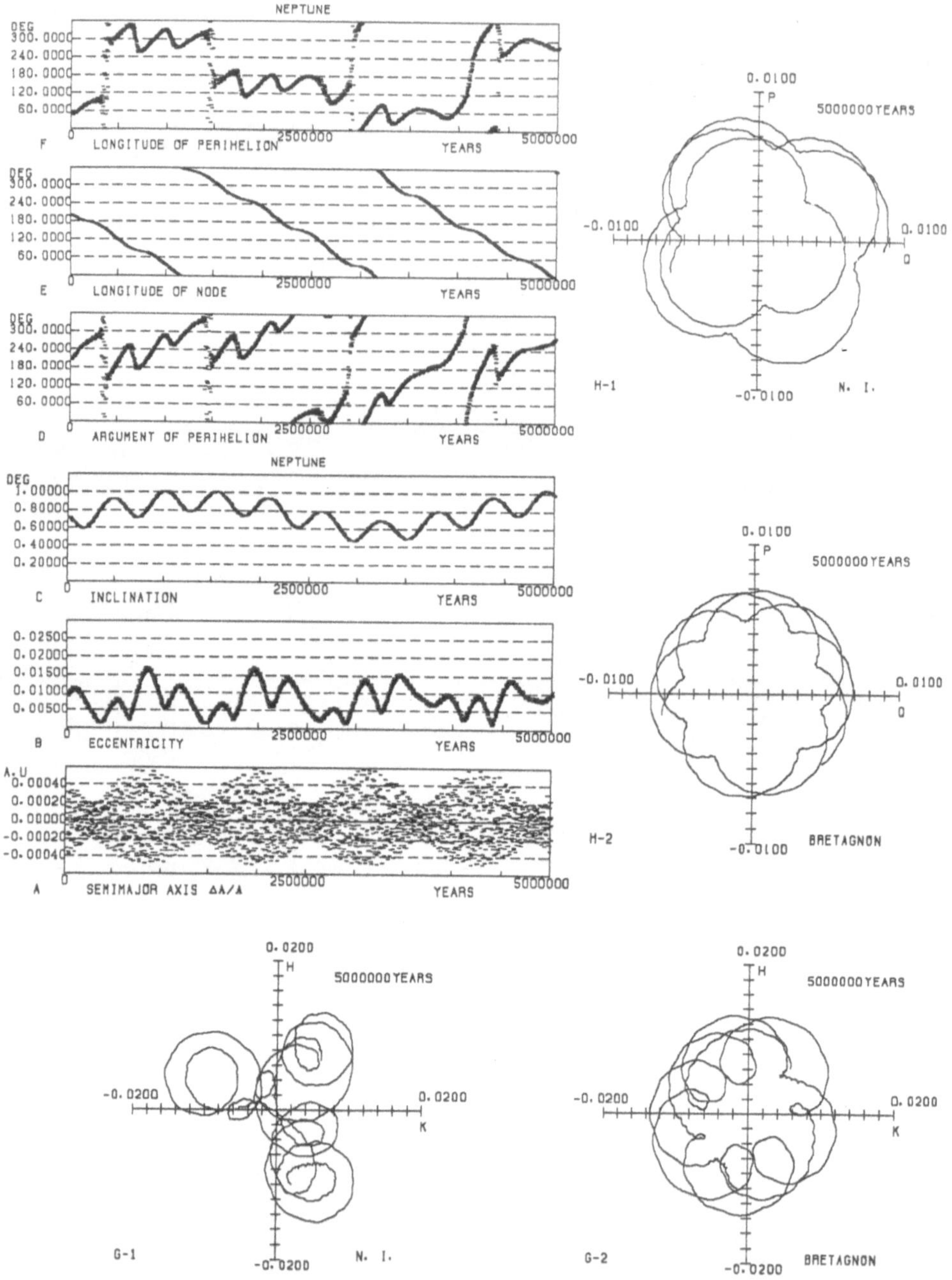

Fig.4 Neptune's Orbital Elements
(see the legend in Fig.1)
$M_{pluto}=2.78\times 10^{-6}M_{sun}$, $a_{mean}=30.106a.u.$

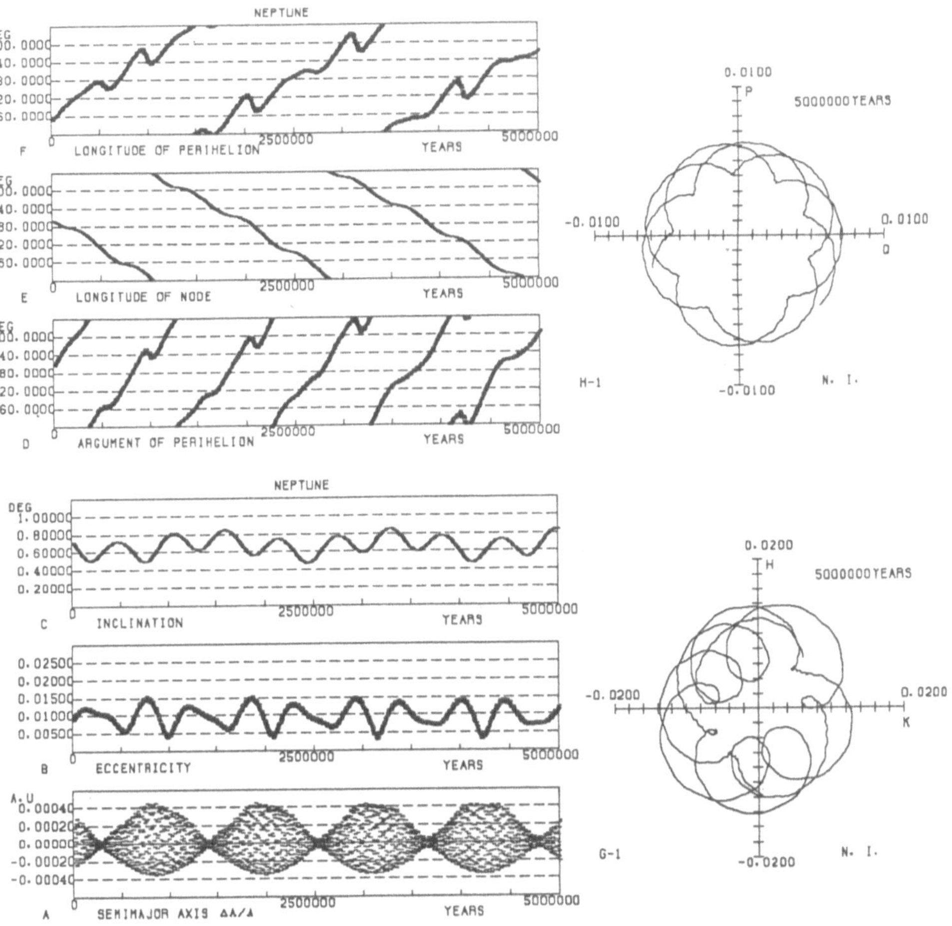

Fig.5 Neptune's Orbital Elements
(see the legend in Fig.1)
M_{pluto}=7.69×$10^{-9}M_{sun}$, a_{mean}=30.103a.u.

These elements are referred to the barycenter of the solar system, instead of the heliocentric coordinates. By choosing the barycentric cooordinates,variations of orbital elements become smooth,especially for Neptune's orbital elements. Our results are in quite good agreement with those by Cohen *et al.* (1973) during the overlapped 500,000 years in both short and long-periodic variations.In addition to the Keplerian elements, $K=e\cos\varpi$.$H=e\sin\varpi$ and $Q=\sin I/2\cos\Omega$,$P=\sin I/2\sin\Omega$ are plotted in the same figures with those obtained from Bretagnon's secular perturbation theory(1974),which is the second-order secular perturbation theory with respect to the masses of disturbing planets. From comparing them,Bretagnon's theory are qualitatively in good agreement with the

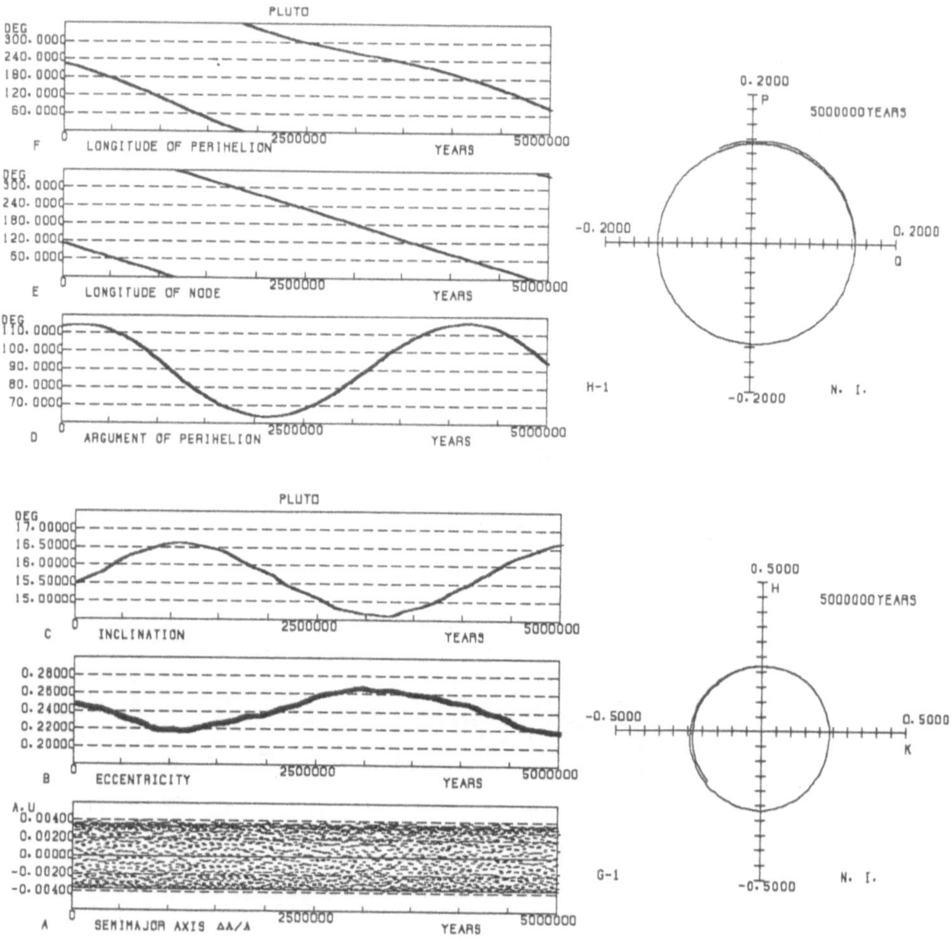

Fig.6 Pluto's Orbital Elements
(see the legend in Fig.1)
$M_{pluto}=2.78\times10^{-6}M_{sun}$, $a_{mean}=39.458$ a.u.

numerical integration except Neptune. Slight deviations in the amplitudes and the phases between them arise mainly from the differences of the adopted planetary masses and initial orbital elements, and the effect due to the mutual perturbation with respect to the square of masses of disturbing planets may be small.

4. Motion of the Perihelion of Neptune

As seen in Figure 4, the motion of Neptune's perihelion derived from the numerical integration is complicated and the Neptune's eccentricity does

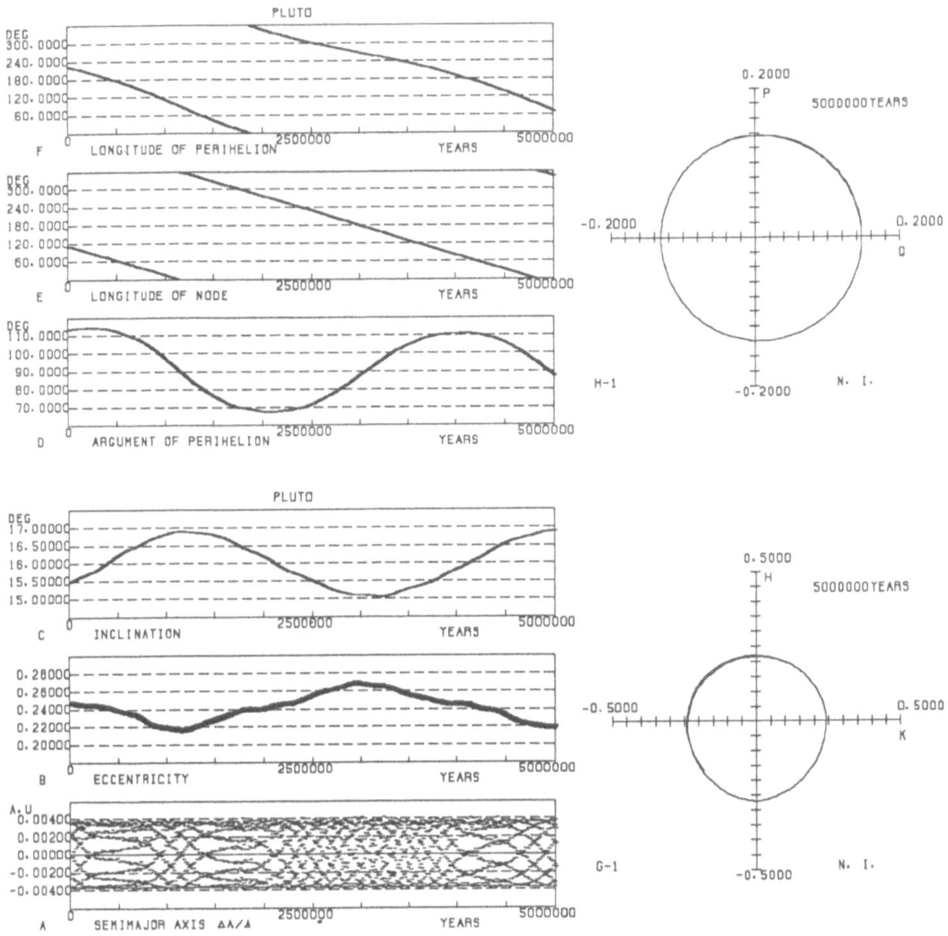

Fig.7 Pluto's Orbital Elements
(see the legend in Fig.1)
$M_{pluto}=7.69\times10^{-9}M_{sun}$, $a_{mean}=39.466a.u.$

not have a minimum,and is quite different from the motion predicted by Bretagnon's theory(1974). Bretagnon(1974) did not include Pluto because of the difficulty in application of conventional perturbation method. On the other hand,the mass of Pluto in our integration($2.78\times10^{-6}M_{sun}$), which is the same one used by Cohen *et al.* (1973),is larger 360 times than the newly determined mass of Pluto ($7.69\times10^{-9}M_{sun}$:see the column C in Table 1). Therefore,we integrated outer planets with use of the new masses used in DE200 and the same initial values. The result is plotted in Figure 5 and is in good agreement with Bretagnon's theory for details. This fact indicate that in the discussion of long-periodic behaiviours of Jupiter, Saturn, Uranus, and Neptune, the dynamical

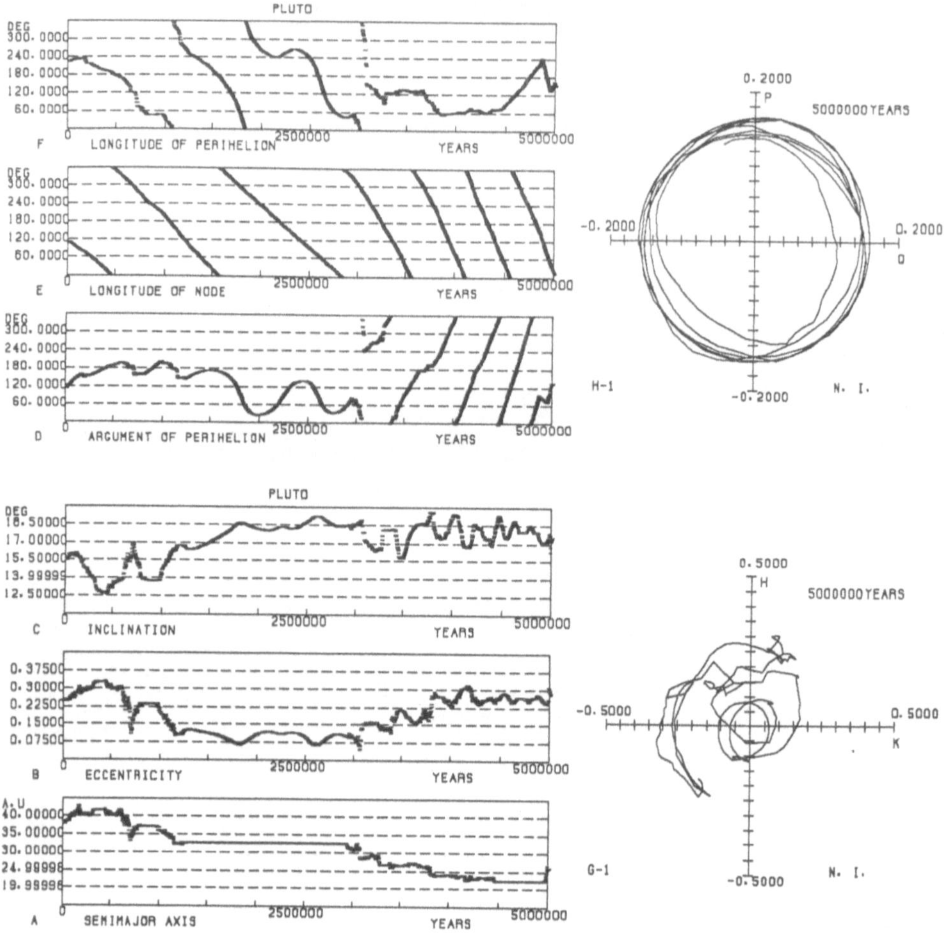

Fig.8 Pluto's Orbital Elements
(see the legend in Fig.1)
$M_{pluto}=7.69\times10^{-9}M_{sun}$, initial osculating semi-major axis = 38.0 a.u., Other initial osculating elements are the same as those in Fig.6.

effect of Pluto could be negligible and Pluto could be treated as a restricted problem. By changing the mass of Pluto, the orbital elements of Pluto itself do not show any essential change due to the indirect effect except a slight change of the semi-major axis (compare Figure 6 with Figure 7).

5. Motion of Pluto's Perihelion

As seen from Figure 6 or 7, the argument of Pluto's perihelion librates about 90 degrees with an aplitude of 23 degrees and the librational

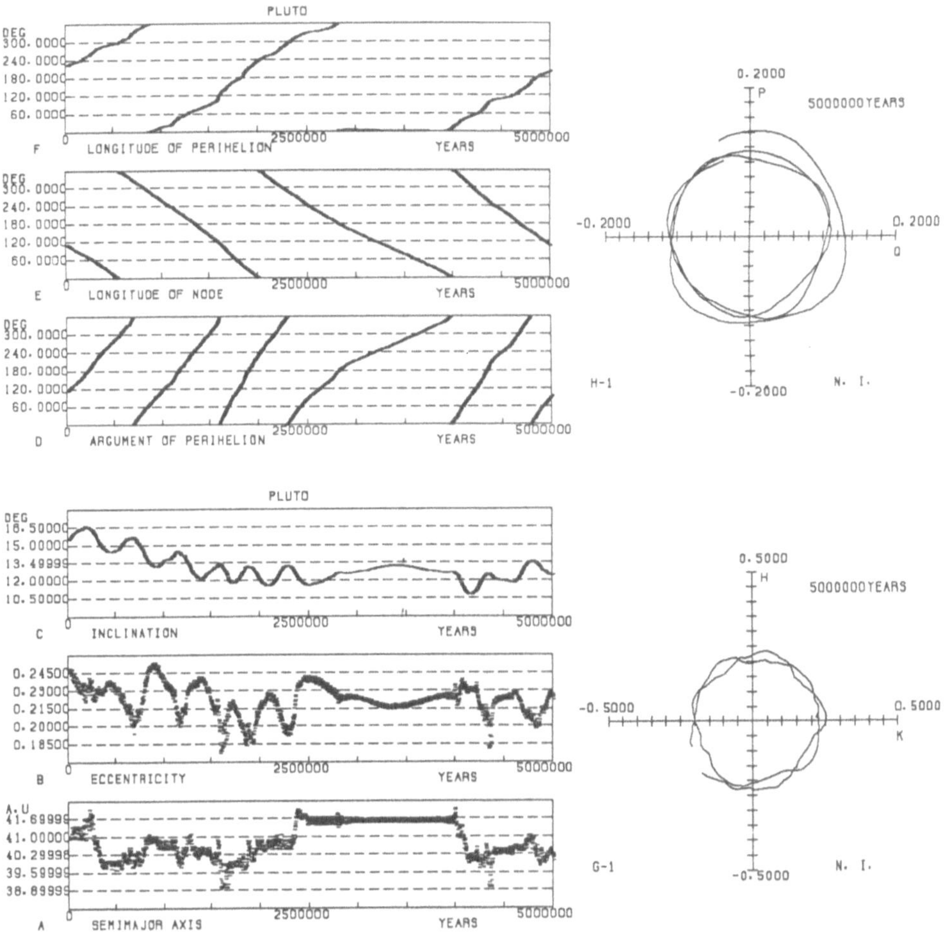

Fig.9 Pluto's Orbital Elements
(see the legend in Fig.1)
$M_{pluto}=7.69\times10^{-9}M_{sun}$, initial osculating semi-major axis = 41.0 a.u., Other initial osculating elements are the same as those in Fig.6.

period is 4.0×10^{6} years for $M_{pluto}/M_{sun}=2.78\times10^{-6}$ and 3.8×10^{6} years for $M_{pluto}/M_{sun}=7.69\times10^{-9}$. This result qualitatively agrees with the semi-analytical theory by Nacozy and Diehl (1978) and with the numerical integration of averaged equation of Pluto's motion as a restricted problem by Williams and Benson(1971). The libration of the argument of Pluto's perihelion means that the aphelion of Pluto does not come in the orbital plane of Neptune and the Neptune-Pluto system tends to avoids a close approach, because the close approach of Pluto to Neptune takes place near the aphelion of Pluto.

In the mechanism of keeping the avoidance of close approach between Pluto and Neptune, the commensurable relation between mean motions between Pluto and Neptune does play an essential role. In order to see how the commensurable relation is important in keeping the stable configuration of Neptune–Pluto system, we carried out a numerical integration with the semi-major axis that is chosen outside of the commensurable region, keeping other initial osculating elements unchanged. Figure 8 is the case of the initial osculating semi-major axis a_0=38 a.u. In this case Pluto repeats frequently a close approach with Neptune and gradually moves inward and crosses apparently the orbit of Uranus after 4.5 million years. The minimum distance of the close approaches is about 2a.u. and is far outside the sphere of influence of Neptune, of which radius is 0.6a.u. The variations of orbital elements are irregular and chaotic. Five million years integration does not tell whether the orbit of Pluto becomes stable in the future or not. The case of a_0=41 a.u. is shown in Figure 9. In this case, Pluto does not cross the orbit of Neptune during this five million years. The orbital change, however, is also irregular and chaotic.

6. Conclusion

Our numerical integrations of outer planets over five million years are in good agreement with the secular perturbation theories ,Nacozy and Diehl(1978) and Bretagnon(1974). This fact indicates that with use of the secular perturbation theory we can discuss long-range and global behaivior of planets. As far as the solar system is assumed to be conservative, judging from the good agreement of the secular perturbation theory with the numerical integration, the present configuration of outer planets might be stable in the age of the solar system. As seen in the last section, however, the orbital change of Pluto, of which the semi-major axis does not lie in the resonance region with Neptune, is large and irregular due to the close approach to Neptune. The question, therefore, is how Pluto is captured in the present resonant state with Neptune. In order to discuss the evolution of Pluto's orbit into the resonance, we have to take into consideration the non-conservative mechanism in the early stage of the solar system.

Acknowledgements

We acknowledge Dr. Oesterwinter for his helpful comments and Mr. Fukushima for informing us of the numerical integration by extrapolation method developped by Gragg(1965).

References

Bretagnon, P: 1974, *Astron. Astrophys.*, 30, 141.
Cohen, C. J. and Hubbard, E. C.: 1973, *Astron. J.*, 70, 10.
Cohen, C. J., Hubbard, E. C., and Oesterwinter, C.: 1973, *Astronomical Papers of* the American Ephemeris), 22, pt 1.
Gragg, W. B.: 1965, *SIAM J. Numer. Anal.*, 2, 384.

Hori,G.and Gicaglia,G.E.O.:1968,*Research in Celestial Mechanics and Differential Equations*(University of Sao Paulo)),1,4.
Nacozy,P.E. and Diehl,R.E.:1978a,*Astron.J.*,83,522.
Nacozy,P.E. and Diehl,R.E.:1978b,*Celest. Mech.*,17,405.
Williams,J.G. and Benson,G.S.:1971,*Astron.J.*76,167.

PROGRESS IN GENERAL PLANETARY THEORY

J. Laskar
Bureau des Longitudes
77 Ave. Denfert-Rochereau
F-75014 Paris
France

ABSTRACT. When on searches for a planetary theory valid over 1 million years, one can leave in the solution the short period terms whose amplitude are small, and compute only the long period terms which give the essential of the solution; but one must take into account the short period terms contribution in the computation of the long period terms.

Among recent works on such theories, Bretagnon (1974) computed a long period solution for the 8 planets, including the contribution of all terms up to order 2 with respect to the masses and degree 3 in the eccentricity-inclination. He improved his solution later on by adding the effects of the relativity and the Moon, and by adjusting the main frequencies of the outer planets to the secular terms of his short period theory VSOP82 (Bretagnon 1984). Duriez (1979) computed an analytical solution for the 4 outer planets up to order 2 of the masses and degree 5 in the eccentricity-inclination (and even 7 for the Jupiter-Saturn couple).

Nowadays, the great development of the Milankovitch astronomical theory of palaeoclimates, resulting from the improvements of geological datation procedures for sediments (Hays, Imbrie and Shackleton 1976), requests a very accurate knowledge of the Earth orbital elements over a very long span of time.

We resumed the study of General Planetary Theory, focussing on the accuracy of the solution. The aim of our study is no longer to have the global features of the solution, but to obtain an accurate solution with known error bounds over 1 million years, in order to know how far we can extend the solution keeping a reasonable precision.

In the first step of the work, we have developed a new method for computing the terms of the autonomous system which gives the secular variations of the 8 planets orbital elements. This system includes all the secular terms and the contribution of the short period terms up to

Celestial Mechanics **34** (1984) 219–221. 0008–8714/84.15

order 2 of the masses and up to degree 5 in the eccentricity-inclination. The method is sufficiently efficient to allow us to kep all the 150 000 monomial terms in the autonomous system. We have also paid special attention to the precision of the computation and all the coefficients of the autonomous system are given with a relative precision of 10^{-6} (Laskar, 1984a).

We then elaborated an algorithm to compute an analytical solution of the autonomous system, following the method of normalization given in Brumberg (1980). We thus convinced ourselves that it was very difficult, if not impossible to obtain a very accurate solution with this method in the case of our 8 planet system, nowithstanding the fact that if the degree 3 solution contains 25 000 terms, the degree 5 solution contains 3 000 000 terms, and may not give a more accurate solution, due to the presence of small divisors which damage very much the solution.

On the contrary, to integrate numerically the autonomous system is rather easy: the main periods are larger than 50 000 years, so we can choose a step of integration of about 1000 years. We thus made a numerical integration of the whole autonomous system over 1 million years using an Adams method with varying stepsize and tolerance of 10^{-12}.

To check the accuracy of the solution, we made an expansion of the solution at the origin in polynomial of the time, in order to compare it with the secular variations given by the highly accurate short period solution VSOP82 (Bretagnon 1982). As the constants of the two theories are the same, we can assume that the comparison of the coefficients of T gives a good estimate of the precision of the solution. We made also a comparison with Bretagnon (1984) long period theory. For the variable $k = e \cos \bar{\omega}$ of the Earth, for example, we obtain a precision of 0.068×10^{-6} rd/ky for our solution against 13.371×10^{-6} rd/ky for the Bretagnon (1984) solution.

If we take then the terms in T^3, T^4, T^5, T^6 given by our general theory and add them to the secular terms of VSOP82, we reduce considerably the discrepancies VSOP82-DE102 over the whole range of DE102, that is 3000 years.

In conclusion, our general planetary theory fulfils a double task:

- It gives a good approximation with given error bounds for the long period variations of the 8 planets orbital elements over 1 million years.
- It can be used over 10 000 years to prolong the highly accurate short period theories.

A complete exposition of the results is to be found in Laskar (1984b).

REFERENCES

Bretagnon, P.: 1974, Astron. Astrophys. 30, 141.

Bretagnon, P.: 1982, Astron. Astrophys. 114, 278.

Bretagnon, P.: 1984, in Berger A.L. and J. Imbrie eds. 'Milankovitch and Climate' proceedings, New York, Nov. 1982.

Brumberg, V.A.: 1980, 'Analytical Algorithms of Celestial Mechanics', Nauka, Moscow (in Russian).

Duriez, L.: 1979, Thèse, Lille.

Hays, J.D., J. Imbrie and N.J. Shakleton: 1976, Science 194, 1121.

Laskar, J.: 1984, 'Accurate methods in general planetary theory', Astron. Astrophys., submitted for publication.

Laskar, J.: 1984, Thèse de 3ème cycle, Paris.

SATELLITE ORBITS AND EPHEMERIDES

S. Ferraz-Mello
Departamento de Astronomia
Universidade de São Paulo
01051-São Paulo, Brazil

ABSTRACT. This review considers theoretical work leading to the actual knowledge of the motions of all planetary satellites, except the Moon. It covers recent calculations of inequalities of satellites' motions and the determination of their orbits on the basis of existing observational data. The results are presented and some of them discussed in views of possible future developments.

1. INTRODUCTION

I would entitle this review as Satellites galore if a similar title had not been used by Marsden in a review by himself sometime ago (Marsden, 1980), and redefine the Solar System as the place where planetary satellites are kept for showing. Indeed, the most striking feature the satellites show is their diversity. One glance at the images transmitted by space probes already allows one to perceive a richness of situations, from active volcanoes to frozen surfaces, from tiny rocks to satellites almost as great as their central planet as in the case of the Moon. We know 44 of them: our Moon, 2 satellites around Mars, 16 around Jupiter, 17 around Saturn, 5 around Uranus, 2 around Neptune and one around Pluto. Their rich dynamics includes satellites shepherding rings, satellites accelerated by tidal interactions with the planet and 14 satellites participating in resonant gravitational intersatellite interactions. The very interesting phenomena exhibited by their motions and the continuous quest for better ephemerides led to an intensification in the researches in this domain. Since the preparation of the last report of IAU Commission 20 Working Group on satellite orbits and ephemerides (Kozai, 1982) new elements were published for the Galilean satellites of Jupiter, for some of the outer satellites of Jupiter, for almost all satellites of Saturn, for the satellites of Uranus and Neptune, and for most of the satellites newly discovered around Jupiter and Saturn. If to these results we add the orbits published after 1975, we have an almost complete revision of the motions of the satellites that leaves untouched only a few satellite orbits. The information on this progress is gathered in Table I, where orbit determinations (from observations) made since 1975 are listed. This table is far from exhausting the contribu-

Celestial Mechanics **34** (1984) 223–241. 0008–8714/84.15

tions to our knowledge of the satellites motions. Indeed, much theoretical work related, directly or indirectly, to the motion of the planetary satellites, was published in this period.

Orbital elements not being quantities directly accessible to observation, their determination is generally made through the least-squares fitting of a model. The models used range from very complete theories, as for the Galilean satellites, to simple Keplerian ellipses in the equatorial plane of the planet.

The Moon has not been considered in Table I and is not included in this review. Indeed, the problem of the motion of the Moon is, for several reasons, very different from the problem of the motion of the other planetary satellites. The main reason arises nowadays from the fact that the observations of the Moon already give results at the decimeter level. In order to reach this level of precision a theory of the Moon must consider, besides the disturbing effects of the Sun, those due to the figure of both, the Earth and the Moon, effects of the planets, the tidal interactions between the Moon and the Earth, and must also include relativistic corrections. Therefore the problem of the motion of the Moon is classically considered separate from the problem of the motion of the other planetary satellites.

2. THE SATELLITES OF MARS

The only theoretical complexity arises from the fact that the oblateness of Mars and the Sun make the orbital plane of the satellite to have a retrograde precession. However, while the oblateness drives this precession along the planet's equator, the Sun drives it along the orbital plane of Mars. The two planes have a mutual inclination of 25°.2 (Martian obliquity). The form of the solution is simplified if one adopts as reference planes the Laplacian planes (one for each satellite since their interaction is too small to be considered). On the other hand Born and Duxbury (1975) have shown that the mean motion of Deimos has a periodic variation of amplitude $0°.86 \times 10^{-4}$ d^{-1} and period 54 yr (the argument is the angle from the ascending node of the Laplacian plane of Deimos on the orbit of Mars to the ascending node of the satellite orbit on this Laplacian plane). They also showed that there is an oscillation of amplitude 4.6 km and period 65 h in the longitude of Deimos due to a resonance with one harmonic (J_{22}) of the Martian potential. While small, these inequalities have to be considered to reproduce the observed positions (50 of Phobos and 30 of Deimos) derived from the Mariner 9 TV images.

The most controversial point regarding the satellites of Mars is the acceleration of their motion (Table II). The value announced by Sharpless in 1945 led to many discussions and hypothesis before more detailed tidal models became available. The idea that some drag could be exercising a brake on the satellites motion led to very high values for their area/mass ratio, and thus to very low satellite density. The curious hypothesis of a hollow satellite became famous at the end of the fifties. However the most favoured opinion was that, in fact, the error of early observations of the satellites was great and gave not enough support to Sharpless' results. From a set of 3107 observations collected

TABLE I - Recent Determinations of Orbital Elements

M I (Phobos)	Shor (1975),	Born & Duxbury (1975)*,	Sinclair (1978)*.
M II (Deimos)	id.	id.	id.
J I (Io)	Lieske (1980),	Arlot (1982).	
J II (Europa)	id.	id.	
J III (Ganymede)	id.	id.	
J IV (Callisto)	id.	id.	
J V (Amalthea)			
J VI (Himalia)	Bordovitsyna & Bykova (1978),	Rocher (1983).	
J VII (Elara)	id.	id.	
J VIII (Pasiphae)	Rocher (1984).		
J IX (Sinope)	id.		
J X (Lysithea)			
J XI (Carme)			
J XII (Ananke)			
J XIII (Leda)	Aksnes (1978).		
J XIV (Thebe)			
J XV (Adrastea)			
J XVI (Metis)			
S I (Mimas)	Chugunov (1983b),	Rapaport (1977)*.	
S II (Enceladus)	id.	id.	Jefferys & Ries (1979a)*.
S III (Tethys)	id.	id.	
S IV (Dione)	id.	id.	id.
S V (Rhea)	id.	id.	
S VI (Titan)	id.		
S VII (Hyperion)	Hatanaka (1979)*.		
S VIII (Iapetus)	Sinclair (1974),	Rapaport (1978)*.	
S IX (Phoebe)	Rose (1979), Bec-Borsenberger & Rocher (1982), Bykova & Shikhalev (1984).		
S X (Janus)	Yoder *et al.* (1983)*.		
S XI (Epimetheus)	id.		
S XII ("Dione B")	Reitsema (1981a)*.		
S XIII (Telesto)	Reitsema (1981b)*.		
S XIV (Calypso)	id.		
S XV (Atlas)			
S XVI	Synnott *et al.* (1983).		
S XVII	id.		
U I (Ariel)	Veillet (1983).		
U II (Umbriel)	id.		
U III (Titania)	id.		
U IV (Oberon)	id.		
U V (Miranda)	id.		
N I (Triton)	Harris (1983).		
N II (Nereid)	Mignard (1981),	Veillet (1982).	
P I (Charon)	Harrington & Christy (1981),	Bonneau & Foy (1980)*.	

* Determination of only some elements or from a particular set of observations.

TABLE II - Acceleration of the satellites of Mars (century^{-2})

Author	Phobos	Deimos
Sharpless (1945)	$+ 18^\circ.8 \pm 1^\circ.7$	$- 2^\circ.7 \pm 1^\circ.6$
Sinclair (1972)	$\left\{\begin{array}{l} + 9^\circ.6 \pm 1^\circ.6 \\ - 8^\circ.3 \pm 1^\circ.9 \end{array}\right.$	
Shor (1975)	$+ 14^\circ.3 \pm 1^\circ.5$	$+ 0^\circ.1 \pm 0^\circ.4$
Sinclair (1978)	$+ 6^\circ.6 \pm 0^\circ.4$	$- 0^\circ.4 \pm 0^\circ.2$

by G. Wilkins, Sinclair (1972) derived the value $+9^\circ.6 \pm 1^\circ.6$ cy^{-2}; however this value was strongly dependent on the use of the positions observed by Asaph Hall between 1877 and 1881. If these early observations were discarded the accelerations became negative as shown in Table II. Sinclair's conclusion was that the observations were not precise enough to show conclusively that $\dot{n} \neq 0$. For Deimos the results, for different sets of observations, ranged in the interval $-0^\circ.6 \pm 0^\circ.4$ to $-4^\circ.0 \pm 0^\circ.6$ cy^{-2} allowing a conclusion in favour of the existence of a small negative acceleration. The existence of a positive acceleration in the motion of Phobos was however confirmed by Shor (1975) from an orbit determination based on 4902 observations made in the period 1877-1971.

The fitting of Shor's orbit to observations leads to σ(O-C) = 0".48. For the best series of observations (Flagstaff, 1969; Pulkovo, 1967; Washington, 1926 and 1928) the r.m.s-value of the (O-C) is 0".2 - 0".3, while for the period before 1925 it is always higher than 0".5 with only one exception. The σ found for Phobos is smaller than that found for Deimos. Shor's orbits fit Mariner 9 TV data within 100 km.

Certainly the best results are those obtained from Mariner 9 TV data, which are assumed to have a precision in the range 3-10 km. However these observations cover only a small time interval (November 1971 to October 1972) and a comparison of ephemerides based on it with old observations is certainly necessary in order to know their actual precision.

3. THE GALILEAN SATELLITES OF JUPITER

The Galilean satellites of Jupiter form with the Moon, Titan and Triton, a family of giant satellites with masses ranging from 5×10^{25} g (Europa) to 15×10^{25} g (Ganymede). The best available model for their motion is obtained using Lieske-Sampson theory (Lieske, 1977). Two sets of elements are available. The E-2 set (Lieske, 1980) was built on the basis of 3405 Earth-based observations: 1753 eclipse timings (636 of which from the Harvard collection, 1878-1903), 170 mutual event timings from the 1973 series and 1482 photographic observations made in the period 1967-1978. The G-5 set (Arlot, 1982) was built on photographic observations only: the total number of observations considered is around 4150 ranging from 1891 to 1978 but only 2980 show a quality good enough to contribute to the results; also, 580 of these observations are incomplete since only the abscissas on a presumed equatorial line have been published. It is necessary to emphasize that the above figures do not coincide with those published by both authors since we adopted a different mode of reckoning. Here, one photographic observation means the relative position of two satellites regardless of the fact that a complete datum contains two independent measures; also, the independent measurement of the position of the planet was not included in the reckoning.

Ephemerides constructed on the basis of these elements give equivalent results and correspond to a precision of around 200 km for the middle of the past decade. Table III shows the r.m.s.-value of the (O-C) in

TABLE III - Comparison of Ephemerides of Galilean Satellites to Photographic Observations

Opposition	σ(O-C) in geoc.distance		Observatory	σ(O-C) in the coordinates §
	G-5	E-2		
1913 I*	0".09	0".13	Cape	0".06
1913 II*	0".07	0".11	Cape	0".05
1914 I*	0".10	0".15	Cape	0".07
1914 II*	0".09	0".14	Cape	0".06
1915 I*	0".10	0".15	Cape	0".07
1915 II*	0".10	0".15	Cape	0".07
1916 I*	0".08	0".12	Greenwich	0".06
1916 II*	0".11	0".16	Greenwich	0".08
1918-19	0".09	0".12	Greenwich	0".07
1922 *	0".10	0".15	Leiden	0".07
1924	0".16	0".19	Cape	0".11
1927 *	0".12	0".14	Johannesburg	0".08
1928 *	0".16	0".21	Johannesburg	0".11
1968	0".07	0".08	L. McCormick	0".05
1973	0".07	0".08	U.S.N.O.	0".05
1974	0".09	0".09	U.S.N.O.	0".06
1975	0".08	0".08	U.S.N.O.	0".06
1976	0".09	0".10	U.S.N.O.	0".06
1977	0".10	0".11	U.S.N.O.	0".07

* Series for which only x-measures are available.

§ σ(O-C) in the satellite coordinates referred to an ideal center in the plate for the elements G-5.

the geocentric distances between satellites when the positions are calculated using either Lieske's elements E-2 (Biancale *et al.*, 1982) or Arlot's elements G-5 (Tsuchida and Zanchi, 1983). All data considered were previously analysed (Ferraz-Mello and De Paula, 1976; Tsuchida *et al.*, 1982) showing standard errors in the range 0".06-0".10. The data before 1928 are means over 6 exposures; the data of the period after 1968 are means of exposures taken in a single night (4 exposures in general). For most of the series before 1928 only the abscissas on a presumed equatorial line have been published. We could expect some major influences of this fact on the results but tests made over other sets considering one or both measures show equivalence. Since results in Table III were derived using only distances between satellites they express only the behaviour of the studied ephemerides in longitude. Their behaviour in latitude was not studied. The last column of Table III gives the results obtained with the elements G-5 converted into residuals in the coordinates referred to an ideal center in the plate (independent of the satellite measures; see Ferraz-Mello, 1983) and are given to make easier future comparisons.

In a set of 845 eclipses of Io observed in the period 1775-1802 Arlot *et al.* (1984) found that elements G-5 lead to an error of 1.0 min in the eclipse timings (the calculated date being earlier than the observed one), while elements E-2 lead to only 0.3 min.

From the qualitative point of view the most interesting elements are those related to the integration constants and to the main inequalities. Some of the parameters of the planar motion are given in Table IV. The errors indicated are standard and are those given by the authors; it is not superfluous to stress that the rules used by Lieske and Arlot were not the same and the results may not be compared (also, the errors originally published by Arlot are probable errors).

TABLE IV - Galilean Satellites

Element	E-2	G-5
Amplitude of the Libration	3'.95 ± 0'.76	5'.16 ± 0'.44
Phase of the Libration	184°.4 ± 20°.3	185°.8 ± 0°.4
Proper Eccentricity of Io	0.00001 ± 4E-6	0.000024 ± 7E-6
id. of Europa	0.00009 ± 2E-5	0.000121 ± 13E-6
id. of Ganymede	0.00147 ± 3E-5	0.001450 ± 10E-6
id. of Callisto	0.00733 ± 3E-5	0.007346 ± 6E-6
Forced Eccentricity of Io	0.00415	
id. of Europa	-0.00937	
id. of Ganymede	0.00059	

It is worth emphasizing that the osculating eccentricities of Io and Europa are almost equal to their forced eccentricities (0.00415 and 0.00937) and not to their proper eccentricities as unfortunately stated in many general books and papers on the Solar System. Their periapsides have a very fast motion and are always on the conjunction line of Io and Europa, at opposite sides. Their relative situation is shown in Figure 1. At the conjunctions Io is at its perijove while Europa is at its apojove; at the oppositions both are at their perijoves. The way in which the proper and the forced eccentricity of Io combine to give rise to the osculating eccentricity is shown in Figure 2. For Europa the situation is similar except that the center is on the negative semi-axis (this is the meaning of the minus sign in Table IV). Some authors call this situation also a libration. But this oscillation is not a true libration, the oscillating character of the angle $\lambda_1-2\lambda_2+\tilde{\omega}_1$ being due only to the

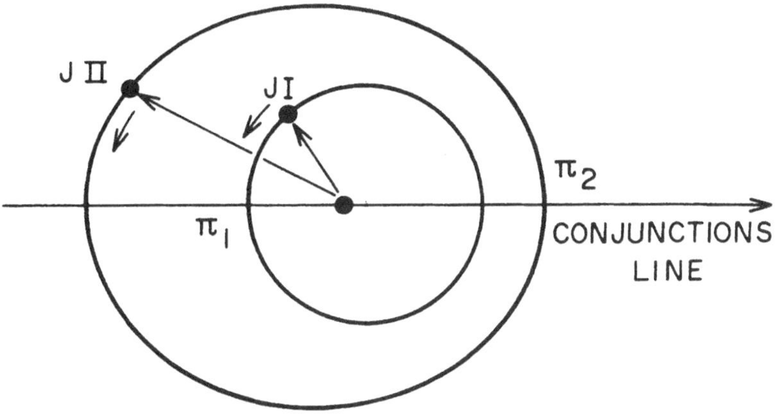

Figure 1 - Relative orbits of Io and Europe in a frame rotating with angular velocity n_1-2n_2. At the conjunctions Io is at its perijove while Europa is at its apojove; at the oppositions both are their perijoves.

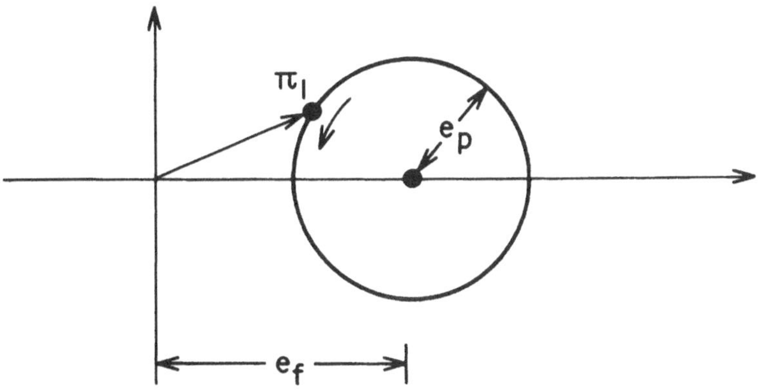

Figure 2 - Motion of the osculating periapsis of Io. The radius vector of π_1 is the osculating eccentricity of Io and the polar angle is the angle $\lambda_1-2\lambda_2+\varpi_1$. The osculating perijove of Io oscillates about the angle $\lambda_1-2\lambda_2$ due to the fact that $e_p<e_f$. If $e_p<e_f$ the angle $\lambda_1-2\lambda_2+\varpi_1$ would circulate.

fact that $e_p<e_f$. If $e_f<e_p$, we would have a circulatory behaviour of the angle $\lambda_1-2\lambda_2+\varpi_1$. Some authors propose to call this oscillation an apocentric libration, but the opposite situation for Io and Europa shows that the word apocentric is not a good choice. In an ideal situation I called these motions paradoxal librations, but for this particular situation the name small-eccentricity libration (Greenberg, 1977) is also appropriate.

At variance with Io and Europa, the osculating eccentricity of Ganymede is dominated by the proper component (0.00147), but it is not the only one; the motion of the periapsis of Ganymede is also affected by a long term component of coefficient 0.00064 and by the forced eccentricity of shorter period and coefficient 0.00059. Here we have $e_f<e_p$ and the result is shown in Figure 3.

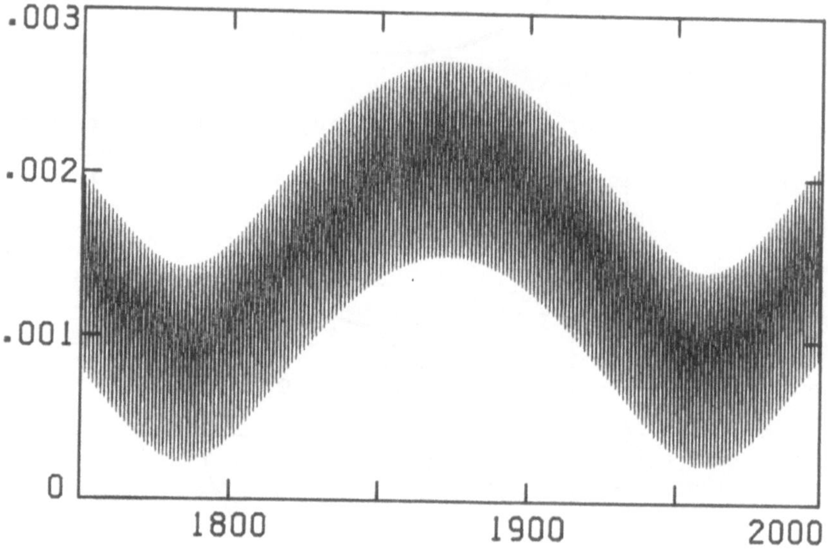

Figure 3 - Variation of the osculating eccentricity of Ganymede. (From Ferraz-Mello, 1979).

The osculating eccentricity of Callisto is almost equal to its proper eccentricity (0.00733).

The immediate future for the study of the motion of the Galilean satellites lies in the construction of a still more accurate theory allowing for a better calculation of the secular and long-period perturbations (Thuillot and Vu, 1983), and on the consideration of the old eclipses from the already published collections, together with the sets of Delisle's and Delambre's old manuscripts, recently found by Lieske (1983) and by Arlot *et al.* (1984).

4. THE OUTER SATELLITES OF JUPITER

The outer satellites of Jupiter lie in two groups of 4 satellites each: one whose motion is direct (satellites with names ending by a; see Table I) and the other whose motion is retrograde (names ending by e). The direct satellites are at an average distance of 11 million kilometers from the planet and have very similar inclinations (∿30°) and eccentricities (0.1-0.2). The retrograde satellites are at an average of 22 million kilometers from the planet. Pasiphae has the distinction of attaining a greater distance from its primary than any other known satellite in the Solar System: 33 million kilometers. Sinope completes its revolution in 2.07 years and has the longest orbital period among known planetary satellites.

The orbits of Himalia and Elara have been determined by Bordovitsyna and Bykova (1978) and by Rocher (1983). Rocher used 422 observations of Himalia and 171 of Elara. The standard deviation of the ephemerides with respect to the observations is around $0^s.2$ in right ascension for both satellites. In declination the standard deviations are 1".8 for Himalia and 2".9 for Elara.

The orbit of Leda has been determined by Aksnes (1978) from 31 observations covering nearly five revolutions about Jupiter. The r.m.s.-value of the residuals is 1".0 in right ascension and 0".7 in declination in 1974-1977.

Rocher (1984) also studied the motions of two retrograde satellites: Pasiphae and Sinope. The orbit of Pasiphae was determined from 287 observations and that of Sinope from 92 observations. The standard deviations of ephemerides and observations are around $0^s.1$ in right ascension and 1"-1".2 in declination. The analysis of the residuals shows the upmost importance of observations made after 1974 at the MacDonald Observatory and allow us to see how crucial is the existence of extended series of observations of faint satellites by great telescopes. It is a pity that those observations were discontinued some years ago because of funding difficulties.

There are reports of an extended work by Bordovitsyna *et al.* (1981) regarding satellites JVI to JXI.

5. THE INNER SATELLITES OF JUPITER

There are no recent works on the motion of the inner satellites of Jupiter other than the determination of the main parameters of the three new satellites Thebe (or 1972 J2), Adrastea (or 1979 J1) and Metis (or 1979 J3) (Veverka *et al.*, 1982). The last published study of the motion of Amalthea is the orbit and observations discussion by Sudbury (1969). That discussion indicates that the elements changed appreciably in the interval 1917-1949, when no observations were made. Among the possible causes Sudbury lists that the orbit may have undergone a single perturbation presumably by the close approach of another body of comparable mass. His results do not confirm the existence of a positive acceleration as formely found by van Woerkom (1950).

6. SATURN'S MAIN SYSTEM

They are 6 satellites and together with Iapetus they are the most bright ($m<12$), and are all the satellites known at the end of the XVIIIth century. Their gathering in one section comes from the fact that there is an old tradition of studying them together and untill 20 years ago they were referred to as the *inner* satellites of Saturn. These satellites have masses ranging from 7×10^{-8} (Mimas) to 2.4×10^{-4} (Titan) of the mass of Saturn and semi-major axes in the interval 3 (Mimas) to 20 (Titan) planetary radii. The other relatively bright satellite, Iapetus, whose relative mass is 3.4×10^{-6}, is too far away to influence the motion of these six satellites (semi-major axis equal to 59 planetary radii). However, Hyperion, whose relative mass is around 2×10^{-7} and circulates quite close to the orbit of Titan, interacts strongly with Titan. But the magnitude of Hyperion is much higher (>14) and this satellite is generally not observed together with the main six.

The orbits of these satellites are generally computed using the theories of G. Struve (1930). The elements and the motions of the nodes and periapsides have been determined from mainly micrometric measurements of the positions of the satellites with respect to the planet or of one satellite with respect to another.

The dynamical consistency and potential accuracy of the theories have been investigated by least-squares analysis of 8200 U.S.N.O. micrometer data (1877-1947) and 920 recent photographic data (1972-1976). Data noise standards determined by fitting short-time intervals were in the range 0".15 to 0".3 (geocentric) i.e. 1000 to 2000 km, for inter-satellite micrometer data, 0".3 to 0".5 for planet relative micrometer data and about 0".2 for photographic data of satellite positions relative to the star background. The errors in the coordinates calculated with the theories of Struve using Kozai (1957) coefficients is less than 2000 km i.e. 0".3 geocentric (Null and Taraji, 1979; Null *et al.*, 1981).

The longitude of the epoch and the mean motions of the first five satellites were redetermined by Rapaport (1977). The residuals given by Rapaport (averages of individual residuals taken in absolute value) indicate individual standard deviations higher than those reported above. In fact, Rapaport's modification of Kozai elements lead to corrections in the satellite positions not higher than 0".1 geocentric, except for Mimas where they reach 0".3. However his modifications to Kozai mean motions and longitudes of the epoch were not confirmed by Chugunov (1983b) with the only exception of the mean motion of Dione.

A complete new set of elements has been recently announced by Chugunov (1983b) using as a basis an extension of Struve's theory. In this extension (Chugunov, 1983a) the perturbations were calculated with a nominal precision of 10^{-6} (or 0".2 on the orbit). In the determination of the constants all observations available were used. They are 28500 measures of the six satellites plus Hyperion. 6% of these observations were discarded mainly because errors in their original publication. 75 unknowns were considered, among them: the mean elements of Mimas and Tethys, the osculating elements of Enceladus, Dione, Rhea and Titan, the amplitude and periods of the libration related inequalities, and the motions of the ascending nodes and periapsides (because of high correlations the motion of the nodes and periapsides of Rhea and Titan were considered equal in absolute value: $\dot{\omega}_i = -\dot{\Omega}_i$). Information on the residuals of the comparison to observations of ephemerides constructed with these elements is not known.

7. ENCELADUS-DIONE

Table V shows Chugunov's determination of some of the parameters related to the resonant pair Enceladus-Dione (the errors indicated are those given by Chugunov). The resonance amongst these satellites is of the

TABLE V - Enceladus-Dione (Chugunov's elements)

Element	Enceladus	Dione
Mean Motion (d^{-1})	$262^\circ.7319372 \pm 10E-7$	$131^\circ.5349733 \pm 4E-7$
Apsidal Motion (d^{-1})	$0^\circ.337927 \pm 14E-6$	$0^\circ.084520 \pm 8E-6$
Eccentricity	$0.004591 \pm 1E-6$	$0.002243 \pm 1E-6$
Period of the 1st. great ineqaulity (yr)		11.02 ± 0.01
Period of the 2nd. great inequality (yr)		3.867 ± 0.001

TABLE IV - Periapsides of Enceladus and Dione

Element	Enceladus	Dione
Proper Eccentricity (e_p)	0.00012	0.00224
Forced Eccentricity (e_f)	0.00459	0.00004
Motion of Proper Periapisis (g)	$0^\circ.4274\ d^{-1}$	$0^\circ.084\ d^{-1}$
$2n_4 - n_2 - g$	$0^\circ.0894\ d^{-1}$	$0^\circ.254\ d^{-1}$
Associated Period	11.0 yr	3.88 yr

TABLE VII - Amplitude of the Great Inequalities

	Kozai (1957)	Jefferys (1979)	Chugunov(1983b)
1st. Inequality			
Enceladus	15'.38	15'.54	14'.82 ± 0'.03
Dione	0'.88	0'.84	0'.81 ± 0'.01
2nd. Inequality			
Enceladus	13'.04	12'.18	12'.62 ± 0'.03
Dione	0'.75	0'.66	0'.66 ± 0'.01

same nature as that verified in the 2:1 coupling of the Galilean satellites of Jupiter. For Enceladus the phenomenon is the same as shown in Figure 2 and we have $e_p \ll e_f$, while for Dione $e_f \sim 0$ and the circle is almost centered at the origin of the plane. If the results for Enceladus and Dione are compared to the classical theory of the Galilean satellites of Jupiter (Ferraz-Mello, 1979) the two inequalities sometimes called *librations* are just the great inequalities in longitude that arise in the integration of the variational equations of Lagrange for the mean longitudes when the periapsides are supposed to be moving. The arguments called V_1 and V_2 in Enceladus-Dione theories (V_2 is the faster one), are then in fact equal to $\pi_2-2\lambda_4+\lambda_2$ and $2\lambda_4-\lambda_2-\pi_4$ where π_i are the proper periapsides. For V_2, since the proper periapsis of Dione coincides with the osculating periapsis we must have $2n_4-n_2-\dot{\varpi}_4-V_2 = 0$. With Chugunov's data we obtain $-0^\circ.0014\ d^{-1}$. The elements of the motion of the periapsides of Enceladus and Dione derived from Chugunov's are given in Table VI where the number of figures has been dropped in order to have satisfied the relation just introduced. The proper eccentricity of Enceladus was derived from the observed value of the first great inequality of the motion of Enceladus (Sato, 1984). The amplitudes of the great inequalities are given in Table VII as determined from the observations by several authors.

8. MIMAS-TETHYS

The Mimas-Tethys resonance is a very particular kind of resonance. It is of the inclination type and the axis along which the conjunctions occur oscillates about the mid-point between the ascending nodes of the two satellites on Saturn's equatorial plane. The motion of the nodes is a retrograde precession determined by the oblateness of Saturn; the relative influence of the satellites on this motion is less than 10^{-3} and may not even alter the uniformity of the precessions. As a consequence the equations for the motion of the orbital planes and the inequalities of the longitude may be solved separately and the average behaviour of the critical angle $W = 4\lambda_3-2\lambda_1-\Omega_3-\Omega_1$ is given by the pendulum equation.

A new theory of the Mimas-Tethys resonance has been completed by Jefferys and Ries (1979b). The inequalities due to the interactions of the two satellites are given in literal form as functions of the orbital elements, the satellite masses and the amplitude of the libration. They show that, besides the libration of W, the 2:1 resonance of the two satellites leads to other inequalities in the longitude of Mimas, the most important of which has argument $\lambda_1-2\lambda_3+\varpi_1$ and coefficient 2.4×10^{-3} (8'.2).

The data determined by Chugunov for the libration of Mimas and Tethys and other orbital data are given in Table VIII. It must be noted that the value published by Chugunov for the semi-major axis of Tethys (45".591 heliocentric) is clearly incorrect and not consistent with the value of the mean motion. The inequalities considered in the theoretical model used by Chugunov (1983a), include terms in t^3 but with a coefficient very small (3'.6 cy^{-3} for Mimas and -1'.7 cy^{-3} for Tethys) unable to give rise to an observable inequality. Chugunov's model also includes

TABLE VIII - Mimas-Tethys (Chugunov's Mean Elements)

Mean Motion (d^{-1})	381°.994552 ± 3E-6	190°.6979506±5E-7
Motions of the Node (d^{-1})	-0°.9994907±14E-7	-0°.1977029±9E-7
Orbital Inclination	1°.5068 ± 3E-4	1°.08035 ±6E-5
Eccentricity	0.02001 ± 1E-5	
Apsidal Motion (d^{-1})	1°.000867 ± 9E-6	
Libration in Longitude	-43°.460 ± 2E-3	2°.0072 ±3E-4
Period of Libration (yr)	71.143 ± 0.007	

Poisson terms in the critical arguments. These terms, as well as the t^3-terms, are not present in the results of Jefferys and Ries (1979b). No attempt was made to confirm the existence of an acceleration in the motion of Mimas (see Kozai, 1957).

9. THE OUTER SATELLITES OF SATURN

They are Hyperion, Iapetus and Phoebe, Hyperion is tied to Titan by a 4:3 resonance and shows a librational motion. Following the quasi-periodic orbit determined by Woltjer (1928) this libration, of period 640.5 d has a relatively large amplitude (see Figure 4). Besides the libration, the periapsis shows an harmonic component of very long period (18.77 yr); we recall that a recent study of the 2:1 planetary resonance (Sessin and Ferraz-Mello, 1984) shows that first-order planetary libra-

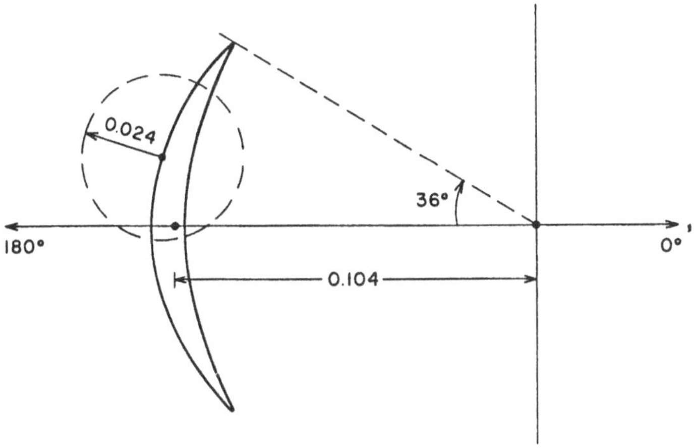

Figure 4 - Motion of the osculating periapsis of Hyperion. The radius vector is the osculating eccentricyty of Hyperion and the polar angle is the angle $3\lambda_6-4\lambda_7+\varpi_7$.

tions are always linked to harmonic oscillations of this kind. Ephemerides founded on Woltjer elements are wrong, by now, by 10,000 km (Null *et al.*, 1981). Chugunov tried to include Hyperion together with the other satellites in his determination but his results for Hyperion were unsatisfactory. The only recent determination of Hyperion's elements was made by Hatanaka (1979) and considered only the observations made by himself.

A new theory of the motion of Iapetus has been derived by Sinclair (1974), the constants being determined by comparison with recent photographic observations made at Herstmonceux. Sinclair's orbit has been compared by Rapaport (1978) to observations made in the period 1917-1973 and the errors found are of the order of 0".4-0".5 for the observations made after 1972, but reach 1" for observations made before 1927. The mean motion and longitude of the epoch were redetermined by him and the corrected elements lead to errors of the order of 0".3 for the modern observations and 0".4-0".5 for the old ones.

The motion of Phoebe has been studied by several authors who derived independent sets of orbital elements. The orbit determined by Rose (1979) is founded on 269 observations, that by Bec-Borsenberger and Rocher (1982) on 182 observations and that by Bykova and Shikhalev (1984) on 324 observations. In both cases the standard deviation of the residuals is of the order of 1".5.

10. THE NEW SATELLITES OF SATURN

They are the co-orbitals Janus and Epimetheus, the Dione's Lagrangian Dione B, the Tethys' Lagrangians Telesto and Calypso, all discovered from ground-based observations at epochs when the Earth crosses the plane of the Saturn's rings, which thus appear edge on to us. They are also Atlas, the shepherd of the outer edge of ring A, and Prometheus and Pandora, the shepherds of ring F, discovered on images transmitted by the space probe Voyager I.

All satellites in the first family are involved in three-body librations, displaying both tadpole and horseshoe orbits (Figure 5). The orbits of the co-orbitals have been thoroughly studied by Yoder *et al.* (1983). They determined the integration constants related to the libration and showed that the two satellites undergo separate oscillations and are prevented from passing each other by their mutual perturbations.

The precision of the observations in the last edge-on period are not enough to allow an identification of the satellites observed in the 1966 edge-on period. However, due to the lower magnitude of Janus, Dollfus and Brunier (1981) identified 1980 S1 with Janus, the satellite observed at the Pic du Midi in 1966, and 1980 S3 with Epimetheus, the satellite later found on some photographic plates.

Dione B is a satellite in stable 1:1 resonance with Dione. Its orbit has been determined by Reitsema (1981a) from 80 positions observed in 1980 and some other from 1981. The motion is a 785-day libration around Dione's L_4. The maximum separation between Dione and Dione B is 137 degrees and was reached on April 20, 1980; the minimum separation is only 13 degrees. The r.m.s.-value of the residuals of the comparison of the observations to the ephemerides is $0^\circ.3$.

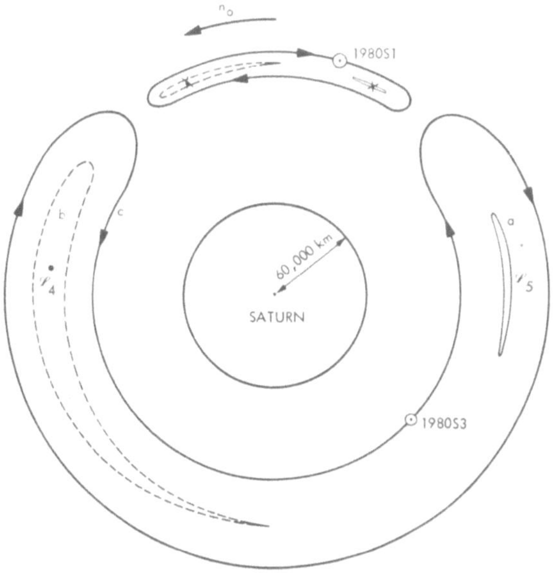

Figure 5 - Relative orbits of two co-orbital satellites in the frame rotating with their average angular velocity n_0 for three separate cases: (a) Small librations about the Lagrange L_5 point. (b) Limiting tadpole orbit about L_4 (dashed curves). (c) Horseshoe orbit. The relative positions of Janus (1980 S1) and Epimetheus (1980 S3) in October 1980 are also shown. (From Yoder *et al.*, 1983).

Telesto and Calypso are in 1:1 resonance with Tethys. Telesto librates around Tethys' L_4 and Calypso around L_5. From 17 observations of these satellites, Reitsema (1982b) obtained a well established libration for Telesto, of $1^\circ.4$ of each side of L_4. The libration of Calypso is still uncertain, some fittings leading to an half-amplitude of 4° and others to 15°-20°.

The shepherds Prometheus and Pandora have neighbouring orbits; the semi-major axis of the inner one is 139350 km and that of the outer is 141700 km. They are one at each side of Saturn's F ring (R = 140185 ± 30 km). Their motion has been determined by Synnot *et al.* (1983) from 25 Voyager 2 images. They also studied the relationship to some features (braids) observed in ring F. The precision of the orbital parameters is still insufficient to allow a study of the gravitational interaction between the satellites and the ring F.

11. THE SATELLITES OF URANUS

Uranus has five satellites that form a very regular system (small eccentricities and small inclinations to the planet's equator). The three inner (and smaller) satellites have mean motions relatively close to a commensurability ($n_5-3n_1+2n_2 = -0°.0785\ d^{-1}$) which produces a great inequality of period 12.6 years, clearly observed in the motion of Miranda. A complete determination of the orbits of these satellites has been finished by Veillet (1983), that leaves residuals of only 0".07 for the very recent observations of Miranda (up to 0".2 for more ancient observations). For the other satellites the residuals after 1948 (time of the discovery of Miranda) are of the same order as for Miranda; for the old observations, the residuals range in the interval 0".3-0".6 for the observations made in the period 1856-1926, and 1"-3" for the observations made before that period. Veillet's determination used around 4500 observations, micrometric and photometric. Among the observations, there were more than 1600 from plates obtained by himself and by G. Ratier at several observatories (Pic du Midi, ESO, CFH) since 1977.

TABLE IX - Inequalities in the longitude of Uranian satellites

	Coeff. of sin θ	Coeff. of sin 2θ
Miranda	1°.4	0°.16
Ariel	0°.21	0°.03
Umbriel	0°.11	0°.01

N.B. $\theta = \lambda_5-3\lambda_1+2\lambda_2$

The great inequality of the longitudes due to the Laplacian commensurability has been recalculated by Lazzaro *et al.* (1984). The values obtained using Veillet's system of masses are shown in Table IX. The period obtained from the integration of the Laplace pendulum-like equation is 12.0 years, and is smaller than the values obtained directly from $n_5-3n_1+2n_2$ that range in the interval 12.47 - 12.57 years. This discrepancy could be due to actual masses a few percent smaller than those used.

12. THE SATELLITES OF NEPTUNE AND PLUTO

The great satellite of Neptune, Triton, had its orbit recently determined by Harris (1983) from sets of observations covering the periods 1899-1910, 1939-1942 and 1977-1981. The ephemerides were obtained, using a simple precessing circular orbit.

Nereid, the small and very eccentric satellite of Neptune (e=0.745) had its orbit redetermined by Mignard (1981) from 44 observations (35 in 1949-1955 and 9 in 1967-1969) and by Veillet (1982) from these 35 observations and 6 new ones (2 from MacDonald and 4 from ESO). The residual of 1" obtained by other authors were confirmed. Veillet's orbit allows us to expect residuals with σ = 0".3-0".4 for the present epoch, but which are becoming worse very rapidly.

Charon, the satellite of Pluto, was discovered in 1978, and has been the object of several orbit determinations. The most complete is the determination by Harrington and Christy (1981) using pre- and post-discovery observations of elongated images of Pluto together with the resolved images obtained by Bonneau and Foy (1980) using speckle interferometry. The orbital period is 6.3871 ± 0.0002 d and the semi-major axis is 19700 ± 300 km. These values correspond to a reciprocal mass of Pluto equal to (1.34 ± 0.07) × 10^8. The present orbit suggests that occultations, transits and eclipses should be visible beginning with the 1984 apparition.

13. CONCLUSIONS

There are no complete reviews, and this one is not an exception. Certainly some important contributions were ommitted. Completeness was impaired by the fact that many results are not widely diffused remaining inside unpublished dissertations, internal reports, or even in undistributed periodicals or in less accessible languages. On the other hand the space and time available oblige us to fix a particular point of view.

In this review the adopted point of view is to privilege recent theoretical work leading to an actual knowledge of the motions of all planetary satellites except the Moon. It covers recent calculations of the inequalities observed in the motion, and determinations of satellite orbits, taking for a basis the existing observational data. Purely theoretical work was excluded. Indeed it is almost impossible to classify many theoretical works as dealing with the motions of planets, natural satellites, artificial satellites, or even with solutions of the three-body problem. Strict frontiers do not exist and because of the diversity of the motions amongst satellites, almost every theory may be sooner or later related to some actual satellite motion.

It is worth emphasizing the importance of ground-based observations for the knowledge of the motions of the planetary satellites. The progress reported here was only possible because of the existence of very fine new observations of many bodies in the Solar System. The precision requested for the planning of future space missions will be attained only if the observational effort continues and is even increased. An observation not made has no substitute.

ACKNOWLEDGEMENTS

This review has been written while the author was at the Bureau des Longitudes, Paris, participating in a CNPq/CNRS sponsored cooperative research project. The author is indebted to the research staff of the Bureau des Longitudes for their help and for many very valuable discussions on the subjects under consideration in this review. Thanks are also due to Dr. Y. Kozai and J. Message for their suggestions.

REFERENCES

Aksnes, K.: 1978, *Astron. J.* **83**. 1249.
Arlot, J.E.: 1982, *Astron. Astrophys.* **107**, 305.
Arlot, J.E., Morando, B. and Thuillot, W.: 1984, *Astron. Astrophys.* (in press).
Bec-Borsenberger, A. and Rocher, P.: 1982, *Astron. Astrophys. Suppl.Ser.* **50**, 423.
Biancale, R., Ferraz-Mello, S. and Tsuchida, M.: 1982, *Celes. Mech.* **26**, 225.
Bonneau, D. and Foy, R.: 1980, *Astron. Astrophys.* **92**, L1.
Bordovitsyna, T.V., Boronenko, T.S., Bykova, L.E. and Chernitsov, A.M.: 1981, in *Referat. Zh. Ser. 51*, No. 12.51.132.
Bordovitsyna, T.V. and Bykova, L.E.: 1978, *Teorii Dvizheniya Efemeridy VI, VII Sputnikov Yupitera na 1979-2000 gody*, Izd. Univ., Tomsk.
Born, G.H. and Duxbury, T.C.: 1975, *Celes. Mech.* **12**, 77.
Bykova, L.E. and Shikhalev, V.V.: 1984, *Celes. Mech.* **32**, 185.
Chugunov, I.G.: 1983a, *Pis'ma b Astron. Zh.* **9**, 368.
Chugunov, I.G.: 1983b, *Pis'ma b Astron. Zh.* **9**, 508.
Dollfus, A. and Brunier, S.: 1981, *Icarus* **48**, 29.
Ferraz-Mello, S.: 1979, *Dynamics of the Galilean Satellites. An Introductory treatise*, Univ. Sao Paulo, Sao Paulo.
Ferraz-Mello, S.: 1983, *An. Acad. Bras. Cienc.* **55**, 219.
Ferraz-Mello, S. and De Paula, L.R.: 1976, *Astron. J.* **81**, 127.
Greenberg, R.: 1977, *Vistas Astron.* **21**, 209.
Harrington, R.S. and Christy, J.W.: 1981, *Astron. J.* **86**, 442.
Harris, A.W.: 1983, *Bull. Amer. Astron. Soc.* **15**, 870.
Hatanaka, Y.: 1979, in R.L. Duncombe (ed.), *Dynamics of the Solar System*, D. Reidel, Dordrecht, p. 167.
Jefferys, W. and Ries, L.M.: 1979a, in R.L. Duncombe (ed.), *Dynamics of the Solar System*, D. Reidel, Dordrecht, p. 171.
Jefferys, W. and Ries, L.M.: 1979b, *Astron. J.* **84**, 1778 and **86**, 640.
Kozai, Y.: 1957, *Ann. Tokyo Astron. Obs. 2nd. Ser.* **5**, 73.
Kozai, Y.: 1982, in *Trans. Int. Astron. Union* **18** A, 205.
Lazzaro, D., Ferraz-Mello, S. and Veillet, C.: 1984, *Astron. Astrophys.* (in press).
Lieske, J.H.: 1977, *Astron. Astrophys.* **56**, 333.
Lieske, J.H.: 1980, *Astron. Astrophys.* **82**, 340.
Lieske, J.H.: 1983, in V.V. Markelos and Y. Kozai (eds.), *Dynamical Trapping and Evolution in the Solar System*, D. Reidel, Dordrecht, p. 51.
Marsden, B.: 1980, *Icarus* **44**, 29.
Mignard, F.: 1981, *Astron. J.* **86**, 1728.
Null, G.W., Lau, E.L., Biller, E.D. and Anderson, J.D.: 1981, *Astron.J.* **86**, 456.
Null, G.W. and Taraji, H.: 1979, *Bull. Amer. Astron. Soc.* **11**, 805.
Rapaport, M.: 1977, Thèse de Doctorat, Bordeaux.
Rapaport, M.: 1978, *Astron. Astrophys.* **62**, 235.
Reitsema, H.J.: 1981a, *Icarus* **48**, 23.
Reitsema, H.J.: 1981b, *Icarus* **48**, 140.

Rocher, P.: 1983, *Astron. Astrophys. Suppl. Ser.* 52, 333.
Rocher, P.: 1984, Private Communication.
Rose, L.E.: 1979, *Astron. J.* 84, 1067.
Sato, M.: 1984, Private Communication.
Sessin, W. and Ferraz-Mello, S.: 1984, *Celes. Mech.* (in press).
Sharpless, H.: 1945, *Astron. J.* 51, 185.
Shor, V.A.: 1975, *Celes. Mech.* 12, 61.
Sinclair, A.T.: 1972, *Monthly Notices Roy. Astron. Soc.* 155, 249.
Sinclair, A.T.: 1974, *Monthly Notices Roy. Astron. Soc.* 169, 591.
Sinclair, A.T.: 1978, *Vistas Astron.* 22, 133.
Struve, G.: 1930, *Veröff. Univ. Stern. Berlin-Babelsberg* Bd. 6.
Sudbury, P.V.: 1969, *Icarus* 10, 116.
Synnott, S.P., Terrile, R.J., Jacobson, R.A. and Smith, B.A.: 1983, *Icarus* 53, 156.
Thuillot, W. and Vu, D.T.: 1983 in S. Ferraz-Mello and P. Nacozy (eds.), *The Motion of Planets and Natural and Artificial Satellites*, Univ. Sao Paulo, Sao Paulo, p. 273.
Tsuchida, M., Ferraz-Mello, S. and Biancale, R.: 1982, *Astron. J.* 87, 924.
Tsuchida, M. and Zanchi, J.: 1983, in *Astron. Din. Matem.* (Annual Report), Sao Paulo.
van Woerkom, A.J.J.: 1950, *Astron. Papers Amer. Ephem.* 13(1).
Veillet, C.: 1982, *Astron. Astrophys.* 112, 277.
Veillet, C.: 1983, Thèse de Doctorat, Paris.
Veverka, J., Thomas, P. and Synnott, S.: 1982, *Vistas Astron.* 25, 245.
Woltjer, J., Jr.: 1928, *Ann. Sterr. Leiden* 16(3).
Yoder, C.F., Colombo, G., Synnott, S.P. and Yoder, K.A.: 1983, *Icarus* 53, 431.

TIDAL EFFECTS AND THE MOTION OF A SATELLITE

J. Kovalevsky
C.E.R.G.A.
Avenue Copernic
F-06130 Grasse
FRANCE

ABSTRACT. The effect of resonant planetary perturbations on the evolution of the orbit of a satellite driven by tidal forces is studied in this paper. The basic equations that govern it are similar to the equations found in orbit-orbit and in spin-orbit couplings. The general form of these equations is:

$$\left.\begin{aligned} \frac{dx}{dt} &= A - Bx + (Q+Q't)\sin\theta \\ \frac{d\theta}{dt} &= 2G(t)x \end{aligned}\right\} \qquad (1)$$

A general treatment of such equations, proposed earlier (J. Kovalevsky, in Dynamical Trapping and Evolution of the Solar system, IAU Colloquium n°74, V.V. Markellos and Y. Kozai, eds., 1983) is sketched.

In particular, the effects of the large long periodic variations of the excentricity e' of the planet are analysed on an example taken from the lunar theory and the Earth's general theory due to Bretagnon.

The argument of the well known planetary term θ=18 V-16T-ℓ due to the tidal friction and quasi-periodic variations due to the presence of e' in the expression of the mean motion of the Moon. Their joint effect, has been to produce in the past resonant situations for this argument that repeated more than 100 times. Every such situation can be treated by equation (1).

Numerical integration, using conditions that might have occurred while θ or similar other arguments were quasi resonant, have produced the following results:

(a) In some cases, the argument becomes temporarily resonant. Between the capture to and the escape from the resonance, the semi-major axis undergoes oscillations, but the tidal secular evolution is stopped.

(b) In other cases, the argument is not trapped into a resonant conditions, but the semi-major axis undergoes a quick change while $d\theta/dt$ is close to zero.

A number of arguments that have been quasi resonant in the past

Celestial Mechanics **34** (1984) 243–244. 0008–8714/84.15

history of the Earth-Moon system has been identified from the Chapront and Chapront-Touzé Lunar Theory. It appears that the phenomena described are frequent features in the evolution of the Lunar orbit.

THE EFFECT OF THE DYNAMICAL PARAMETERS IN THE MOTION OF THE GALILEAN SATELLITES OF JUPITER

W. THUILLOT
Bureau des Longitudes
Equipe de Recherche Associée au C.N.R.S.
77 avenue Denfert-Rochereau
F-75014 Paris
France

ABSTRACT. The construction of an analytical theory of the motion of the Galilean satellites of Jupiter requires that we keep track of the dynamical parameters, that is, the masses of the satellites, and the harmonic coefficients of the potential of the planet J_2 and J_4. This is realized here. But as in other theories the solution becomes partly numerical from the resolution of an autonomous system. The aim of this paper is to present a method to obtain developped solutions of this autonomous system. In these solutions the proper motions of the pericenters and nodes are obtained as short series developped in the neighbourhood of a numerical solution. We have used these results to obtain complementary terms in the general solution which give a complete representation of the motions with respect to the dynamical parameters.

1. INTRODUCTION

The construction of a theory of the motion of the Galilean satellites of Jupiter requires taking into account various effects in the development of the equations. It is necessary to introduce the effects of the interactions between the satellites, the oblateness of the planet, the action of the Sun, and the motion of the jovian equator. Because of this fact, the choice of an analytical formulation for these developments permits us to keep control of the order of each term. Furthermore the use of such a solution, with any set of parameters, and the eventual calculation of partial derivatives with respect to them, are easy.

The method used (Sagnier, 1981) permits us to realize this type of development. The dynamical parameters, the masses of the satellites and the zonal harmonic coefficients J_2 and J_4, are the small parameters in powers of which the developments are ordered.

In this theory, the motions of the orbital planes come from the resolution of a system independant of the time: the autonomous system.

This is an eigenvalue problem, and its solutions introduce the frequencies of the pericenters and nodes in the angles of the different periodic terms. From this step of the construction, the solution becomes partly numerical.

Celestial Mechanics **34** (1984) 245–253. 0008–8714/84.15

Here we have applied a method which permits us to keep track of each dynamical contribution in the solutions of the autonomous system. The present application is limited to the Galilean satellites, but its application is easily extended to any other analogous system. In a planetary case, similar results have recently been obtained by Duriez (1983) on the basis of a similar method. In the present work, the contributions to each periodic term generated by the analytical solution of the autonomous system in the general solution have been computed. These results could be helpful for the analysis of the results of Sampson and de Sitter as suggested by Ferraz-Mello (1976).

2. THE VARIATIONAL SYSTEM

The construction of the analytical theory of the Galilean satellites following Sagnier's method is composed of two main steps. An intermediary solution gives a first representation of the motions, then a variational system completes it to form the general solution of the theory.

A step necessary for the complete representation of the motions of the satellites is accomplished when the motion of the proper pericenters and nodes are obtained. The solution of the variational system gives a representation of these motions.

In this work, all the developments are ordered with respect to the small parameters: the zonal harmonic coefficients J_2 and J_4, and the satellite masses m_1, m_2, m_3, and m_4. A track of the origin of each term is kept and this allows control of the order with respect to

$$\mu = \{J_2, J_4, m_1, m_2, m_3, m_4\}$$

The variables used are the osculating Keplerian variables: the mean longitudes l_j, the mean motion n_j, a complex variable in eccentricity and longitude of the pericenter p_j and a complex variable in inclination and longitude of the node q_j, defined by:

$$p_j = e_j \exp i \varpi_j$$

$$q_j = I_j \exp i \Omega_j$$

and their conjugates $\bar{p}_j$ and $\bar{q}_j$, where j indicates the number of the satellite.

The vector

$$X = \text{column } (l_1,\dots,l_4,n_1\dots n_4,p_1\dots p_4,\bar{p}_1,\dots\bar{p}_4,q_1\dots q_4,\bar{q}_1\dots\bar{q}_4)$$

may be written as

$$X = \text{column } (X_1,X_2,X_3)$$

where X_1= column$(l_1,\dots,l_4)$, X_2 = column$(n_1,\dots,n_4)$ and X_3 = column$(p_1,\dots,\bar{q}_4)$ since the space of these vectors may be split into three subspaces.

If ξ denotes the intermediary solution, the variations with respect to ξ are defined to the first approximation by

$$Y = X - \xi \tag{1}$$

The variations Y are obtained from the variational system

$$\dot{Y} = G\,Y \tag{2}$$

where $$G = \begin{bmatrix} 0 & 1 & G_1 \\ 0 & 0 & G_2 \\ 0 & 0 & G_3 \end{bmatrix}$$

The submatrices G_1, G_2, and G_3 are deduced from the equations of the motions and at this approximation they are series of periodic terms which depend only on the independent variable t.

We can isolate the system

$$\dot{Y}_3 = G_3\,Y_3 \tag{3}$$

and integrate it. Here we use the Krasinskij-Brumberg normalization method for linear system (Brumberg, 1970),limited to the first order. The solution of eqn.(3) is sought in the form

$$Y_3 = (1 + S^1 + \dots)\,Z, \tag{4}$$

where Z is an auxiliary matrix obtained from

$$\dot{Z} = K\,Z, \tag{5}$$

and K and S are matrices satisfying the conditions

$K = K^1$, a constant matrix, and
$S = 1 + S^1$.

This solution leads to the relations

$$\dot{S} = G\,S - S\,K$$

which become, when limited to first order,

$$\dot{S}^1 = G_3^1 - K^1.$$

In order to maintain the quasi-periodic character of Y, the non-periodic part of S^1 must be made equal to zero, which is possible if K^1 is chosen to be equal to the constant part of G_3^1. We then have the complete solution of (4) and (5).

The system (5) is a system of differential equations with constant coefficients, which can be split to the first approximation in two independent systems, and whose general solution is

$$z_p^j = \sum_{k=1}^{4} M_k^j C_k \exp i (g_k t + \beta_k) \quad \text{for the p variables} \tag{6}$$

$$z_q^j = \sum_{k=0}^{4} N_k^j E_k \exp i (h_k t + \gamma_k) \quad \text{for the q variables} \tag{7}$$

In equations (6) and (7), j is the number of the satellite, k=0 corresponds to the variables related to the jovian equator, C and E are arbitrary diagonal matrices the elements of which are the proper eccentricities and inclinations. C and E are to be determined by comparison with observations, as well as the phases β and γ, and in this paper we have used the same values as Sampson (1921) in order to allow comparison to Lieske's (1977) E-0 series. The matrices M and N are formed by the eigenvectors of K and g and h are the corresponding eigenvalues. The quantities g and h are the motions of the proper pericenters and nodes. Y_3 introduces long period inequalities in the solutions Y_2 and Y_1 which are obtained subsequently.

A general solution where we have used such a variational system completed by the resolution of a critical system and an intermediary solution previously described (Vu, 1981), has already been obtained (Thuillot and Vu, 1983). Here we show the possibility of keeping track of the dynamical parameters in M, N, g and h. The method used allowed us to obtain a complete analytical solution of the autonomous system, and to compute the corresponding general solution.

3. THE METHOD

A method proposed by Sagnier (1981) has been applied in order to get analytical eigenvalues and eigenvectors, which are obtained as analytical quantities developped in the neighbourhood of a numerical solution.

The developments are made with respect to small variations $\delta\mu$ around the initial vector μ as determined by Null (1976) (see Table I).

TABLE I: Values of the dynamical parameters

Authors	m_1	m_2	m_3	m_4	J_2	J_4
	(10^{-8} mass of Jupiter)				(10^{-6})	
Sampson (1921)	4497	2529	7988	4504	14848.5	-810.7
Null (1976)	4684 ± 22	2523 ± 25	7803 ± 30	5661 ± 19	14733 ±4	-587 ±7

Let the eigenvector matrix T, the eigenvalue matrix D, and the matrix K be written as

$$T = T^0 + T^1 + \dots + T^n,$$

$$D = D^0 + D^1 + \dots + D^n,$$

$$K = K^0 + K^1 + \dots + K^n,$$

where D^i are diagonal matrices and where the upper index denotes the order with respect to $\delta\mu$.

The substitution of K, T, D in the fundamental equation

$$K\,T - T\,D = 0, \tag{8}$$

gives the equation of order n :

$$D^0\,\Sigma^n - D^n - \Sigma^n\,D^0 = \Psi^n. \tag{9}$$

Here Σ^n is an auxiliary matrix defined by $\Sigma^n = \overset{-1}{T^0}\,T^n$ and Ψ^n a function of known quantities given (for n > 1) by

$$\Psi^n = \overset{-1}{T^0}\,\{-K^n\,T^0 - (K^1\,T^{n-1} + \dots + K^{n-1}T^1) + $$
$$+ (T^1\,D^{n-1} + \dots + T^{n-1}D^1)\},$$

and $\Psi^1 = -\overset{-1}{T^0}\,K^1\,T^0$.

The identification of the diagonal and non diagonal parts of the two members of eqn. (9) gives the following relations relative to the elements of column i and line j :

$$(\Sigma^n)_{ij} = \frac{(\Psi^n)_{ij}}{(D^0)_{ii} - (D^0)_{jj}}, \tag{10}$$

and

$$(D^n)_{ii} = -(\Psi^n)_{ii}. \tag{11}$$

It is noteworthy that the diagonal elements of Σ^n are not given by these equations and this missing datum can generate a divergence in the process as shown by some numerical experiments. On the other hand we have, since the begining, adopted the normalization rule diag (T) = I.Since diag (T^0) = I we assumed diag (T^n) = 0 for every n ≠ 0 and we obtained a rule for determining the diagonal of the matrices Σ^n.

Using the relation diag (T^n) = diag ($T^0\,\Sigma^n$) it is easy to verify that the diagonal of T is

$$\mathrm{diag}\,(T) = I + \mathrm{diag}\,(\Sigma^1 + \overset{''}{T}{}^0\,\overset{''}{\Sigma}{}^1) + \dots + \mathrm{diag}\,(\Sigma^n + \overset{''}{T}{}^0\,\overset{''}{\Sigma}{}^n),$$

where I is the unit matrix, and $\overset{''}{\Sigma}{}^i$ and $\overset{''}{T}{}^i$ represent the matrices (Σ^i - diag Σ^i) and (T^i - diag T^i). In order to obtain a normalized

TABLE II. First-order developments of the eigenvalues g, h, and the eigenvectors M and N. Coefficients are followed by the power of 10. Units for g and h are rad/day.

	order 0	δJ_2	δJ_4	δm_1	δm_2	δm_3	δm_4
M_1^j	1						
	-2.699-2	1.897	-1.561-1	-5.968+2	-2.684+1	-9.100	-6.854-1
	-1.292-2	7.570-2	-5.521-3	-2.755+1	6.539	1.832-1	-1.197-1
	-7.713-5	4.362-3	-3.096-4	-1.642	2.617-1	7.554-2	1.795-3
M_2^j	1.824-2	-1.278	1.055-1	1.432+1	7.051+2	5.163	4.194-1
	1						
	-2.759-2	1.484	-4.882-2	7.002+1	-1.133+3	5.204+1	-1.052+1
	-4.342-4	-3.762-3	1.863-4	-3.984-1	-1.650+1	6.119	2.571-1
M_3^j	5.379-3	-4.428-1	2.941-2	-3.649	6.179+1	6.541+1	7.562-1
	1.075-1	-5.767	1.890-1	-2.765+2	1.529+2	1.176+3	3.855+1
	1						
	-1.097-1	6.026	-8.024-2	1.509+2	6.365+2	-1.629+3	3.174+2
M_4^j	7.951-4	-9.086-2	4.465-3	-1.124	3.542	7.454	1.249+1
	1.295-2	-1.211	2.728-2	-4.313+1	-5.408+1	1.236+2	1.992+2
	1.052-1	-5.745	7.581-2	-1.431+2	-6.256+2	2.124+2	1.553+3
	1						
g_1	23.2046-4	1.527-1	-1.094-2	1.933-2	1.745	2.241-1	3.299-2
g_2	5.7818-4	3.004-2	-8.532-4	1.334	2.182-2	8.556-1	7.807-2
g_3	1.2562-4	5.876-3	-6.687-5	1.420-1	6.048-1	1.019-3	2.684-1
g_4	0.3188-4	8.710-4	-7.811-7	1.707-2	5.240-2	1.750-1	-1.580-2
h_1	-23.2439-4	-1.526-1	1.094-2	-3.523-2	-1.769	-2.245-1	-3.297-2
h_2	-5.7906-4	-3.002-2	8.548-4	-1.317	-4.303-2	-8.669-1	-7.846-2
h_3	-1.2609-4	-5.869-3	6.801-5	-1.417-1	-5.550-1	-7.410-3	-2.854-1
h_4	-0.3072-4	-9.366-4	4.516-6	-1.867-2	-5.709-2	-1.571-1	3.326-2
h_0	-3.33 -8	1.363-6	-2.407-8	3.017-5	6.786-5	3.662-5	-1.208-4

	order 0	δJ_2	δJ_4	δm_1	δm_2	δm_3	δm_4
N_1^j	1						
	-3.592-2	2.516	-2.060-1	-8.018+2	3.566+1	-1.145+1	-9.015-1
	-2.659-3	1.583-1	-1.135-2	-5.783+1	1.190+1	3.521-1	-2.285-1
	-2.759-4	1.599-2	-1.105-3	-6.023	8.323-1	2.284-1	5.809-3
	-1.419-3	9.283-2	-6.648-3	1.829-1	1.073	1.394-1	2.037-2
N_2^j	2.329-2	-1.629	1.373-1	2.091+1	9.398+2	6.167	6.117-1
	1						
	-3.722-2	2.009	-6.657-2	8.879+1	-1.532+3	7.048+1	-1.376+1
	-1.136-3	8.294-3	-1.001-4	-4.474-1	-4.417+1	1.246+1	5.350-1
	-8.816-4	5.424-2	-2.062-3	1.847	-4.441	1.252	1.248-1
N_3^j	7.096-3	-6.429-1	5.278-2	-9.508-1	1.229+2	1.223+2	7.433
	1.414-1	-7.602	2.537-1	-3.521+2	2.099+2	1.588+3	5.049+1
	1						
	-1.613-1	8.861	-1.159-1	2.210+2	8.507+2	-2.387+3	4.926+2
	-3.410-3	2.059-1	-4.212-3	5.185	1.113+1	-7.991	7.414
N_4^j	-1.365-3	2.681-3	9.538-3	3.057	2.665+1	3.222+1	-2.297-1
	2.290-2	-2.188	5.181-2	-7.843+1	-7.763+1	2.525+2	3.735+2
	1.510-1	-8.187	1.051-1	-2.032+2	-8.031+2	3.055+2	2.224+3
	1						
	-3.818-3	2.745-1	-4.032-3	6.184	1.636+1	1.047+1	-3.923+1
N_0^j	0.99936	7.073-2	-3.377-3	7.519-1	-2.271	-2.274	-4.668
	0.99388	5.846-1	-1.178-2	1.879+1	2.008+1	-2.299+1	-4.616+1
	0.96918	2.002	-1.996-2	4.587+1	1.625+2	7.235+1	-2.378+2
	0.85967	4.356	-2.135-2	8.688+1	2.612+2	6.660+2	-1.062+2
	1						

eigenvector, the following relation has to be verified:

$$\operatorname{diag}(\Sigma^n) = -\operatorname{diag}(\overset{\prime\prime}{T}{}^0\ \overset{\prime\prime}{\Sigma}{}^n). \qquad (12)$$

4. RESULTS

Table II gives the results obtained for the eigenvectors M and N, and for the eigenvalues g and h. The column labelled "order 0" contains the numerical results, in the classical manner, of the resolution of the autonomous system, and the others give the increment of these values for each variations $\delta\mu$, expressed in the same units as in Table I. Each line corresponds to a development of the solution limited to the order 1.

Numerical experiments have been made to appreciate the gain of precision when a solution of higher order is used. The analytical solution has been compared for several values of the variations $\delta\mu$ to a numerical solution. This solution of reference has been computed with the use of a matrix K affected by the same variations $\delta\mu$ before its diagonalization. The gain of precision is 10^{-2} when we use an order 2 solution instead of an order 1 solution. These results justify the criterium of normalization described in section 3.

5. EFFECTS ON THE GENERAL SOLUTION

The analytical solution of order 1 for the autonomous system has been used to calculate complementary terms to the general solution. Following the construction explained in section 2, the variational system has been integrated, and all the contributions of $\delta\mu$ in the solutions Y_1, Y_2, Y_3 have been obtained. We have taken into account corrections due to the integration factors generated by arguments where eigenvalues appear.
A general term under the form:

$$C \exp i\,\{a\, l_0 - b\,(g\,t + \beta)\},$$

where $g = g^0 + \sum g^1 \delta\mu$, gives an integrated term approximated by:

$$\frac{C}{a\,n_0 - b\,g^0}\left(1 + \frac{b \sum g^1\ \delta\mu}{a\,n_0 - b\,g^0}\right) \exp i\,\{a\, l_0 - b\,(g\,t + \beta)\}.$$

After the general solution has been obtained, the classical Keplerian formulae give the series of periodic terms in true longitude L, in the radius vector R and in the latitude B. In Table III are gathered the complementary terms in $\delta\mu$ for the quantities $L - l_0$, $R/A - 1$, tan B (l_0 is the observed mean longitude, A is the semi major axis of the orbit). This table is limited to the largest terms and shows the effect of variations of the dynamical parameters on the coefficients of the main inequalities.

From these results, it is also possible to isolate some equations which have already been calculated by Sampson (1921). He used an itera-

TABLE III. First-order increments of the coefficients of the main inequalities. Coefficients are in radian, δμ are to be expressed in units of the Table I.

(R/A − 1) SATELLITE 1 (cosine)		
Coef.		Argument
6.504−4	δJ2	L1−PI3
6.662−4	δJ2	L1−PI4

(R/A − 1) SATELLITE 2 (cosine)		
Coef.		Argument
−1.701−1	δm4	3L2−4L3+PI4
5.896−4	δJ2	3L2−4L3+PI4
8.492−3	δJ2	L2−PI3
−1.728	δm3	L2−PI3
4.062−1	δm1	L2−PI3
−2.246−1	δm2	L2−PI3
3.163−1	δm1	L2−PI4
3.966−1	δm2	L2−PI4
8.879−3	δJ2	L2−PI4
−9.062−1	δm3	L2−PI4
−1.461	δm4	L2−PI4

(R/A − 1) SATELLITE 3 (cosine)		
Coef.		Argument
1.042−1	δm2	L3−PI2
4.213−2	δJ2	L3−PI4
−5.559−4	δJ4	L3−PI4
1.049	δm1	L3−PI4
4.587	δm2	L3−PI4
−1.557	δm3	L3−PI4
−1.139+1	δm4	L3−PI4

(R/A − 1) SATELLITE 4 (cosine)		
Coef.		Argument
−2.217−1	δm1	L4−PI3
2.393	δm3	L4−PI3
−9.350−1	δm2	L4−PI3
−4.663−1	δm4	L4−PI3
−8.852−3	δJ2	L4−PI3

(L − LO) SATELLITE 1 (sine)		
Coef.		Argument
−2.615	δm3	L1−2L2+PI3
−1.374	δm3	L1−2L2+PI4
−2.243	δm4	L1−2L2+PI4
1.297−1	δm2	L1−PI2
−1.301−3	δJ2	L1−PI3
1.815−1	δm2	L1−PI3
−1.333−3	δJ2	L1−PI4
1.093−1	δm3	L1−PI4
1.832−1	δm4	L1−PI4
1.922−1	δm3	L1−PI3

(L − LO) SATELLITE 2 (sine)		
Coef.		Argument
−1.146−3	δJ2	3L2−4L3+PI4
−1.374−1	δm2	3L2−4L3+PI4
3.309−1	δm4	3L2−4L3+PI4
−6.158−4	δJ2	L2−2L3+PI2
5.385−4	δJ2	L2−2L3+PI3
1.096−1	δm2	L2−2L3+PI3
−1.281	δm1	L2−2L3+PI3
5.557	δm3	L2−2L3+PI3
4.102	δm2	L2−2L3+PI4
1.235	δm1	L2−2L3+PI4
−8.619	δm4	L2−2L3+PI4
2.042	δm3	L1−2L2+PI4
3.300	δm4	L1−2L2+PI4
−1.694−2	δJ2	L2−PI3
−8.125−1	δm1	L2−PI3
4.493−1	δm2	L2−PI3
3.456	δm3	L2−PI3
1.133	δm4	L2−PI3
−1.776−2	δJ2	L2−PI4
−6.326−1	δm1	L2−PI4
−7.932−1	δm2	L2−PI4
1.812	δm3	L2−PI4
2.922	δm4	L2−PI4

(L − LO) SATELLITE 3 (sine)		
Coef.		Argument
−1.206−1	δm4	2L2−3L3+PI4
−1.421	δm3	L2−2L3+PI3
−8.778−4	δJ2	L2−2L3+PI4
−1.380	δm2	L2−2L3+PI4
2.204	δm4	L2−2L3+PI4
−2.084−1	δm2	L3−PI2
−8.428−2	δJ2	L3−PI4
1.112−3	δJ4	L3−PI4
−2.099	δm1	L3−PI4
−9.176	δm2	L3−PI4
2.278+1	δm4	L3−PI4
3.115	δm3	L3−PI4

(L − LO) SATELLITE 4 (sine)		
Coef.		Argument
1.770−2	δJ2	L4−PI3
4.433−1	δm1	L4−PI3
1.870	δm2	L4−PI3
−4.786	δm3	L4−PI3
9.326−1	δm4	L4−PI3

(TAN B) SATELLITE 2 (sine)		
Coef.		Argument
−3.812−1	δm1	L2−OM1
1.196−3	δJ2	L2−OM1
−2.364−2	δJ2	L2−OM3
−1.095	δm1	L2−OM3
6.526−1	δm2	L2−OM3
4.937	δm3	L2−OM3
1.570−1	δm4	L2−OM3
−3.720−1	δm1	L2−OM4
−3.682−1	δm2	L2−OM4
1.198	δm3	L2−OM4
1.772	δm4	L2−OM4
3.165−2	δJ2	L2−OM0
1.017	δm1	L2−OM0
1.087	δm2	L2−OM0
−1.245	δm3	L2−OM0
2.449	δm4	L2−OM0

(TAN B) SATELLITE 1 (sine)		
Coef.		Argument
1.704−1	δm1	L1−OM2
7.657	δm2	L1−OM2
3.821−1	δm2	L1−OM3
3.803−1	δm3	L1−OM3
3.821−1	δm2	L1−OM3
1.264−1	δm2	L1−OM4
1.528−1	δm3	L1−OM4
−2.527−1	δm4	L1−OM0
−1.231−1	δm3	L1−OM0

(TAN B) SATELLITE 3 (sine)		
Coef.		Argument
1.634−2	δJ2	L3−OM2
7.234−1	δm1	L3−OM2
−1.248+1	δm2	L3−OM2
5.742−1	δm3	L3−OM2
−1.121−1	δm4	L3−OM2
−3.883−2	δJ2	L3−OM4
−9.637−1	δm1	L3−OM4
−3.809	δm2	L3−OM4
1.449	δm3	L3−OM4
1.060+1	δm4	L3−OM4
1.084−1	δJ2	L3−OM0
2.483	δm1	L3−OM0
8.795	δm2	L3−OM0
3.917	δm3	L3−OM0
1.287+1	δm4	L3−OM0

(TAN B) SATELLITE 4 (sine)		
Coef.		Argument
−3.599−1	δm2	L4−OM2
1.015−1	δm3	L4−OM2
6.872−1	δm1	L4−OM3
2.505	δm2	L4−OM3
−7.422	δm3	L4−OM3
1.532	δm4	L4−OM3
2.358−1	δJ2	L4−OM0
4.703	δm1	L4−OM0
1.414+1	δm2	L4−OM0
3.605+1	δm3	L4−OM0
−5.749	δm4	L4−OM0

tive process to get this type of term expressed as a function of the corrections to the masses and to the coefficient J_2, in order to deduce their values from a comparison with observations.

The correction to free oscillation in longitude of Ganymede, the argument of which is L3 - PI4, is given in the Table III by:

$$- 4''.33 \times 10^5\ \delta m_1 - 18''.93 \times 10^5\ \delta m_2 + 6''.43 \times 10^5\ \delta m_3 +$$

$$+ 46''.93 \times 10^5\ \delta m_4 - 0''.17384 \times 10^5\ \delta J_2 + 0''.229 \times 10^5\ \delta J_4,$$

where the coefficients are expressed in jovicentric arcseconds. Sampson's result is:

$$- 4''.07 \times 10^5\ \delta m_1 - 16''.87 \times 10^5\ \delta m_2 + 50''.25 \times 10^5\ \delta m_4 -$$

$$- 0''.17195 \times 10^5\ \delta J_2,$$

which shows a good agreement. We must keep in mind that Sampson used a different set of parameters (Table I).

6. CONCLUSION

The results obtained show the possibility of obtaining developments of the secular elements related to the pericenters and the nodes of the Galilean satellites orbits with respect to variations of the dynamical parameters. This method has been applied to calculate the effects of variations of these parameters in the most important inequalities of the satellites motion. Several applications of these results can be considered: comparison of eigenvalues and eigenvectors due to different authors use of a unique general solution of the theory with other sets of dynamical parameters, and computation of partial derivatives of the main inequalities with respect to them.

REFERENCES

Brumberg, V. A.: 1970, *Periodic Orbits, Stability and Resonances*, G. Giacaglia Ed., D. Reidel, Dordrecht, 410.

Duriez, L.: 1983, *Compte-rendu du Colloque du GS5*, Paris, octobre 1983, Y. Reqieme Ed., Université de Bordeaux I.

Ferraz-Mello, S.: 1976, *Science*, 192, 1127.

Lieske, J.H.: 1977, *Astron. and Astrophys.* 56, 333.

Null, G.W.: 1976, *Astron. J.* 81, 1153.

Sampson, R. A.: 1921, *Mem. Roy. Astron. Soc.* 63, 1.

Sagnier, J.-L.: 1981, *Thèse de doctorat*, Université Paris VI.

Thuillot, W. and VU, D.T.: 1983, *The motion of Planets and Natural and Artificial Satellites*, S.Ferraz-Mello and P.E. Nacozy Ed. Université de Sao Paulo, 273.

Vu, D.T.: 1981, *Astron. and Astrophys.* 94, 140.

LIBRATION OF LAPLACE'S ARGUMENT IN THE GALILEAN SATELLITES THEORY

J. Henrard
Department of Mathematics
Facultés Univesitaires de Namur
B-5000 Namur Belgium

ABSTRACT. In the Galilean satellites motion, the Laplace argument $\lambda_1 - 3\lambda_2 + 2\lambda_3$ librates around the value π. The amplitude of libration is very small so that the classical theories have not been set up to take into account large librations. On the other hand large librations have to be considered when we describe possible scenarii of capture into resonance by tidal effects. The aim of this paper is to present a new way of applying Hamiltonian perturbation methods to the problem of the Galilean satellites in such a way that the theory is valid for large librations. Preliminary results from such a theory are discussed.

1. THE HAMILTONIAN FUNCTION

This paper does not pretend to develop a complete theory of the Galilean satellites. We shall thus consider only a simplified problem which we hope retains the essential features and difficulties of the full problem.

We take into account the mutual gravitational attraction of Jupiter (m_0), Io (m_1), Europa (m_2) and Ganymède (m_3) considered as point masses and moving in the same plane. We add to this Hamiltonian function the secular effects of Jupiter oblateness and the attraction of Callisto and the Sun.

Let us introduce the modified Delaunay's elements for the Jacobi coordinates :

$$\begin{aligned} \lambda_i &: \text{ mean longitude of satellite (i)}, \\ -p_i &: \text{ argument of perijove for satellite (i)}, \\ L_i &: m_i\,[G\,m_0\,a_i\,\mu_{i-1} / \mu_i]^{1/2}, \\ P_i &: L_i\,[1 - \sqrt{1-e_i^2}], \end{aligned} \tag{1}$$

where (a_i, e_i) are the osculating elliptic elements of satellite (i) in Jacobi coordinates and

$$\mu_i = m_0 + m_1 + \ldots + m_i . \tag{2}$$

Celestial Mechanics **34** (1984) 255–262. 0008–8714/84.15

We introduce also the osculating mean motion N_i by the formula

$$N_i^2\, a_i^3 \;=\; G\, m_0\, \mu_i \,/\, \mu_{i-1} \,. \qquad (3)$$

With these notations, the zero order Hamiltonian function describing the secular effects is given by

$$H_0 = -\sum_{i=1}^{3} \frac{N_i L_i}{2} \Big\{1 + J_2 \left(\frac{R_J}{a_i}\right)^2 [1 - e_i^2]^{-3/2} - \frac{3}{4} J_4 \left(\frac{R_J}{a_i}\right)^4 [1 - e_i^2]^{-5/2} -$$

$$- \frac{1}{4} \frac{M_\Theta}{m_0} \left(\frac{a_i}{A_\Theta}\right) [1 - \frac{3}{2} e_i^2] - \frac{1}{4} \frac{m_4}{m_0} \left(\frac{a_i}{a_4}\right)^3 [1 - \frac{3}{2} e_i^2] + H_S \Big\} \,. \qquad (4)$$

The secular effects of the Sun and Callisto have been truncated at the second order in excentricities and we have added the contribution H_S of the secular perturbation of the function H_P describing the first order (in masses) of the mutual perturbations of the satellites

$$H_P \;=\; \sum_{j=1}^{2} \; \sum_{i=j+1}^{3} \frac{m_j}{m_0} L_i\, N_i\, a_i \left\{\frac{1}{r_{j,i}} - \frac{(\vec{r}_j | \vec{r}_i)}{r_i^3}\right\} \,. \qquad (5)$$

The Hamiltonian function we are considering is thus

$$H \;=\; H_0 + H_1 \qquad \text{with} \qquad H_1 = H_P - H_S \,.$$

2. RESONANT VARIABLES AND THE EQUILIBRIUM

In order to bring forward the combinations of angular variables which are found to be librating in the Galilean satellites problem, we introduce the canonical variables

$$\begin{aligned}
q_1 &= p_1 - \lambda_1 + 2\,\lambda_2 \,, & Q_1 &= P_1 \,, \\
q_2 &= p_2 - \lambda_1 + 2\,\lambda_2 \,, & Q_2 &= P_2 \,, \\
q_3 &= p_3 - \lambda_2 + 2\,\lambda_3 \,, & Q_3 &= P_3 \,, \\
r_1 &= \lambda_1 - \lambda_3 \,, & 3\,R_1 &= 3\,L_1 + L_2 + P_1 + P_2 + P_3 \,, \\
r_2 &= -\lambda_1 + 3\,\lambda_2 - 2\,\lambda_3 \,, & 3\,R_2 &= L_2 - 2\,P_1 - 2\,P_2 + P_3 \,, \\
r_3 &= \lambda_3 \,, & R_3 &= L_1 + L_2 + L_3 - P_1 - P_2 - P_3 \,.
\end{aligned} \qquad (7)$$

The variable r_2 is the well-known Laplacian angle. The usual symmetries governing the expansion of the perturbation function insure that r_3 does not appear in this expansion. This is the direct consequence of the fact that R_3 represents the total angular momentum of the system.

The angular variable r_1 has a frequency of the order of the mean motion of the satellites and can be averaged out without problem. We shall assume that this has been done and that the perturbation H_1 depends only on the remaining angular variables (q_i, r_2) .

We have thus reduced the problem to a four degrees of freedom problem depending upon two parameters R_1 and R_3 .

To find the equilibria of such a system, we introduce the Poincaré-like variables

$$x_i = \sqrt{2 Q_i} \sin q_i \quad , \quad y_i = \sqrt{2 Q_i} \cos q_i \ . \tag{8}$$

The equilibria are located at $x_i = 0$ and $\sin r_2 = 0$ for y_i and R_2 solutions of

$$\frac{\partial H}{\partial y_i} = \frac{\partial H}{\partial R_2} = 0 \ . \tag{9}$$

These solutions are functions of the parameters R_1 , R_3 . The ratio of these parameters is the only important factor as the value of one of them can always be normalized by a proper choice of units of time or distance. This ratio is a function of the ratio of the frequency of the angular variable $u_i = \lambda_1 - 2\lambda_2$ over the frequency of λ_1 . As the latter ratio is more meaningful physically, we shall use it as the basic parameter rather than R_1/R_3 .

The equilibrium $r_2 = \pi$ corresponding to the symmetries of the Galilean satellites is stable when the above ratio of frequencies is larger than approximatively 0.001 (Yoder and Peale, 1981; Henrard, 1983). The actual value of this ratio for the Galilean satellites is 0.0034 .

The equilibria of the reduced four degrees of freedom system correspond in a proper rotating coordinate system to periodic orbits studied by Hadjidemetriou and Michalodimitrakis (1981).

3. ZERO POINT OF THE EXPANSIONS

The Hamiltonian function depends upon the momenta Q_i , R_i and we wish to expand it in power series. While the Q_i are intrinsically small (of the order of the square of the eccentricities), the R_i are not. We shall thus expand the Hamiltonian function around nominal values R_i^* in powers of the undimensional β_i defined by

$$R_i = R_i^* + L_1^* \beta_i \ , \tag{10}$$

where $L_1^* = R_1^* - R_2^*$. The nominal values R_i^* are chosen in order to adjust to observed values the mean motions of the angular variables $u_1 = \lambda_1 - 2\lambda_2$ and $u_2 = \lambda_2 - 2\lambda_3$ as computed from a reduced problem (zero order Hamiltonian H_0 evaluated at zero eccentricities).

More specifically, let us take as observed values the values $0.7395^{o/day}$ for the mean motions of u_1 and u_2 and $203.489^{o/day}$ for the mean motion of λ_1 . On the other hand, we compute that the latter in the reduced problem is equal to $1.001251\ N_1^*$ where N_1^* is the Keplerian mean motion associated with L_1^* by (1) and (3) .

We then impose that the mean motions of u_1 and u_2 in the reduced problem are equal to

$$3.63865\ 10^{-3}\ N_1^* = \frac{0.7395}{203.489} \times 1.001251\ N_1^* \ . \tag{11}$$

Of course, these frequencies will be adjusted again at the end of the computation when we give numerical values to the β_i .

With these definitions, we obtain nominal values for the ratios of Keplerian mean motions

$$N_2^* = 0.498543\ N_1^* \quad , \quad N_3^* = 0.247519\ N_1^* \quad , \tag{12}$$

and the ratios of the Keplerian semi-major axis

$$a_1^* = 0.628741\ a_2^* \quad , \quad a_1^* = 0.394216\ a_3^* \ . \tag{13}$$

These values will be used for the computations of the Laplace coefficients and their derivatives necessary to expand the perturbation H_P .

The Hamiltonian function can now be written as

$$H = H_0(\beta_i, Q_i) + H_1(\beta_i, \sqrt{2\ Q_i}; q_1, q_2, q_3, q_4, r_2) \tag{14}$$

where H_1 is proportional to the mass ratios m_i/m_0 of the order of 10^{-5} and the derivatives of H_0 are given by

$$\begin{aligned} \nu_1 &= \partial H_0/\partial Q_1 = -4.216\ 10^{-3} + f_1(\beta_i, Q_i) \ , \\ \nu_2 &= \partial H_0/\partial Q_2 = -3.726\ 10^{-3} + f_2(\beta_i, Q_i) \ , \\ \nu_3 &= \partial H_0/\partial Q_3 = -3.659\ 10^{-3} + f_3(\beta_i, Q_i) \ , \\ n &= \partial H_0/\partial R_2 = f_4(\beta_i, Q_i) \ . \end{aligned} \tag{15}$$

The functions f_j vanish with β_i and Q_i .

4. NORMALIZATION : A SIMPLIFIED MODEL

We wish to study the general solution of the Hamiltonian (14) in the vicinity of the equilibrium described in section 2. The size of this vicinity is measured by the amplitude of libration of the Laplace argument and the amplitudes of the free eccentricities.

For the orbits of the Galilean satellites as we see them now, all of these amplitudes are small. But if we take into account probable scenarii of the capture into resonance of these satellites by tidal effects (Yoder, 1979; Yoder and Peale, 1981; Henrard, 1983), we have to consider that while the amplitudes of free eccentricities have always been small, the amplitude of libration of the Laplace argument may have been large in the past.

For this reason, we would like to develop a theory without assuming that this latter amplitude is small. This could be achieved if we were able to eliminate from the Hamiltonian (14) the terms depending upon the angular variables q_i . We would be left with a one-degree of freedom, pendulum-like Hamiltonian describing libration as well as circulation of the Laplace argument r_2 .

We shall indeed achieve this goal when we assume that the free eccentricities vanish. We believe that in a further step it is possible to extend the result and make allowance for small but non-zero values of free eccentricities but we have not investigated this completely.

In order to prepare the normalization which will lead to our result, let us discuss a simplified model.

Let us consider the following Hamiltonian function

$$H = [\nu + Q]\,Q + \gamma\sqrt{2\,Q}\cos q \qquad (16)$$

which remains the principal feature of the dependance of the Hamiltonian (14) with respect to one of the couple angle-action (q_i, Q_i). This Hamiltonian is actually what we have called in a previous publication (1983) the second fundamental model for resonance. The contant factors γ and ν are assumed to be small (actually of the order of 10^{-5} and 4.10^{-3}).

When γ^2/ν^3 is small, this Hamiltonian possesses an equilibrium close to the origin localed at approximatively $Q^* = \gamma^2/\nu^2$ and well isolated from any other equilibria which, when they exist (when γ and ν have the same sign), are located at approximatively $Q^{**} = \nu >> Q^*$. If we want to include the equilibrium Q^* but exclude the other ones from consideration, we scale the momenta by

$$Q = \frac{\gamma^2}{\nu^2}\,Q' \qquad (17)$$

which leads after a change of time variable $\tau = \nu\,t$ to the new Hamiltonian

$$H' = \frac{\nu}{\gamma^2}\,H = [1 + \frac{\gamma^2}{\nu^3}\,Q']\,Q' + \sqrt{2\,Q'}\cos q\,. \qquad (18)$$

We see that H_1 is of the same order of magnitude than H_0 and no averaging seems possible, which is indeed understandable as the angular variable q can librate around the equilibrium Q^*.

But if we now introduce a translation

$$\sqrt{2Q'}\sin q' = \sqrt{2S}\sin s \quad ; \quad \sqrt{2Q'}\cos q' = -1 + \sqrt{2S}\cos s \qquad (19)$$

which brings the origin $S = 0$ approximatively to the equilibrium, the new Hamiltonian function becomes

$$H'' = [1 + \frac{\gamma^2}{\nu^3}\,S]\,S + \frac{\gamma^2}{\nu^3}\,\{(1 - 2\,S)\sqrt{2S}\cos s + S\cos 2\,s\} \qquad (20)$$

which can now be averaged. The small parameter which measures the size of H_1 with respect to H_0 is γ^2/ν^3.

But the two operations, the translation and the averaging can both be performed by the Lie transform method. The small parameter associated with these operations are respectively γ/ν and γ^2/ν^3 which for the Galilean satellites have about the same size as $\gamma \sim \nu^2$.

It seems natural and economical to perform both operations at the same time by a unique Lie transformation. This Lie transformation eliminates all the terms depending upon the angular variable from the Hamiltonian function. The elimination of the terms linear in $\sqrt{2\,Q}$ are responsible for the translation, the elimination of the others for the averaging.

The algebra involved in the double operation which we call a normalization is identical to the algebra involved in a simple averaging. Putting it in a paradoxical form, we can say that in this way we are able to average over a librating angular variable.

5. NORMALIZATION

To complete the normalization, we should eliminate at the same time the three angular variables q_i . This presents some difficulties as the frequencies ν_i are similar and we can expect some resonances between them.

To avoid this difficulty, we shall at the present time eliminate only from the Hamiltonian function (14) the terms linear in $\sqrt{2 Q_i}$. It amounts to performing only the translation part of the normalization. In a more complete solution, which should take into account the free eccentricities, we would eliminate from the Hamiltonian function all periodic terms except resonant ones which would necessitate a particular treatment.

We have performed this operation starting from a perturbation H_p truncated at the fourth power in eccentricities and computing all the terms up to order 6 in $\sqrt{Q_i}$, $\sqrt{\beta_i}$, ν_i , $\sqrt{m_i/m_0}$ (as the ν_i appear in the denominator, this actually includes terms up to power 8 in the mass ratios).

The resulting Hamiltonian function can be divided in two parts

$$H = K_L(\bar{\beta}_2,\bar{r}_2;\beta_1,\beta_3) + K_F(\bar{\beta}_i,\bar{r}_2;\bar{Q}_i,\bar{q}_i) \quad . \tag{21}$$

The function K_F contains only terms of degree at least two in $\sqrt{Q_i}$. Particular solutions corresponding to zero free eccentricities can be obtained by considering only the first part K_L which we call the Libration Hamiltonian.

After adjustment of the parameters β_1 , β_3 so that the stable equilibrium of K_L corresponds to $\bar{\beta}_2 = 0$ and so that the mean motion of $u_1 = \lambda_1 - 2\lambda_2$ computed at the equilibrium is equal to his observed value, we have performed a last scaling

$$R = 10^4 \bar{\beta}_2 \quad , \quad \tau = 10^{-4} t \quad . \tag{22}$$

The coefficients K_{ij} of the new Hamiltonian

$$K_L(R,r) = \sum_{i,j} K_{ij} R^i \cos j r \tag{23}$$

are given in Table I .

Table I. the Libration Hamiltonian.

	R^0	R^1	R^2	R^3
-	-	0.2507	-11.1664	-0.2394
cos r	-4.5669	0.3596	-0.3717	0.1911
cos 2 r	-0.2286	0.0822	-0.0944	...
cos 3 r	0.0098	0.0268	...	
cos 4 r	-0.0008	...		

Each coefficient K_{ij} in Table I is the sum of a truncated series in the powers of ν_i , $\sqrt{m_i/m_0}$, the convergence of which is illustrated in Table II for the two main terms underlined in Table I .

Table II. Convergence of the coefficients of Table I.

	Order 0	Order 1	Order 2	Order 3
coeff. of R^2	-11.9888	1.1263	-0.3039	...
coeff. of cos r	-5.1847	0.7085	-0.1122	0.0210

6. COMPARISON WITH OTHER THEORIES

Other authors have computed Libration equations of the form

$$\ddot{r}_2 = \Sigma A_j \sin (j\, r_2) . \quad (24)$$

From our Libration Hamiltonian, we can compute a similar pendulum-like equation but its right hand member will be more involved. It will take the form

$$\ddot{r}_2 = \Sigma A_{ij} R^j \sin (j\, r_2) \quad (25)$$

or equivalently

$$\ddot{r}_2 = \Sigma A'_{ij} \dot{r}_2^i \sin (j\, r_2) . \quad (26)$$

It is not clear to us at present how to compare (24) with (25) or (26) . It seems to us that somewhere the previous authors have made the assumption that the libration in r_2 is small in order to get rid of the terms in R or $\dot{r}_2$ but it is difficult to point out exactly where and the consequence of this upon the coefficients of (24) .

In any case we can compare the period of the infinitesimal librations. For the values of the mass ratios given by Null (1976), we find 2014 days. Ferraz-Mello (1979) after introducing mass corrections in Sampson theory finds 2012 days while in his Ephemrris E-2 , Lieske (1980) gives 2074 days. Brown (1977) finds 2032 days for slightly different values of the mass-ratios.

REFERENCES

Brown, B. : 1977, 'The Long Period Behaviour of the Orbits of the Galilean Satellites of Jupiter', Celest. Mech. **16**, 229-259.

Ferraz-Mello, S. : 1979, 'Dynamics of the Galilean Satellites', Univ. Sao Paulo, IAG.

Hadjidemetriou, J.D., and Michalodimitrakis, M. : 1981, 'Periodic Planetary-Type Orbits of the General 4-Body Problem with an Application to the Satellites of Jupiter', Astron. Astrophys. **93**, 204-211.

Henrard, J. : 1983, 'Orbital Evolution of the Galilean Satellites : Capture into Resonance', Icarus **53**, 55-67.

Henrard, J., and Lemaître, A. : 1983, 'A Second Fundamental Model for Resonance', Celest. Mech. **30**, 197-218.

Null, G.W. : 1976, Astron. J. **81**, 1153.

Lieske, J.H. : 1980, 'Improved Ephemerides of the Galilean Satellites', Astron. Astrophys. **82**, 340-348.

Yoder, C.F. : 1979, 'How Tidal Heating in Io drives the Galilean Orbital Resonance Locks', Nature **279**, 767-770.
Yoder, C.F., and Peale, S.J. : 1981, 'The Tides of Io', Icarus **47**, 1-35.

PLANETARY PERTURBATIONS ON THE LIBRATION OF THE MOON

M. MOONS
Department of Mathematics
Facultés Universitaires N.D. de la Paix
Rempart de la Vierge, 8
B-5000 Namur
BELGIUM

ABSTRACT. A theory of the libration of the Moon, completely analytical with respect to the harmonic coefficients of the lunar gravity field, was recently built (Moons, 1982). The Lie transforms method was used to reduce the Hamiltonian of the main problem of the libration of the Moon and to produce the usual libration series p_1 , p_2 and τ . This main problem takes into account the perturbations due to the Sun and the Earth on the rotation of a rigid Moon about its center of mass. In complement to this theory, we have now computed the planetary effects on the libration, the planetary terms being added to the mean Hamiltonian of the main problem before a last elimination of the angles. For the main problem, as well as for the planetary perturbations, the motion of the center of mass of the Moon is described by the ELP 2000 solution (Chapront and Chapront-Touze, 1983).

1. THE PHASE SPACE AND THE HAMILTONIAN

1.1. Hamiltonian

To describe the libration of the Moon, we construct an Hamiltonian which is composed of two parts : the first one corresponds to the rotation of a rigid Moon about its center of mass and the second one is due to the attraction of other bodies (Earth, Sun and Planets), neglecting their shape.

This second part, denoted by V , has the following form :

$$V = \sum_{C^*} [-G\, E^* \int_{Moon} d\mu \,/\, r^*] ,$$

where C^* is a perturbating body and E^* its mass,
$d\mu$ is an element of mass in the Moon,
G is the universal gravitational constant, and
r^* is the distance between $d\mu$ and the body.

At present, we thus neglect the elasticity of the Moon and the shape of the other bodies, although we know that the flattening of the Earth has some importance and has still to be taken into account.

Celestial Mechanics **34** (1984) 263–273. 0008–8714/84.15

1.2. Phase Space

First of all, we have to choose appropriate variables to express the Hamiltonian.

We shall use the modified Andoyer's elements :

$$\lambda_1 = \mu_1 + \mu_2 + \mu_3 , \qquad \Lambda_1 = \|\vec{L}\| ,$$

$$\lambda_2 = -\mu_3 , \qquad \Lambda_2 = \|\vec{L}\| \ (1 - \cos b) ,$$

$$\lambda_3 = -\mu_1 , \qquad \Lambda_3 = \|\vec{L}\| \ (1 - \cos I) ,$$

which are related to the angular momentum of the Moon, $\vec{L}$, and, as shown in Figure 1, describe the position of a moving reference frame $(\vec{f}_1,\vec{f}_2,\vec{f}_3)$ with respect to a fixed one $(\vec{e}_1,\vec{e}_2,\vec{e}_3)$, both systems being centered at the center of mass of the Moon.

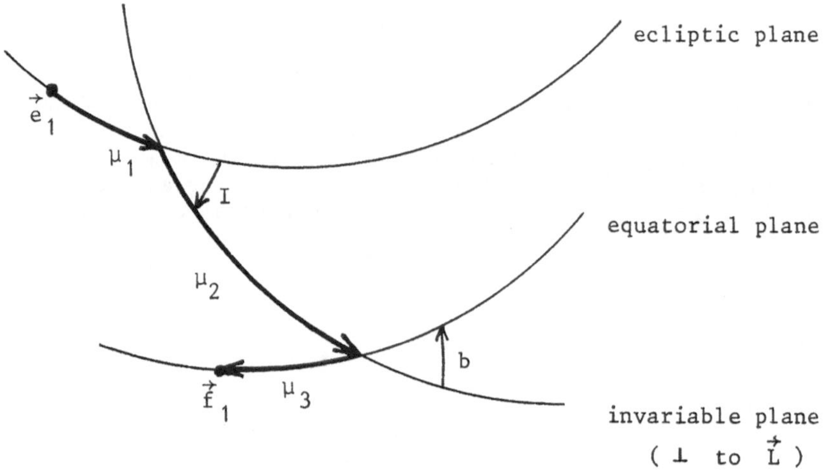

Fig. 1.

More precisely,

- $(\vec{e}_1,\vec{e}_2,\vec{e}_3)$ is inertial, $(\vec{e}_1,\vec{e}_2)$ being the ecliptic 2000 plane with $\vec{e}_1$ towards the perigee of the Sun;
- $(\vec{f}_1,\vec{f}_2,\vec{f}_3)$ is the system of the principal axes of inertia of the Moon for which the momenta of inertia matrix has the form

$$\begin{bmatrix} A & 0 & 0 \\ 0 & B & 0 \\ 0 & 0 & C \end{bmatrix} \quad \text{with} \quad A < B < C .$$

Let us notice (Henrard, 1974) that these canonical variables do not introduce virtual singularities since their only singularities are of polar-coordinates type ($\Lambda_i = 0$ when λ_i is undefined for $i=2,3$) and the Hamiltonian presents the d'Alembert characteristic with respect to the couples $(\lambda_2 , \sqrt{2 \Lambda_2})$ and $(\lambda_3 , \sqrt{2 \Lambda_3})$.

1.3. Form of the Hamiltonian

In the modified Andoyer's variables, the Hamiltonian has the form

$$H = \frac{1}{2}\frac{\Lambda_1^2}{C} + 2\frac{\Lambda_1^2}{C}\cos^2\frac{b}{2}\sin^2\frac{b}{2}\left[\frac{C-A}{A}\sin^2\lambda_2 + \frac{C-B}{B}\cos^2\lambda_2\right] +$$

$$+ \sum_{C^*}\left\{-G\,E^*\frac{M}{r^*}\sum_{n\geq 2}\left(\frac{R}{r^*}\right)^n\sum_{m=0}^{n} P_{nm}(\sin\gamma^*)\left[C_{nm}\cos m\lambda^* + S_{nm}\sin m\lambda^*\right]\right\}$$

where M is the mass of the Moon and R its equatorial radius,
(γ^*,λ^*) are the latitude and the longitude of C^* in $(\vec{f}_1,\vec{f}_2,\vec{f}_3)$,
P_{nm} are the associated Legendre polynomials, and
C_{nm}, S_{nm} are the harmonic coefficients of the lunar gravity field.

In a first step, we can solve a so-called "main problem" (denoted by MP) which takes into account the perturbation due to the Earth (with $n=2$ and 3) and the Sun (with $n=2$) and where (r^*,γ^*,λ^*) are given, for the Earth, by the main problem of lunar theory and, for the Sun, by an elliptical orbit about the Earth-Moon center of mass.

Any other contribution can be handled as perturbation to this main problem.

We already computed by this way the fourth order perturbation due to the Earth (Moons, 1981). We shall see here how to introduce the indirect planetary perturbations, i.e., the planetary perturbations on (r^*,γ^*,λ^*) . The direct effect of the planets is less important and has not yet been computed. The motion of the ecliptic plane is also neglected and will be introduced later.

But, before going further about planetary perturbations, we shall briefly recall the way of solving the main problem.

2. HOW TO SOLVE THE MAIN PROBLEM

2.1. Cassini's Laws

Cassini's laws (Tisserand, 1898) provide us with a good approximation of the motion of the Moon and can be summarized as follows :
- uniform rotation of the Moon about its polar axis with a period equal to the mean sidereal period of its rotation about the Earth;
- constant inclination (near $1°32'$) of the lunar equator on the ecliptic;
- coincidence of the ascending node of the lunar orbit and the descending node of the lunar equator on the ecliptic.

In our phase space, they are expressed by

$$\lambda_1 = \lambda + \pi\ , \quad \Lambda_1 = n\,C\ , \quad I = \text{constant}\ , \quad b = 0\ , \quad \lambda_3 = -h$$

where λ is the mean longitude of the Moon,
n its mean motion, and
h is the longitude of its ascending node on the ecliptic.

Graphically, these laws correspond to the following configuration :

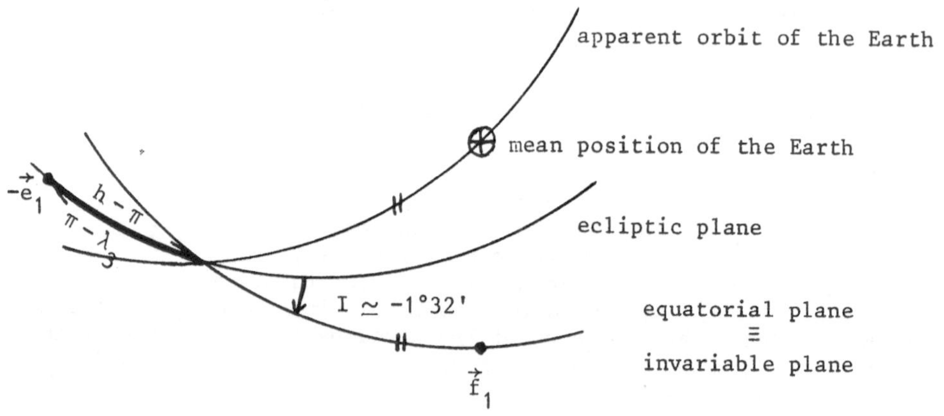

Fig. 2.

2.2. Expansion around mean values

Let us perform a canonical transformation which introduces the action-angle variables (p,P), (q,Q) and (r,R) as deviations with respect to Cassini's motion :

$$\alpha_1 \sqrt{2P} \sin p = \lambda_1 - \lambda - \pi , \qquad \sqrt{2P} \cos p = \alpha_1 (\Lambda_1/nC - \nu) ,$$

$$\alpha_2 \sqrt{2Q} \sin q = \sqrt{2\lambda_2/nC} \sin \lambda_2 , \qquad \sqrt{2Q} \cos q = \alpha_2 (\sqrt{2\Lambda_2/nC} \cos \lambda_2) ,$$

$$\alpha_3 \sqrt{2R} \sin r = \sqrt{2\lambda_3/nC} \sin(\lambda_3+h) , \quad \sqrt{2R} \cos r = \alpha_3 [\sqrt{2\Lambda_3/nC} \cos(\lambda_3+h) - 2\mu].$$

The quantities $(\sqrt{2P} \sin p , \sqrt{2P} \cos p)$, $(\sqrt{2Q} \sin q , \sqrt{2Q} \cos q)$ and $(\sqrt{2R} \sin r , \sqrt{2R} \cos r)$ are cartesian-like dimensionless coordinates centered at a mean equilibrium (i.e., μ and ν are computed so as to cancel the mean linear terms of the Hamiltonian). The constants α_1, α_2 and α_3 are determined so as to reduce the mean quadratic part of the Hamiltonian to three harmonic uncoupled oscillators :

$$H_{MP} = n_1 P + n_2 Q + n_3 R + F(p,q,r,P,Q,R,t)$$

and we get the following approximation of the frequencies of libration :

$$n_1 = 0.025950015853*n , \; n_2 = 0.000994243436*n , \; n_3 = -0.00309927617*n .$$

The Hamiltonian is depending on the time through the angles ℓ (mean anomaly of the Moon), ℓ' (mean anomaly of the Sun), $F = \lambda - h$

and $D = \lambda - \lambda'$ (λ' is the mean longitude of the Sun) describing the position of the Earth with respect to the Moon.

2.3. Elimination of the angles

In the new canonical variables, the Hamiltonian can be expressed in a form suitable for applying a perturbation method :

$$H_{MP} = H_0 + H_1$$

with

$$H_0 = n\,L + n'\,L' + n_g\,G + n_h\,H + n_1\,P + n_2\,Q + n_3\,R$$

and where $H_1(\ell,\ell',F,D,p,q,r,P,Q,R)$ is of the order of 10^{-3} with respect to H_0 and is expanded in powers (up to the fourth order) in $\sqrt{2P}$, $\sqrt{2Q}$ and $\sqrt{2R}$.

The perturbation method used is the Lie transforms method and the Hamiltonian H_{MP} is reduced to an averaged Hamiltonian $\overline{H}_{MP}$ which is independent of the angular variables (i.e., the new momenta are constant and the new angles are linear functions of the time).

The elimination of the angles is performed in several steps due to the large number of terms in H_1 (more or less 23 000 terms only for the first order in $\sqrt{2P}$, $\sqrt{2Q}$, $\sqrt{2R}$). There are two reasons for this : the loss of accuracy on the long period terms and the number of physical parameters (7 trigonometric variables corresponding to the angles ℓ, ℓ', F, D, p, q, r and 12 analytic variables correponding to $\sqrt{2P}$, $\sqrt{2Q}$, $\sqrt{2R}$, $\delta = (2C-A-B)/C$, $\gamma = (B-A)/C$, C_{30}, C_{31}, S_{31}, C_{32}, S_{32}, C_{33}, S_{33}) since our solution is analytical with respect to the harmonic coefficients of the lunar gravity field.

3. HOW TO INTRODUCE THE PLANETARY PERTURBATIONS

3.1. Modified Hamiltonian

Now the main problem is solved, let us see how to take into account the influence of the planets. The total Hamiltonian is

$$H = H_{MP} + \delta H$$

where δH represents the indirect planetary perturbations.
More precisely,

$$\delta H = \frac{\partial H_{MP}}{\partial r^*}\,\delta r^* + \frac{\partial H_{MP}}{\partial \gamma^*}\,\delta\gamma^* + \frac{\partial H_{MP}}{\partial \lambda^*}\,\delta\lambda^*$$

where $(\delta r^*, \delta\gamma^*, \delta\lambda^*)$ denote the planetary perturbations on the center of mass of the Moon.

The Hamiltonian H_{MP} of the main problem being already transformed into an averaged Hamiltonian $\overline{H}_{MP}$, we have first to transform δH

with the generator S_{MP} of this transformation in order to have an expression of δH in the new canonical variables. Fortunately, $(\delta r^*, \delta\gamma^*, \delta\lambda^*)$ being independent of our phase space, we can compute this transformation in three distinct parts :

$$[\frac{\partial H_{MP}}{\partial r^*}, S_{MP}] \quad , \quad [\frac{\partial H_{MP}}{\partial \gamma^*}, S_{MP}] \quad \text{and} \quad [\frac{\partial H_{MP}}{\partial \lambda^*}, S_{MP}] \ .$$

3.2. Integration

The last step is then the averaging of K which is equal to

$$\overline{H}_{MP} + [\frac{\partial H_{MP}}{\partial r^*}, S_{MP}]\ \delta r^* + [\frac{\partial H_{MP}}{\partial \gamma^*}, S_{MP}]\ \delta\gamma^* + [\frac{\partial H_{MP}}{\partial \lambda^*}, S_{MP}]\ \delta\lambda^* \ .$$

We first rewrite K under the form

$$K = K_0 + K_1$$

with

$$K_0 = n\, L + n'\, L' + n_g\, G + n_h\, H + \sum_{j=1}^{8} np_j\, LP_j + n_1^*\, P + n_2^*\, Q + n_3^*\, R$$

where $(np_j\ ,\ j=1,\ldots,8)$ denote the mean motions associated to the 8 planets with $(LP_j\ ,\ j=1,\ldots,8)$ the corresponding momenta and where

$$n_i^* = n_i + \text{polynome in the analytical physical parameters} \quad (\ i=1,\ldots,3) \ .$$

In practice, it is impossible to compute the whole K_1 and we have to construct the generator S_{PP} of the planetary perturbations term by term, by multiplication and simultaneous integration.

3.3. Series of Libration

The generator of the planetary perturbations can be applied to the usual series of libration p_1 , p_2 and τ (Eckhardt, 1965), previously transformed by S_{MP} , in order to obtain the planetary perturbations on the libration :

$$(p_1,p_2,\tau) \xrightarrow[S_{MP}]{} (p_1,p_2,\tau)_{MP} \xrightarrow[S_{PP}]{} (p_1,p_2,\tau)_{MP+PP} \ .$$

4. DATA, RESULTS AND COMPARISONS

4.1. Data

In our theory, the motion of the center of mass of the Moon, as well as the planetary perturbations of this motion, are described by the ELP 2000 solution (Chapront and Chapront-Touze, 1983) of the lunar theory.

4.2. Results : Forced Libration

The theory of libration is obtained as Poisson series for p_1 , p_2 and τ , i.e., multiple Fourier series in the mean longitudes of the planets with analytical expressions as coefficients. These series are available on magnetic tape. We present here in Table 1 the terms greater than 0.1 second of arc (i.e., more or less 1 m) when the physical parameters take the value used in Eckhardt's solutions 500 and 501 (Eckhardt, 1981).

The planetary terms are denoted by * and, in p_1 and in τ , we have an independent term (denoted by **) coming from the fourth order perturbation due to the Earth.

In the planetary perturbations, the terms in ℓ' have been transformed in terms in T , the mean longitude of the Earth, in the following way :

$$\ell' = T + \pi - 282°56'14.42753'' .$$

4.3. Results : Free Libration

The planetary effects on the free libration are almost neglectible. The most important of them give the following four terms

$$-0.0043\ 10^5 \sqrt{2P} \cos(8\ T - 16\ Ma + 4\ J + 5\ S \pm p) ,$$

$$-0.0014\ 10^5 \sqrt{2P} \sin(8\ T + 16\ Ma + 4\ J + 5\ S \pm p) .$$

As well as for the main problem (Moons, 1982), their magnitudes depend on unknowns : the amplitudes and phases of the free libration angles p , q and r .

4.4. Comparisons

The general agreement between our theory and the Eckhardt's 500 theory (Eckhardt, 1981) is very good for the main problem (Moons, 1982).

We present here in Table 2 the comparison of the two theories (Eckhardt, 1982) for the terms greater than 0.″01 in the planetary perturbations on τ . The amplitudes are tabulated in Columns 1 , 2 (Eckhardt's solution) and 3 . The corresponding combinations (sine terms) are in column 6 . In columns 1 and 3 , we have two different versions of our theory : the complete solution and the solution without perturbations in parallax. In column 4 , we can find the differences between our theory and the one of Eckhardt but, if we want to get a real comparison with Eckhardt's theory, we have to drop the perturbations in parallax that Eckhardt does not take into account. As we can see in Column 5 , the corresponding portion of our solution is in very good agreement with Eckhardt's solution.

Finally, let us notice that Eckhardt's solution gives two terms with amplitudes -0.274 and 0.228 that do not appear in our solution. The periods of these two terms are very close to the free libration period of p . Such resonant terms are always difficult to compute accurately although an analytical method such as ours is usually more stable in this circumstance.

5. TABLES

5.1. Table 1

Most important terms in the 500 series :

For p_1 :

5562.459	sin F	
124.483	sin(ℓ-F)	
-80.644		
5.752	cos F	
2.910	sin(F-2D)	
-2.678	sin(ℓ+F-2D)	
1.575	sin(ℓ+F)	
1.244	sin(ℓ'+F)	
1.021	sin(ℓ'-F)	
-0.824	cos ℓ	
-0.715	sin ℓ	
0.379	sin(2ℓ-F)	
-0.350	sin(ℓ-F-2D)	
0.307	sin(ℓ'-F+D)	
-0.267		**
0.232	sin(2ℓ-3F)	
0.231	sin(2ℓ-F-2D)	
0.170	sin(18V-16T-ℓ+F)	*
0.170	sin(18V-16T-ℓ-F)	*
0.152	sin(2ℓ-2D)	
0.131	cos(T+D)	*
0.128	sin(ℓ-ℓ'-F)	
-0.123	cos(ℓ-F)	
0.119	sin(F+2D)	
0.118	sin(ℓ'-F+2D)	
0.116	sin(ℓ+ℓ'-F)	
0.116	sin(F-D)	

For p_2 :

5540.331	cos F	
-75.433	cos(ℓ-F)	
-5.775	sin F	
-3.198	cos(F-2D)	
-1.612	cos(ℓ+F-2D)	
1.284	cos(ℓ'+F)	
-1.059	cos(ℓ'-F)	
0.835	sin ℓ	
-0.721	cos ℓ	
0.483	cos(2ℓ-F)	
0.390		
0.299	cos(ℓ+F)	
-0.233	cos(2ℓ-3F)	
-0.218	cos(2ℓ-F-2D)	
0.192	cos(2ℓ+F-2D)	
0.170	cos(18V-16T-ℓ+F)	*
-0.169	cos(18V-16T-ℓ-F)	*
-0.161	cos(ℓ+ℓ'-F)	
-0.137	cos(ℓ-ℓ'-F)	
-0.131	sin(T+D)	

For τ :

214.187		
90.704	sin ℓ'	
17.020	sin(2ℓ-2F)	
-16.795	sin ℓ	
12.748	sin(18V-16T-ℓ)	*
9.941	sin(2ℓ-2D)	
-6.597	cos(ℓ-F)	
6.368	cos(18V-16T-ℓ)	*
4.130	sin(ℓ-2D)	
-3.463	sin(ℓ-D)	
-2.456		**
1.646	sin(2F-2D)	

-1.394	sin(ℓ-F)	
-1.152	sin(ℓ-ℓ'-D)	
1.084	cos F	
0.954	sin(2ℓ-ℓ'-2D)	
-0.867	cos(4T-8Ma+3J)	*
-0.729	cos(2V-3T)	*
-0.673	cos(3T-5Ma)	*
0.504	sin(10V-3T-ℓ)	*
0.494	cos(2ℓ-2F)	
-0.487	sin 2D	
-0.445	sin 2ℓ	
-0.426	sin(ℓ-2F)	
0.408	sin(2ℓ-2ℓ'-2D)	
0.401	sin(3T-5Ma)	*
0.361	sin(V-T)	*
0.281	sin(2T-3J-2ℓ+2D)	*
-0.253	cos(10V-3T-ℓ)	*
0.253	sin(2T-2J-2ℓ+2D)	*
0.251	sin(T-2Ma)	*
0.248	sin(4T-8Ma+3J)	*
-0.248	cos(V-2T)	*
0.231	sin(ℓ+ℓ'-2D)	
-0.225	cos(T-2Ma)	*
0.226	sin 2ℓ'	
-0.187	cos(8V-13T)	*
-0.164	sin(ℓ-ℓ')	
0.161	sin(2ℓ-5S)	*
0.155	sin(2ℓ+ℓ'-2D)	
0.142	sin(ℓ'-2F+2D)	
-0.127	sin(8V-13T)	*
0.107	cos(20V-21T+ℓ-2D)	*
-0.106	sin(T-J)	*
-0.104	cos(3V-4T-ℓ+D)	*
0.102	sin(ℓ+ℓ')	
0.101	cos(26V-29T-ℓ)	*
0.100	sin D	

6.2. Table 2

Comparisons

Moons (1)	Eckhardt	Moons (2)	(1)-E	(2)-E	
14.250	14.403	14.385	-0.153	-0.018	18V-16T-ℓ
-0.902	-0.936	-0.936	0.034		4T-8Ma+3J
-0.784	-0.847	-0.818	0.063	0.029	3T-5Ma
-0.729	-0.750	-0.754	0.021	-0.004	2V-3T
-0.564	-0.579	-0.569	0.015	0.010	10V-3T-ℓ
0.361	0.372	0.373	-0.011	0.001	V-T
-0.336	-0.347	-0.349	0.011	-0.002	T-2Ma
0.285	0.320	0.321	-0.035	0.001	2T-3J-2ℓ+2D

-0.255	-0.269	-0.263	0.014	0.006	V-2T
0.253	0.159	0.159	0.094		2T-2J-2ℓ+2D
-0.226	-0.233	-0.234	0.007	-0.001	8V-13T
-0.165	-0.175	-0.176	0.010	-0.001	2J-5S
-0.107	-0.108	-0.108	0.001		20V-21T+ℓ-2D
0.106	0.112	0.111	-0.006	-0.001	2T-3Ma
-0.106	-0.110	-0.111	0.004	-0.001	T-J
0.105	0.105	0.106		0.001	26V-29T-ℓ
-0.105	-0.109	-0.109	0.004		2T-4Ma
-0.104	-0.109	-0.110	0.005	-0.001	3V-4T-ℓ+D
-0.094	-0.098	-0.098	0.004		3V-5T
-0.092	-0.099	-0.100	0.007	-0.001	J
0.080	0.023	0.023	0.057		3V-3T-2ℓ+2D
0.063	0.064	0.066	-0.001	0.002	2Ma+D+F
0.060	0.066	0.066	-0.006		6V-8T-2ℓ+2D
-0.052	-0.053	-0.054	0.001	-0.001	5V-6T+2D-2F
-0.043	-0.043	-0.043			20V-20T+ℓ-D-F
0.037	0.037	0.037			21V-21T-ℓ
-0.035	-0.035	-0.035			12V-8T+ℓ-2D
0.034	0.034	0.036		0.002	T-2J
0.032	0.033	0.033	-0.001		3V-7T+4Ma
0.032	0.032	0.033		0.001	3V-4T
-0.031	-0.031	-0.032		-0.001	4T-7Ma
0.031	0.031	0.032		0.001	2T-2Ma
0.030	0.029	0.030	0.001	0.001	3Me-T+ℓ-2D
0.028	0.027	0.028	0.001	0.001	15V-13T+ℓ-2D
0.027	0.027	0.027			5V-8T
-0.026	-0.032	-0.026	0.006	0.006	19V-18T+ℓ-D-F
0.026	0.026	0.026			8T-15Ma
0.025	0.026	0.026	-0.001		T+J+D-F
-0.025	-0.025	-0.026		-0.001	2V-2T
-0.024	-0.024	-0.025		-0.001	3T-6Ma
0.023	0.018	0.016	0.005		T-J-ℓ+D
0.023	0.023	0.024		0.001	4V-6T
-0.021	-0.023	-0.023	0.002		6T-8Ma-2ℓ+2D
-0.019	-0.019	-0.020		-0.001	S
0.019	0.019	0.019			T-2J+D-F
0.018	0.019	0.019	-0.001		5V-7T+D-F
0.018	0.018	0.018			3T-7Ma+2ℓ-D-F
-0.017	-0.013	-0.014	-0.004	-0.001	5T-6Ma-2ℓ+2D
0.016	0.016	0.016			18V-18T+ℓ-2D
-0.016	-0.019	-0.016	0.003	0.003	4Me-3T-ℓ
0.015	0.015	0.015			15V-12T-D
-0.014	-0.014	-0.015		-0.001	3T-4Ma-ℓ+D
-0.014	-0.015	-0.014	0.001	0.001	4V-5T+D-F
0.013	0.013	0.013			23V-25T-D
-0.012	-0.013	-0.013	0.001		T-Ma
0.012	0.012	0.012			4V-4T+2D-2F
-0.011	-0.011	-0.011			5T-9Ma
-0.011	-0.011	-0.011			2T-3J+S-2ℓ+2D
-0.011	-0.011	-0.012		-0.001	2J

REFERENCES

Chapront, J., Chapront-Touze, M. : 1983, Astron. Astrophys. **124**, 50.
Eckhardt, D.H. : 1965, The Astronomical Journal **70**, 466.
Eckhardt, D.H. : 1981, The Moon and the Planets **25**, 3.
Eckhardt, D.H. : 1982, High-Precision Earth Rotation and Earth-Moon Dynamics, Ed. O. Calame, D. Reidel, 193.
Henrard, J. : 1974, Celestial Mechanics **10**, 437.
Moons, M. : 1981, 'Libration Physique de la Lune', Thèse de Doctorat, Facultés Universitaires de Namur.
Moons, M. : 1982, The Moon and the Planets **27**, 257.
Tisserand, D. : 1898, Traité de Mécanique Céleste, Gauthier-Villars, Paris.

STABILITY OF L_4 AND L_5 AGAINST RADIATION PRESSURE

F. MIGNARD
C.E.R.G.A.
Avenue Copernic
06130 Grasse
France

ABSTRACT. The restricted three-body problem is generalized with the inclusion of solar radiation pressure. For small particles (typically 1 μm to 1 mm) the familiar equilibrium triangular points L_4 and L_5 no longer exist. However libration orbits are not completely destroyed, although an effect of resonance causes their amplitude to be very large, for a particle initially at rest at either of the triangular point. Finally the results of a study of the linearized equations of motion, supplemented by a numerical integration, rule out the possibility of an accumulation of dust at the Earth-Moon lagrangian triangular points.

INTRODUCTION

It is well known that when two bodies orbit about each other, a massless particle can be at rest in a rotating coordinate frame, at five particular points. Three of them are the so-called lagrangian points L_1, L_2, L_3 aligned with the two primaries, while the last two are the triangular points L_4, L_5. These equilateral points are linearly stable, provided the mass ratio of the primaries is small enough. This follows from a local linearization of the equations of motion in the rotating frame. Stability of the triangular points occurs in spite of the fact that the potential energy has a maximum rather than a minimum at L_4 and L_5 . The stability is actually achieved through the influence of the Coriolis forces (Szebehely, 1967).

The proper modes of oscillation about L_4 and L_5 are ellipses centered at the equilibrium points, described with a characteristic frequency for each mode. In the Earth-Moon case, the short period oscillation is completed in about 1.05 the sideral period of the Moon while the long period oscillation is about three times larger. The shape of the ellipses are respectively $b/a = 0.5$ and $b/a = 0.2$ for the two modes, where b and a are the small and long axis of the ellipses.

These classical studies in the restricted three body problem can be extended in many respects, in particular with the introduction of additional forces to the gravitational attraction of the primaries. The solar induced gravitational force is the first that comes into mind. Both

Celestial Mechanics **34** (1984) 275–287. 0008–8714/84.15

analytical (A.A. Kamel and J.V. Breakwell, 1970) and numerical studies (B.E. Schutz and B.D. Tapley, 1970) were carried out in the recent past. Those investigators found that while the periodicity of the proper oscillation is generally destroyed, a particle can nonetheless be kept in the vicinity of the triangular points for several years, unless the particle experiences a lunar encounter which destroys the libration orbit.

When small particles (say 1 μm to 1 mm in diameter) are considered, the radiation forces become very relevant. D. Schuerman (1980) investigated the problem with two radiating stars as primaries ; he demonstrated that the two equilibrium points L_4 and L_5 are no longer confined at their familiar triangular position. Depending on the particle sensitivity to the radiation pressure, ultimately on its size, Schuerman concluded that L_4-L_5 can be located anywhere on a family of circles centered at the primaries. Whereas the location of the equilibrium points changes drastically with the inclusion of the radiation, it has little effect on the condition on the mass ratio for these points to be stable. However the Poynting-Robertson transverse component causes a small particle displaced from its equilibrium point to oscillate with a growing amplitude, and eventually to be lost. Typically at the Sun-Jupiter triangular points a particle as large as 1 cm would not stay there for more than 10 million years. The lifetime becomes shorter as the size of particle becomes smaller.

In the problem taken up by Schuerman at least one of the primaries is also the source of radiation, which is equivalent to decreasing its mass when computing the gravitational attraction onto the dust grain. Quite a different situation must be considered when dealing with a planet-satellite system, with the Sun as source of radiation. Within this framework we will have time-dependent additional force directed along the Sun-particle line, whose direction will change regularly over the synodic period of the satellite. We will show that this supplementary force makes the stability of the triangular points questionable for sufficiently small particles. For larger particles, radiation forces are taken over by the solar gravitational influence and we recover the study by Schutz and Tapley mentioned above. As a consequence such large excursions about $L_{4,5}$ caused by radiation forces make the existence of concentration of dust in the neighborhood of the triangular points very unlikely. This agrees with the failure experienced by Rosen (1968) and Rosen and Wolf (1969) in their attempt to confirm the alleged discovery of diffuse light at the Earth-Moon lagrangian points reported by Kordylewski (1961). At the moment there is still not a consensus about the existence or non-existence of accumulation of dust in the vicinity of the Earth-Moon triangular points. The calculation to be developed below may show for one time that "the absence of evidence is probably an evidence for an absence".

1. RADIATION FORCES IN THE RESTRICTED THREE BODY PROBLEM

1.1. Qualitative approach

The planet and the satellite have respectively the mass M_P and M_S and move about their common center of mass on a circular orbit. As usual the radius of this orbit is the unit of distance and the unit of time is such that the orbital angular velocity is n=1. Then the sideral period of revolution is precisely 2π. μ denotes the mass ratio $M_s/(M_p + M_s)$

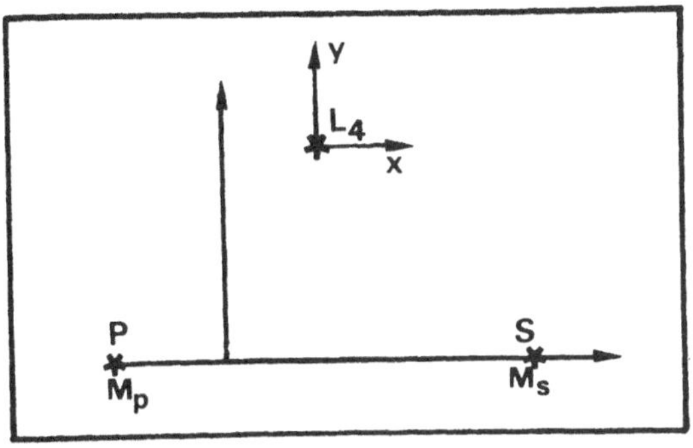

Figure 1. The rotating reference frame with the planet at P and the satellite at S. G is the center of mass of the two bodies.

The linearized equations of motion of a massless particle around one of the triangular point are,

$$\ddot{x} - 2\dot{y} = \frac{3}{4}x + L\frac{3\sqrt{3}}{4}(1 - 2\mu)y,$$
$$\ddot{y} + 2\dot{x} = L\frac{3\sqrt{3}}{4}(1 - 2\mu)x + \frac{9}{4}y, \qquad (1)$$

where $L = +1$ at L_4 and $L = -1$ at L_5. The origin of the coordinate system is at the triangular points and directed as the main system in fig. (1). From Eqs.(1) it is clear that the forces involved are of the order of the departure of a particle from its equilibrium point. Let us introduce now the radiation forces. It is common to characterize the sensitivity of a particle to radiation by the ratio β of the radiation force to that due to the gravitational attraction of the Sun, namely with F_r for the force of radiation and F_g for the attraction,

$$F_r = \beta F_g = \beta G M_o/a_o^2 \qquad (2)$$

where a_o is the planet's semi major-axis, M_o the Sun's mass and G the gravitational constant. The precise value of β depends primarily on the particle size and to a lesser extent on its surface properties for the interaction with the incoming photons. Numerical computation of β for various material of interest in the solar system were carried out by Burns et al. (1979). For the following discussion we will restrict to the approximation of the geometrical optics,

$$\beta = 0.2/r \qquad (r = \text{particle's radius in } \mu m) \tag{3}$$

The latter approximation fails completely for very small particle, $r < 0.1\ \mu m$, when the Rayleigh scattering becomes prominent.

With the units employed in the problem the force of radiation can be written,

$$F_r = \beta \frac{M_o}{M_p} \frac{a^2}{a_o^2} = \bar{\beta} \tag{4}$$

Values of $\bar{\beta}/\beta$ are listed in table (1) for several bodies of the solar system.

	$\bar{\beta}/\beta$	$r_m (\mu m)$
Earth-Moon	2.21	100
Jupiter-Io	$3.0\ 10^{-4}$	35
Jupiter-Europa	$7.8\ 10^{-4}$	20
Jupiter-Ganymede	$2.0\ 10^{-3}$	14
Jupiter-Callisto	$6.1\ 10^{-3}$	10
Saturn-Titan	$2.6\ 10^{-3}$	20

TABLE 1. Multiplying factor to compute the force of radiation (col.2) and particle size beyond which the tidal force exerted by the Sun becomes larger than the radiation force (col.3).

Because of the small size of the planet-satellites system compared to the distances of the planets from the Sun, radiation force expressed by Eq.(4) has a fairly constant magnitude over the region covered by any satellite about its planet. Let us consider a displacement of the order of 0.1 from L_4. The centrifugal force is about 0.1 while the radiation force is $\bar{\beta}$. For a particle with $\beta = 0.1$ this force is of the same order as the centrifugal force and we can expect radiation forces to play an important role in the dynamics of small grains around L_4 and L_5. In any case L_4 and L_5 are no longer equilibrium points. In fact gravitational and centrifugal forces balance exactly at the triangular points, therefore the net force is equal to the radiation force and is different from zero for small particles. Nor can any other point be a new lagrangian point because of the time dependence of the radiation forces.

We have previously mentioned that the radiation force would cease to be the major source of disturbing force when the particle is too large. For a certain particle's size the tidal force exerted by the Sun would become comparable in magnitude to the radiation force. For a particle at the distance r from the center of mass, the tidal force is equal to

$$F_t = \frac{GM_o}{a_o^2} \frac{r}{a_o} , \tag{5}$$

which yields the ratio at L_4 or L_5 ,

$$F_r/F_t = \beta(a_o/a) \tag{6}$$

Then as long as $\beta(a_o/a) > 1$, we can restrict the perturbing action of the Sun to that resulting from the radiation forces and neglect its gravitational pull on the particle. In other words we need not investigate a restricted four-body problem for particle smaller than a limiting radius r_m given by

$$r_m = 0.2 \ (a_o/a) \qquad \text{(in } \mu\text{m)} . \tag{7}$$

Relevant values of r_m are also listed in table (I). With some anticipation, it can be said that the values listed are lower limits, because resonance effects maintain the predominant role of radiation forces even for particles several times larger than r_m.

1.2. Equations of motion in presence of radiation

By setting t = 0 when the Sun is along the Earth-Moon line, in the direction of the Earth we have linearized equations of motion for a grain,

$$\begin{aligned} \ddot{x} - 2\dot{y} &= \frac{3}{4}x + L\,\frac{3\sqrt{3}}{4}\,(1-2\mu)y + \bar{\beta}\cos\omega t, \\ \ddot{y} - 2\dot{x} &= L\,\frac{3\sqrt{3}}{4}\,(1-2\mu)x + \frac{9}{4}y - \bar{\beta}\sin\omega t, \end{aligned} \tag{8}$$

where $\omega = 1 - n_o$ is the mean motion of the Sun in the rotating frame. The apparent motion of the Sun in this frame is retrograde which accounts for the signs of the radiation terms in Eqs.(8), because ω is positive. Eqs. (8) represent a linear system driven by an harmonic external force. Whether the forced solution would have a large amplitude depends on both the magnitude of the excitation, i.e. on $\bar{\beta}$, and on the proximity of the external frequency to one of the proper frequencies of the free oscillations. Since $n_o \gg 1$, ω is close to 1 ; therefore the relevant proper frequency is the largest, corresponding to the short period motion. This largest frequency is the largest root of the quartic equation,

$$\omega^4 - \omega^2 + \frac{27}{4}\,\mu\,(1-\mu) = 0.$$

Eventually, for $\mu \ll 1$, we obtain for the relevant proper frequency

$$\omega^4 \sim 1 - \frac{27}{8}\,\mu\,(1-\mu) \tag{9}$$

For the same set of bodies as before we give in table (II) μ , ω_4 , ω , along with the quantity $\varepsilon = \omega_4 - \omega \sim n_0 - 3.375\ \mu(1-\mu)$ which is an indicator of the closeness of the resonance.

The resonance would be exact for $\varepsilon = 0$ and this case would occur were the Moon or Titan only twice as heavy as they actually are. A quick comparison of $\bar{\beta}/\beta$ and ε in tables (I) and (II) indicates that the relative smallness of $\bar{\beta}$ for all the satellite but the Moon, which let us think of a negligible effect of radiation forces, should be largely compensated by a near exact resonance situation, since the frequency of the driving force is nearly the same as the proper frequency of the short period mode of oscillation.

	$1/\mu$	ω_4	ω	ε
Earth-Moon	82.3	0.954500	0.9252	$2.94\ 10^{-2}$
Jupiter-Io	21300	0.999841	0.999592	$2.45\ 10^{-4}$
Jupiter-Europa	39000	0.9999134	0.999180	$7.33\ 10^{-4}$
Jupiter-Ganymede	12700	0.999734	0.99834	$1.39\ 10^{-3}$
Jupiter-Callisto	17700	0.999809	0.99614	$3.67\ 10^{-3}$
Saturn-Titan	4200	0.999200	0.99852	$6.80\ 10^{-4}$

TABLE II. Mass ratio for relevant planet-satellites systems. ω_4 is the proper frequency of the mode with the shortest period while ω denotes the synodic mean motion of the Sun. $\varepsilon = \omega_4 - \omega$ is an indicator for the closeness of resonance.

2. SOLUTION OF THE LINEARIZED EQUATIONS OF MOTION

2.1. Exact solution

The complete solution of the differential system of equations (8) is a superposition of free oscillations plus the forced solution, whose amplitude in the near resonance situation governs more or less the global behavior of the orbit of the particle. The forced solution for an harmonic driving force is of the form

$$x = A\ e^{i\omega t}$$
$$y = B\ e^{i\omega t} \qquad (10)$$

By inserting Eqs. (10) in the differential equations we obtain the linear system in A and B,

$$(\omega^2 + 3/4)\ A + (2i\omega + 3/4\gamma)\ B = -\bar{\beta}$$

$$(3/4\gamma - 2i\omega)\ A + (\omega^2 + 9/4)\ B = -i\bar{\beta}$$

with $\gamma = \sqrt{3}(1-2\mu)L$.

Then the complex forced amplitudes A and B are

$$A = -\ \bar{\beta}\ \frac{(\omega^2 + 2\omega + 9/4) - 3i\gamma/4}{\omega^4 - \omega^2 + 27\mu(1-\mu)\ /4} \quad , \qquad (11)$$

$$B = \bar{\beta}\,\frac{3/4\gamma - i(\omega^2 + 2\omega + 3/4)}{\omega^4 - \omega^2 + 27\mu\,(1-\mu)/4}\;. \tag{12}$$

Let's use expressions in real number by putting

$$\begin{aligned} x &= |A|\,\cos(\omega t + \phi) \\ y &= |B|\,\cos(\omega t + \psi) \end{aligned} \tag{13}$$

where $|A|$, $|B|$, ϕ , and ψ are respectively the modulus and the phase of the complex A and B . Since we are only dealing with small mass-ratio and near resonant situations these phases and modulus can be expressed through an expansion to the first order of ε and μ.

We obtain after elementary algebraic computations,

$$\begin{aligned} |A| &= \frac{3}{2}\sqrt{13}\,\left(1 - \frac{28}{39}\,\varepsilon + \frac{1491}{104}\,\mu\right)\frac{\bar{\beta}}{2\varepsilon}\;, \\ |B| &= \frac{3}{2}\sqrt{7}\,\left(1 - \frac{20}{21}\,\varepsilon + \frac{753}{56}\,\mu\right)\frac{\bar{\beta}}{2\varepsilon}\;, \end{aligned} \tag{14}$$

$$\begin{aligned} \sin\phi &= -\frac{\sqrt{3}}{2\sqrt{13}}\,\left(1 + \frac{28}{39}\,\varepsilon + \frac{7}{13}\,\mu\right) L\;, \\ \cos\phi &= \frac{7}{2\sqrt{13}}\,\left(1 - \frac{4}{91}\,\varepsilon - \frac{3}{91}\,\mu\right)\;, \end{aligned} \tag{15}$$

$$\begin{aligned} \sin\psi &= \frac{5}{2\sqrt{7}}\,\left(1 - \frac{4}{35}\,\varepsilon - \frac{6}{35}\,\mu\right)\;, \\ \cos\psi &= -\frac{\sqrt{3}}{2\sqrt{7}}\,\left(1 + \frac{20}{21}\,\varepsilon + \frac{10}{7}\,\mu\right) L\;. \end{aligned} \tag{16}$$

The fact that we find $|A|$ and $|B|$ in $1/\varepsilon$ comes from the presence of the quartic polymonial in ε in the denominator of A and B in Eqs. (11-12) and the tiny difference existing between its root ω_4 and the forced frequency ω . The phases ϕ and ψ do not depend very much on the small parameters and we can generally use

$$\phi = 346°$$

$$\psi = 109°$$

Table (III) contains the values of the phases and modulus of A and B. The figures result from an exact numerical computation and may slightly differ from those that would otherwise be obtained from the first order expansion, especially for the Moon. The values for $|A|$ and $|B|$ correspond to a particle of 1 μm in radius. For any other size it suffices to divide the numbers of the table by the particle's radius in μm.

	A (r=1μm)	φ	B (r=1μm)	ψ
Earth-Moon	53.52	345°66	38.51	110°14
Jupiter-Io	0.65	346.10	0.48	109.11
Jupiter-Europa	0.58	346.09	0.42	109.12
Jupiter-Ganymede	0.77	346.09	0.57	109.14
Jupiter-Callisto	0.90	346.06	0.67	109.18
Saturn-Titan	2.08	346.09	1.52	109.13

TABLE III. Modulus and phases for the forced mode of oscillation. A, B and ϕ, ψ respectively refer to an oscillation along the X and Y axis.

Eqs.(13) indicate that the trajectory of a particle evolving according to the forced solution is an ellipse. The center of the ellipse is at L_4 or L_5 and its size and orientation are uniquely known from A and B. For small ε and μ we have the following approximate solution

$$b/a = \frac{1}{2}(1 - \frac{2}{3}\varepsilon - \frac{9}{8}\mu) , \quad (17)$$

$$\chi = (-\frac{\pi}{6} + \frac{\sqrt{3}}{4}\mu) L , \quad (18)$$

where the meaning of χ is made clear in Fig.(2).

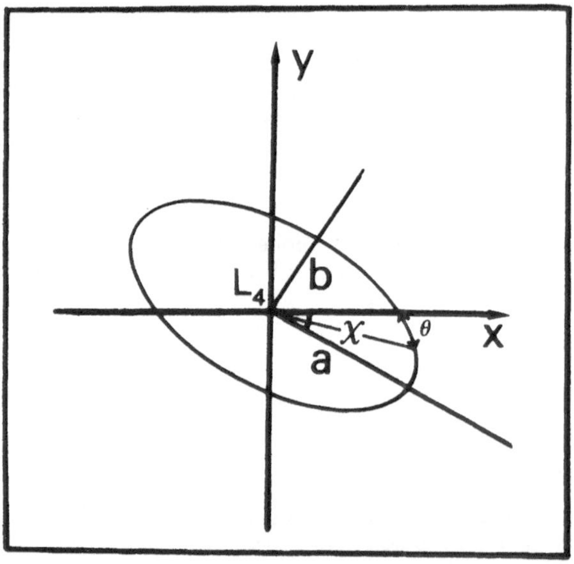

Figure 2. The elliptical trajectory for the forced oscillation about L_4. It is similar in shape and size to the free mode of oscillation with the short period.

At the exact resonance, the long axis is twice the short axis and is directed on the line (-30°, 150°). We can also locate the particle at t = 0 when the Sun is along the X-axis in the negative region, by the angle θ made by the particle's vector radius with the X-axis (fig.2).

We have

$$\tan\theta = |B| \cos\psi \; / \; |A| \cos\psi$$

or
$$\theta \sim -\frac{\sqrt{3}}{7} (1 + \frac{16}{21} \varepsilon + \frac{4}{7} \mu) \sim -14° \; .$$

The elliptical path is quite similar to the one associated with the short period proper mode of oscillation, apart from the amplitude determined by the magnitude of the driving force instead of the initial conditions. For $\varepsilon = 0$ equation (17) reduces to that which is obtained when a similar expansion is performed for the short period mode. As for the orientation expressed by Eq.(18) it is the same as the short period ellipse about L_4 or L_5 . This similarity implies that a particle initially released at either of the triangular point will subsequently move on an ellipse of shape and orientation prescribed by Eqs.(17-18), but contracting and expanding on a timescale comparable to $1/\varepsilon$, the duration of a beat between the forced and free oscillation. This point will be fully established below with the help of a numerical integration.

The modulus of A and B tells us information about the distance a particle of specified size may travel from the equilibrium point and eventually should provide a convenient criterion to assess the long-time stability around the triangular points. Instead of A and B it is more exact to refer to the true dimension of the ellipse given by its semi-axes a and b. With the same limitations of validity as before, we have for these axes

$$a = 3 \frac{\bar{\beta}}{\varepsilon} (1 - \frac{2}{3} \varepsilon + \frac{114}{8} \mu) \; , \qquad (19)$$

$$b = \frac{3}{2} \frac{\bar{\beta}}{\varepsilon} (1 - \frac{4}{3} \varepsilon + \frac{105}{8} \mu) \; , \qquad (20)$$

where $\bar{\beta}$ is numerically obtained from table (I)

$$\bar{\beta} = (0.2/r) \; (\bar{\beta}/\beta) \qquad \text{with } r \text{ in } \mu m.$$

Values of a,b and χ are collected in table (IV) and result from an exact numerical evaluation, instead of Eqs.(19-20).

Are a and b really representative of the removal from L_4-L_5 which a grain would experience as a consequence of radiation forces ? Whenever a grain is released from the triangular point with a zero-velocity in the rotating frame, these initial conditions completely specify its subsequent motion, for a given model of forces. The actual motion is a combination of the forced oscillation with the two free modes. In principle the four unknowns involved in the general solution (the two complex amplitudes of the free modes) could be obtained by solving a linear system of order four. Analytically it results in clumsy and unuseful formulas and I would not go further in this direction. The only point of interest lies in the fact that the amplitude of the free oscillations is of the same order of magnitude as that of the forced oscillation. Hence the size of the ellipse of the forced oscillation gives us an acceptable order of magnitude of the distance a grain wanders around the classical equilibrium points.

	a (r=1μm)	b (r=1μm)	χ
Earth-Moon	59.41	28.60	-29°.69
Jupiter-Io	0.722	0.361	-30.00
Jupiter-Europa	0.639	0.319	-30.00
Jupiter-Ganymede	0.855	0.427	-30.00
Jupiter-Callisto	1.007	0.502	-30.00
Saturn-Titan	2.307	1.152	-30.00

TABLE IV. Semi-major and minor axis and orientation of the ellipse generated by the linear forced oscillation.

Let us put the limit of validity of the linearized theory to excursions around L_4- L_5 no larger than 0.1. Then any particle smaller than a certain radius r_{min} such that

$$a/r_{min} > 0.1 \qquad (r_{min} \text{ in } \mu m)$$

will be considered unstable in the vicinity of L_4- L_5. In the previous equation, a is the semi-major axis of the ellipse for a particle of 1 μm in radius, as given in table(IV). The first column of table (IV) contains a for the same set of planets and satellites we have considered so far. We can see that particles smaller than 600 μm ∿ 1 mm cannot accumulate at the Earth-Moon lagrangian points because of their elimination by radiation forces. Conditions are less stringent for the other satellites since the minimal radius for stability drops to 10 μm. It must be pointed out that these limits are very close to the point when solar induced gravitational perturbations take over the radiation forces as the most prominent additional force to the attraction of the primaries (see r_m in table I)

In any case, for particle smaller than r_{min}, the solar influenced motion of a grain tends to be very large and the linearized solution is only an indication of the grain tendency to move away far from the lagrangian triangular points. Within the validity of a linear theory, it accounts well for the absence of faint glow at the Earth-Moon lagrangian point ; nonetheless it is desirable to supplement our computation by either a non-linear theory or a numerical integration. The latter being the easier, it was selected.

3. NUMERICAL INTEGRATION OF THE EXACT EQUATIONS OF MOTIONS

3.1. Outline

In this section the exact equations of motion of the restricted circular three-body problem are retained. Obviously the radiation forces are also included. The numerical integration used a fourth order Runge-Kutta integrator implemented on a CDC 7600 using 14 digits single precision arithmetic, an overabundant accuracy for the present work.

At first, initial conditions were so determined as to generate the pure free modes of oscillation in absence of radiation force to check the routine. Then cm-sized particles were chosen on which radiation forces fall in the linear regime to generate a pure forced oscillation undisturbed by interference with either of the free modes. The ellipse predicted in the linear solution was recovered exactly in size, shape and orientation. The program has then been considered proper to show the effects of non-linearity on smaller particles.

3.2. Results

The different results obtained for the Earth-Moon system are shown in Fig.(3) for the motion about L_4 and in Fig.(4) for L_5.

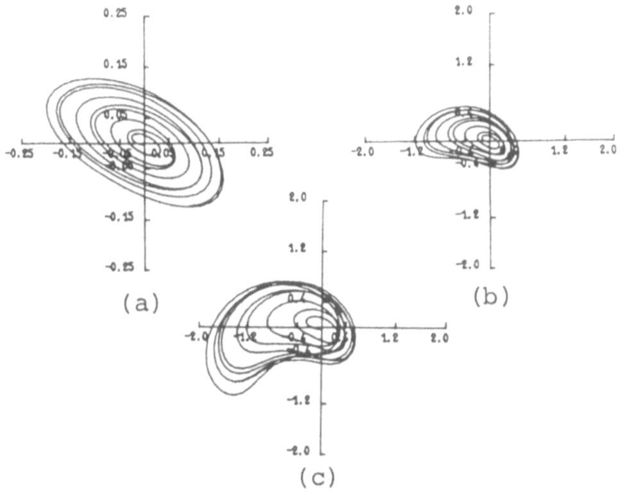

Figure 3. Libration about L_4 for particle located at L_4 at t=0. Radii are 500 μm (a), 100 μm (b) and 50 μm (c). The scale is such that the Earth-Moon distance is precisely equal to 1. The Earth's coordinates are y= -0.866, x= -0.5, while the Moon's are y= -0.866, x= +0.5.

Consequences of the non-linearity appear clearly as the particle's size decreases from 500 μm to 50 μm. The linear libration pattern changes gradually from a more or less elliptical shape to a kidney-shaped orbit for small particles. Likewise there is a manifest evolution in the amplitude of the motion reflected by the different scales used in the figures. The plots of the integration have been intentionally limited to a fraction of a complete beat to facilitate the readability of the diagrams. Numerical integration over 3000 days shows the successive contractions and expansions of an instantaneous ellipse with a period of about 900 days or 33 times the sideral period of the Moon. This last figure is in agreement with our previous analysis which predicted a duration of the order of $1/\varepsilon$ for the period of the beat, that is to say 34 times the Moon's sideral period. The duration of a beat decreases as the particle gets smaller and non-linear effects pervade the solution ; for a radius of 100 μm the beat period drops to only 630 days. As a result of the interference between free and forced oscillations, the particle goes farther out of L_4 and L_5 than expected if the sole linear forced modes were present. This increase is about by a factor 2 in the linear regime ($r > 500$ μm) and larger for smaller particles. To illustrate this point, I found that a particle of 500 μm in radius actually ventures to distance beyond 0.26 from the triangular point ; while according table (IV), it should not be more than 59.4/500 = 0.12. For smaller grains, excursions around the triangular points may yield a fatal close encounter either with the Earth or the Moon. Such an escape has occured for a particle with $r = 50$ μm after only 200 days. With radius of 100 μm distances from L_4 larger than the Earth-Moon separation occur frequently. We cannot consider in this case that the triangular points are stable against radiation pressure.

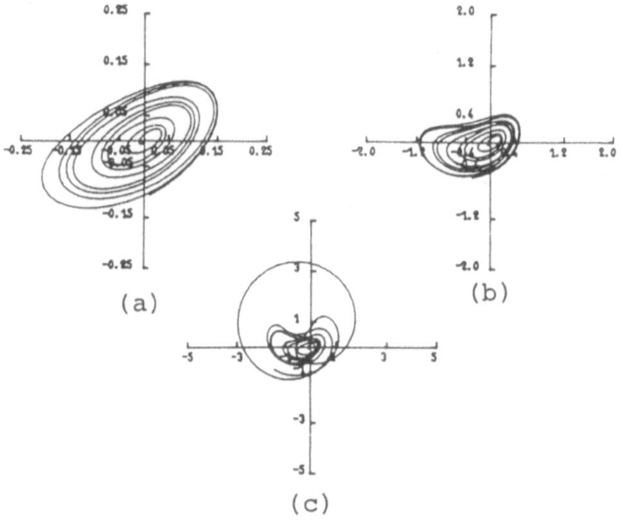

Figure 4. Libration about L_5 for particle located at L_5 at t=0. Radii are 500 μm (a), 100 μm (b) and 50 μm (c). The scale is such that the Earth-Moon distance is precisely equal to 1. The Earth's coordinates are x= -0.5, y= +0.866 and the Moon's x= 0.5, y= 0.866.

In addition, as can be expected from a close encounter, the final outcome is very sensitive to the initial conditions. A non-zero, but still small, velocity at the instant of release at L_4 or L_5 may put the particle into an escaping or collision orbit.

A comparison between Fig.(3c) and (4c) shows the effect of the asymmetry of the Coriolis force between L_4 and L_5.

CONCLUSION

In conclusion, I would stress the limitations of the previous investigation. I have shown that no micron-sized particle can stay in the neighborhood of the triangular points of the Earth-Moon system nor at the similar points of other planet-satellites systems. Nor can these grains be trapped in a stable libration. However this does not prevent discrete particles or rocks to be located there. The detailed numerical integration carried out by Schutz and Tapley, in which they found an ejection of the particle after 579 days, is a unique experiment conducted over a limited duration.

Large stable libration orbits have been found stable by Schechter (1968) and Kamel (1969) when solar attraction is taken into account. A search for discrete objects associated with these librations orbits was undertaken by Valdes and Freitas (1983) up to 17-19th magnitude and no objects were found.

Then both theoretical calculations and observations give great support to the fact that neither dust grains nor larger chunks can stay for long at the Earth-Moon triangular points surroundings.

REFERENCES

Freitas, R.A., Valdes, F., 1980, 'A search for natural or artificial objects located at the Earth-Moon libration points', Icarus, 42 , 442-447.

Kamel, A.A., Breakwell, J.V., 1970, 'Stability of motion near Sun-perturbed Earth-Moon triangular points', in Periodic Orbits, Stability and Resonances, G.E.O. Giacaglia Ed., Reidel.

Kordylewski, K., 1961, 'Photographic investigation of the libration point L_5 in the Earth-Moon system, Acta Astronomica, 11, 165-169.

Roosen, R.G., 1968, 'A photographic investigation of the gegenschein and the Earth-Moon libration point L_5', Icarus, 9, 429-439.

Roosen, R.G., Wolff, C.L., 1969, 'Are the libration clouds real ?', Nature, 224 , 571.

Schutz, B.E., Tapley, B.D., 1970, 'Numerical studies of solar influenced particle motion near the triangular Earth-Moon libration points', in Periodic Orbits, Stability and Resonances, G.E.O. Giacaglia Ed., Reidel.

Szebehely, V., 1967, 'Theory of orbits', chap. 4 & 5, Academic Press.

Valdes, F., Freitas, R.A., 1983, 'A search for objects near the Earth-Moon lagrangian points', Icarus, 53, 453-457.

THE FORMATION OF THE PLANETARY SYSTEM

W. M. Tscharnuter
Institut für Astronomie
Türkenschanzstraße 17
A-1180 Wien, Austria

1. INTRODUCTION

One of the most challenging problems of natural science is the question of how our solar system was formed some 4,6 billion years ago. The main difficulties encountered in trying to find any definite answer to this problem are twofold. In the first place we cannot design direct experiments which could give us clues to the relative importance of the various physical processes. In fact, we are restricted to try to gather all available pieces of information we can get from the present day solar system about its origin. Secondly, we could look for an entire sample of different objects in the universe with similar properties, which again could provide us with important insights into the physical processes relevant for the formation of planetary systems.

Such an astrophysical method has been successfully applied to stellar structure and evolution problems. However, the sensitivity of our instruments is still too low in order to detect other planetary systems or their precursors. Thus, in order to arrive at a coherent picture of the origin of our solar system the setup of a sufficiently complex and elaborate theory is necessary which does not only account for the astrophysical and cosmochemical data available but also gives constraints to future observations.

As a matter of fact, the basic ideas concerning solar system formation were developed by Kant (1755) and Laplace (1796) whose starting point was the socalled nebular hypothesis. Due to its self-gravitation a rather structureless dispersed "cloud" consisting of gas and dust would contract and, in doing so, become denser, finally forming a relatively compact central sun that is surrounded by material of high angular momentum. In such a "solar nebula" the planets are able to form, since the dust grains will coagulate, thus leading to macroscopic solid bodies within the nebula. Particularly in this respect Kant's theory is surprisingly "modern", however, there have been justly raised severe objections to the reality of this otherwise quite attractive conception of a consistent cosmogonic theory for almost two centuries.

The basic difficulty of the nebular hypothesis is the well-known

Celestial Mechanics **34** (1984) 289–296. 0008–8714/84.15

angular momentum problem. How is it possible that there is such an non-equilibrated distribution of mass and angular momentum in our solar system? This crucial question can be answered, only if there are mechanisms that are able to separate mass and angular momentum during the contraction of the cloud; otherwise it would be impossible to arrive at a system, where almost all the mass is contained in the central sun and most of the angular momentum is contained in the planets.

But the great advantage of the nebular hypothesis is that many regularities, e.g. prograde motions of all planets and asteroids in almost coplanar orbits, can be explained by reducing the number of ad hoc assumptions to a minimum. Planetary formation is considered as a part of star formation, which then should be a rather common occurrence, since stars are observed to form continuously in the galaxy. Observations in the radio and infrared region strongly support the nebular hypothesis to be applicable to stellar and planetary system formation, provided the angular momentum problem can be solved in some way or another. We may list three possibilities:

i) magnetic fields via Alfvén waves which can transport angular momentum from the contracting cloud fragment into the external medium (see e.g. Mestel, 1977; Mouschovias, 1978; Dorfi, 1982),
ii) turbulent friction (see e.g. Morfill, 1984; Morfill et al. 1984),
iii) gravitational torques exerted by high amplitude spiral or bar-like density waves in the nebula (Mestel, 1977; Larson, 1984; Boss, 1984).

Any transport process listed above can and presumably will become important as the evolution of the protostellar cloud proceeds.

2. INITIAL CONDITIONS

A necessary condition for the onset of gravitational collapse is the socalled Jeans-criterion. It determines the minimum mass M_J of an interstellar cloud with temperature T and density ρ given.

$$\frac{M_J}{1M_\odot} \simeq 1.25 \times \left(\frac{T}{10K}\right)^{3/2} \left(\frac{\rho}{10^{-19}\ g/cm^3}\right)^{-1/2} \tag{1}$$

Since dense and cool molecular clouds in which stars are observed to form exhibit fairly uniform temperatures, typically around 10K, we find a minimum density ρ_J for a $1M_\odot$-fragment to become Jeans-unstable:

$$\rho_J = 1.6 \times 10^{-19}\ g/cm^3 \tag{2}$$

Typical numbers for cloud densities are 10^{-20} - 10^{-17} g/cm^3 or even higher. Thus, for a given temperature of 10K the initial value for the density ought to be in between 10^{-19} and 10^{-18} g/cm^3.

Too high a density for a given total mass of the fragment, i.e.

smaller Jeans-masses, could promote further fragmentation, which would lead to a binary or multiple stellar system rather than to a single star with a planetary system. The Jeans-radius R_J corresponding to $M_J = 1\ M_\odot$ and ρ_J in (2) is about 1.4×10^{17} cm.

We now roughly estimate the amount of angular momentum that has to be adopted in order to arrive at a centrifugo-thermal equilibrium configuration extending to about the size of our planetary system (several 10^{14} cm) after the collapse phase. The ratio χ of the centrifugal force and gravitation at the outer boundary in the equatorial plane of a homogeneous, rotating cloud is

$$\chi = \frac{R^3\Omega^2}{GM} \tag{3}$$

where R is the distance of the outer boundary from the center, Ω the angular velocity, M the total mass and G the gravitational constant. The specific angular momentum j of a mass element at the boundary then is $R^2\Omega$. Assuming, for the sake of simplicity, j to be a conserved quantity during collapse, we have

$$\chi\ R = \frac{j^2}{GM} = \text{const.} \tag{4}$$

The product χ R is thus an invariant. $\chi = 1$ denotes the point where centrifugal forces catch up with gravity, the corresponding minimum radius $R_{min} = \chi_{initial}$ R. If we now choose the semi-major axis of Saturn's orbit as respresentative for R_{min} we get, putting R equal to the Jeans-radius $R_J = 1.4 \times 10^{17}$ cm, $\chi_{initial} = 10^{-3}$. From eq.(3) the initial angular velocity $\Omega_{initial} = 7 \times 10^{-15}\ s^{-1}$ is easily obtained. The total angular momentum $J = 0.4 \times MR^2\Omega$ of a homogeneous, uniformly rotating cloud is found to be 1.1×10^{53} g cm^2 s^{-1}, the rotational energy $E_{rot} = 0.5\ J\Omega$ and the gravitatonal energy $E_{grav} = 0.6\ G\frac{M^2}{R}$ are 3.8×10^{38} erg and $- 1.14 \times 10^{42}$ erg, respectivELY. The initial ratio β of rotational energy and the absolute value of the gravitational energy is given by

$$\beta_{initial} = \frac{E_{rot}}{|E_{grav}|} = \frac{1}{3}\frac{R^3\Omega^2}{GM} = \frac{1}{3}\ \chi_{initial} \tag{5}$$

i.e. $\beta_{initial} = 3.3 \times 10^{-4}$, which is indeed a very small number. From these rough estimates we conclude that a very slowly rotating, Jeans-unstable fragment of an interstellar cloud containing only a few solar masses (1-3 $M_\odot$, say) would be a reasonable candidate as the precursor of the solar nebula. This is also in perfect accordance with the results of numerical experiments (see e.g. Bodenheimer, 1980 and references cited therein)which show an increasing tendency of the cloud fragment to break up more easily with increasing initial β. Only a low angular momentum fragment has the chance to survive as an identity and to eventually form a central condensation that might then be regarded as the solar embryo imbedded in the primitive solar nebula.

3. NUMERICAL COLLAPSE MODELS

Early studies of the axisymmetric collapse of interstellar clouds (Tscharnuter, 1978) led to the conclusion that without proper redistribution of angular momentum even for a small initial $\beta = E_{rot}/|E_{grav}|$ the formation of a single central object is inhibited. However, recent 3-dimensional calculations including radiative transfer in the Eddington approximation (Boss, 1984) indicate that very slowly rotating clouds with $\beta \simeq 10^{-3}$ can avoid fragmentation and do indeed form single quasi-static cores with densities of about 10^{-10} g/cm^3. These cores are surrounded by a bar-like structure over the scale of several 10^{14} cm. It couples with differential rotation, resulting in gravitational torques that produce rapid outward transfer of angular momentum. Thus, by building up non-axisymmetric ordered structures, the center of the primitive solar nebula could avoid further fission or fragmentation and reach pre-main-sequence densities.

As mentioned in the introductory section turbulence, i.e. random eddy motions and the associated frictional process, will also give rise to a substantial outward angular momentum transport. Although the driving mechanism for turbulence is not known, the advantage of including this process by means of numerical experiments is that they can be performed with 2-dimensional (axisymmetric) models. This is particularly favourable with regard to the notorious accuracy problems of the hydrodynamical codes that work also in case of collapse flows (Tscharnuter and Winkler, 1979; Winkler et al., 1984).

Of course, both the "stochastic" mechanism (turbulence) and the aforementioned "deterministic" mechanism (gravitational torques) may be responsible for an effective redistribution of angular momentum; one could even speculate that there might exist phases in the evolution during which the importance of the two processes alternate. In other words, turbulence would then exhibit a transmittent behavior. If this were true, such a scenario would have most important consequences on the sedimentation of the solid particles toward the equatorial plane in order to form planetesimals (Safronov, 1969; Goldreich and Ward, 1973) and, finally, planets.

Magnetic fields are certainly important for stabilizing molecular clouds with 10^4 - 10^5 $M_\odot$ and for solving the angular momentum problem to a high degree already in the dispersed cloud stage (Dorfi, 1982). They are also responsible for braking the sun's rotation by coupling the ions of the solar wind to the field lines. The magnetic field is presumably maintained by turbulent dynamo effects.

However, during the overall collapse the degree of ionization becomes so low that the magnetic field decouples from the gas and should therefore hardly play any significant dynamical role. This is a period of evolution which is critical because the optically thick core takes shape that is always in danger of fission: during the collapse the core has spun up approaching the critical $\beta \simeq 0.27$ where a dynamical instability sets in (Ostriker and Bodenheimer, 1973). Only turbulence and/or gravitational torques, as discussed above, will prevent the core from breaking apart!

In the sequel we will briefly discuss the main results of preliminary collapse calculations including turbulent friction. For more details we refer to a forthcoming paper by Morfill et al. (1984). The starting parameters are:

total mass [$M_\odot$]	3
specific angular momentum J/M [cm^2/s]	1.1×10^{20}
density ρ[g/cm^3]	10^{-20}
radius R [cm]	5.2×10^{17}
temperature T [K]	8.3
angular velocity Ω[sec^{-1}]	10^{-15}
$E_{rot}/\lvert E_{grav}\rvert$ β	1.2×10^{-4}
free-falltime t_{ff} [yr]	7×10^{15}
mean molecular weight μ	2.73

This set of initial values has been adopted from earlier calculations (Tscharnuter, 1978) and fits well into the general discussion given in section 2, but it should be emphasized that we have not tried to choose the "best" parameters model of the solar nebula proper.

Turbulent friction is expressed by assuming a simple form of the kinematic viscosity

$$\nu = \frac{1}{3} \alpha \quad c_s \quad l \tag{6}$$

where c_s is the speed of sound, l the length scale of the longest turbulent eddies, e.g. the half-thickness of a turbulent disk, and α was assumed to be equal to 0.3. The set of equations that have to be solved consists of the usual hydrodynamical equations (mass, momentum, energy) augmented by the moment equations (radiation energy, radiative flux) for radiative energy transport. The Eddington approximation has been used, which, in the strict sense, applies only to optically thick layers where the radiation field is fairly isotropic (Tscharnuter and Winkler, 1979).

Since the intial β is extremly low, the collapse flow develops almost spherically symmetric. But due to the nonhomology of the flow centrifugal forces, which are otherwise negligibly small, become dominant in the central regions first. The amount of viscosity according to (6) is sufficient to suppress the excitation of the toroidal density wave in the core (Tscharnuter, 1978); the excess angular momentum is quickly removed from the central parts of the cloud and an almost spherical innermost pressure-supported core region forms. Similar to the spherically symmetric models, due to dissociation of H_2 molecules above 2000 K, a further collapse sets in which leads to almost stellar densities of about 10^{-4} g/cm^3 in the center. The outer parts of the core extend to several 10^{14} cm in the equatorial plane and to about 10^{14} cm in the vertical z-direction. Still further outside the envelope is practically in free-fall. After a period of approximately one free-fall time ($\simeq 7 \times 10^5$ yr) the partly collapsed cloud fragment can already be regarded as the precursor of a primitive preplanetary nebula that surrounds a tiny "embryonic" protosun with a few 10^{-2} $M_\odot$ and is continuously fed by infalling material from the envelope.

The calculations give some indication that, contrary to what we know from strictly spherical accretion (zero angular momentum), the further evolution of the nebula undergoes non-stationary phases at a time scale of several thousand years. Unfortunately, it is not quite clear, whether this unexpected behavior is realistic or an artifact of the numerical models. Among other explanations (Morfill et al. 1984) it can be argued that we grossly underestimate the amount of energy which is transported by convection in optically thick layers where the viscous energy generation is high. There the thermal pressure is steadily increased and, finally, an outward directed bulk motion is driven, thus interrupting the "quiet" quasi-stationary accretion.

Because of limitations on computer time it is not possible to follow the high amplitude oscillations which are the result. However, if we are not so much interested in the detailed structure of the central stellar core, we can overcome non-stationary phases (for details see Morfill et al., 1984). After a period of very effective redistribution of angular momentum within a region of about 10^{14} cm, lasting about 3×10^4 yr, a quasi-stationary, disk-like nebula with Keplerian rotation around a rapidly spinning 0.5 $M_\odot$ "protosun" takes shape.

Although the approximations introduced in order to make the problem at least partially tractable are very crude, the numerical models, nevertheless resemble a semi-quantitative evolutionary picture of what we may call a "preplanetary" nebula. One of the most important results relates to the existence of a viscous accretion disk period for protostellar evolution. In the absence of extended numerical parameter studies because of computer time limitations it is therefore the next non-trivial, well-defined step to construct simpler accretion disk models (see e.g. Verbunt, 1982). In such models it is then possible to incorporate the physical and chemical processes and to adjust the parameters that are given by the constraints derived from our own solar system.

4. DUST AND VAPORS IN TURBULENT DISKS

We have seen that turbulent friction removes the excess angular momentum from the central parts of the protostellar cloud so that a central star can form; material of higher angular momentum is left in a turbulent protostellar accretion disk. From these basic results one can draw important consequences on the chemistry of the dust grains. Since there is a large scale radial temperature gradient of the disk, small dust particles that are tied to the random turbulent motions will diffuse to hot interior regions where they sublimate. The reverse process, namely diffusion of their vapors in the outer disk regions, will give rise to a recondensation on cooler grains. The result is a large scale chemical fractionation. In this way the turbulent transport theory may serve as a strong link between astrophysical models of stellar fromation and cosmochemical evidences in the present solar system.

Turbulence, which was originally considered as a mechanism to solve the angular momentum problem, meanwhile has turned out to have striking consequences on the growth and chemical composition of solids (coagulation, cooperation, recondensation) in turbulent disks. Analytical solutions of the set of the coupled, second-order transport equations so far obtained are in remarkably good agreement with cosmochemical measurements: large scale variations in abundance ratios, the importance of condensation, refractory inclusions and their incorporation into meteorites, rims surrounding inclusions and chondrules (Morfill, 1983,1984; Morfill and Völk, 1984; Morfill et al., 1984).

The collapse models predict - at least as long as one starts out with masses of the order of one Jeans-mass - relatively low temperatures for the accumulation of the solids we observe in the form of meteorites today. The temperatures of the various particles carrying exactly the amount of specific angular momentum that corresponds to the different planets in our solar system hardly rise above 10^3 K and they remain fairly constant over long periods of time. Nevertheless, due to the extensive thermal reprocessing of material, completely "presolar" material is not expected to have survived in the inner solar system, even if the temperatures never exceeded 400-500 K in the vicinity of the Earth (Morfill and Völk, 1984).

Another conclusion suggested by the analytical solution of the transport equations concerns the location of the regions where planet formation is promoted. The processes of sublimation, turbulent transport and recondensation lead to a systematic and significant enhancement of the dust particle sizes just outside the sublimation boundary of major chemical constituents. This result is of particular importance, since larger particles can more easily decouple from turbulence and move toward the equatorial plane. There a thin dust sub-disk is built up out of which, due to local gravitational instabilities, km-sized planetesimals are produced (Safronov, 1969; Goldreich and Ward, 1973) and, through further gravitational accumulation, large planetary bodies as well as the cares of the giant gas planets are finally formed (Safronov, 1969).

The last point refers to a basic difficulty of naively interpreted turbulent accretion disk models. It turns out that particle drift and diffusion times in the inner disk are much shorter than the time scale of mass accretion from the protostellar envelope (≃ free-falltime). This means that for steady and uniform accretion from the surrounding envelope and strictly stationary turbulent protostellar disks any recording of chemical fractionation due to turbulent mass transport is lost, unless turbulence is turned off or is intermittent on time scales short compared to diffusion and drift times. But if so, the information about the physico-chemical state of the turbulent disk is conserved in the larger bodies of our solar system whose properties can be analyzed by means of well-established cosmochemical methods.

REFERENCES

Bodenheimer,P., 1980, IAU Symp. 93, "Fundamental Problems in the Theory of Stellar Evolution", eds. D.Sugimoto, D.Q.Lamb, and D.N.Schramm, Reidel Dordrecht, p.5

Boss,A.P., 1984, Icarus, in press

Dorfi,E.A., 1982, Astron.Astrophys. 114, 151

Goldreich,P., and Ward,W.R., 1973, Astrophys.J. 183, 1051
Kant,I., 1755, "Allgemeine Naturgeschichte und Theorie des Himmels"
Laplace,P.S.de, 1796, "Exposition du Système du Monde"
Larson,R.B., 1984, Mon.Not.R.astr.Soc. 206, 197
Mestel,L., 1977, IAU Symp.75, "Star Formation", eds. T.de Jong and A.Maeder, Reidel, Dordrecht, p.213
Morfill,G.E., 1983, Icarus 53, 41
Morfill,G.E., 1984, Proceedings of Summer School, Les Houches, "Birth and Infancy of Stars", eds. A.Omont and R.Lucas
Morfill,G.E., and Völk,M.J., 1984, preprint
Morfill,G.E., Tscharnuter,W.M., and Völk,M.J., 1984, in: "Protostars and Planets", ed. T.Gehrels, University of Arizona Press (to appear)
Mouschovias,T.Ch., 1978, "Protostars and Planets", ed. T.Gehrels, University of Arizona Press, p.209
Ostriker,J.P., and Bodenheimer,P., 1973, Astrophys.J. 180, 171
Safronov,V.S., 1969, "Evolution of the Protoplanetary Cloud and Formation of the Earth and Planets", NASA TT-F-677
Tscharnuter,W.M., 1978, Moon and Planets 19, 229
Tscharnuter,W.M., and Winkler,K.-H.A., 1979, Computer Physics Communications 18, 171
Verbunt,F., 1982, Space Sci.Rev. 32, 379
Winkler,K.-H.A., Morman,M.L., and Mihalas,D., 1984, "Implicit Adaptive-mesh Radiation Hydrodynamics", submitted to Computational Techniques Series, eds. J.U.Brackbill and B.I.Cohen, (preprint)

DYNAMICS OF RING-SATELLITE SYTEMS AROUND SATURN AND URANUS

N. BORDERIES
Observatoire du Pic du Midi et de Toulouse
14 avenue Edouard Belin,
31400 Toulouse, FRANCE.

Abstract : We first recall the observations concerning the opaque rings of Saturn and Uranus. Then we describe a model which represents the kinematics of these rings. Finally we review how we treat collisions, self-gravity and satellite perturbations and what are the basic effects of these forces.

I - INTRODUCTION

This paper deals with the dynamics of planetary rings of high optical depth ($\tau \approx 10^{-1}-1$) which have their brightness dominated by large particles so that electromagnetic effects are small. The rings of Uranus and the major rings A,B,C of Saturn, as well as the Cassini division, belong to this category. These rings contain many sharply defined radial structures which are shaped by gravitational interactions with satellites and by collective effects due to mutual collisions and self-gravity.

These opaque rings can be contrasted with the tenuous Jovian ring and the E and G rings of Saturn. These rings have a low optical depth ($10^{-4}-10^{-6}$), and they are mostly made of micron-sized particles. Their morphology is affected by electromagnetic forces. The ethereal rings of Jupiter and Saturn have recently been reviewed by Burns, Showalter and Morfill (1984).

The F ring of Saturn, also studied by Burns and his co-workers, is somewhat intermediate with an optical depth in the narrow core of the order 10^{-1} - 1 and an optical brightness due to particles with a radius of one micron or smaller. However, the small spacing between particles make the electromagnetic forces unimportant. Since most characteristics of the F ring can probably be explained in terms of gravitational forces, we will include it in the present review.

The paper is divided into four parts. In the next section we review the data which are relevant to the dynamics of the Saturnian and Uranian rings and recall how they have been interprated. Sections 3 and 4 are devoted to the kinematics and to the dynamics.

II - DATA

2-1 - The rings of Uranus

Celestial Mechanics **34** (1984) 297–327. 0008–8714/84.15

The rings of Uranus were discovered on 10 March 1977, during an occultation of a star by the planet. That night, several groups of astronomers were observing this occultation by measuring the intensity of the starlight as the star passed behind Uranus. The presence of rings around the planet produced attenuations of the light before and after the planetary occultation. Elliot, Dunham and Mink (1977) recognized in their discovery data five rings called $\alpha, \beta, \gamma, \delta$ and ϵ. In 1978, Elliot et al (1978) identified four more rings : η, 4, 5 and 6. At present these comprise the nine known rings around the planet. The first occultation data also revealed the two following unusual features :

(i) the extinction of the signal due to the ring material was abrupt, indicating that the rings are narrow, opaque and have sharp edges ;

(ii) the ϵ ring appeared clearly non-circular or inclined.

After the discovery of the rings of Uranus, the first question which arose was the one of their shape. The outermost ϵ ring, which was the widest, was consequently the one which was the best studied. After about a year, when data from several occultations were available, it appeared that the width of the ϵ ring was variable, ranging approximately from 20 to 100 km ; the edges were always sharp and reproducible features were observed in the different occultation profiles. The observations suggested that the ring behaves as a unit.

Nicholson et al (1978) discovered that there exists a linear relation between the width of the ring and its distance to the planet. Their interpretation was that the linear relation is "consistent with an elliptical ring, coplanar with the other rings, whose inner and outer boundaries are two keplerian ellipses with slightly different semi-major axes and eccentricities". Colombo (1978) demonstrated that a keplerian ellipse which lies in the satellites' orbital plane, and whose apsidal line precesses due to the oblateness of Uranus, is a reasonable representation of the observations. This was the first evidence for non-circular rings. The existence of a non-circular ring was surprising because it was expected that differential rotation of the particles and differential precession of the periapses of their orbits would broaden and smear the rings in a few hundred years.

After it had been demonstrated that the ϵ ring is elliptical, Nicholson et al (1981, 1982) and Elliot et al (1981a, 1981b) made kinematic models of the Uranus rings, fitting the free parameters by least squares to the occultation timings. Six of the nine rings were found to be ellipses precessing under the influence of the oblateness of Uranus. The coefficients J_2 and J_4 of Uranus' gravity field were determined from the precession rates.

The models became more and more sophisticated, and in 1982, French, Elliot and Allen introduced two more free parameters for each ring : an inclination to a common plane assumed to be the equatorial plane and a longitude of the nodal line. French, Elliot and Allen (1982) established firmly that seven of the nine rings have significant inclinations. A general trend of the inclinations to decrease with increasing semi-major axis was noted. Remarkably, the ϵ-ring has zero inclination, despite the fact that almost all the other rings of Uranus are inclined and that the ε ring has the largest eccentricity. Differences of inclination between the inner and outer boundaries of inclined rings have not yet been detected although they are expected theoretically (see § 4-2). The nodes precess under the influence of the zonal harmonics of the Uranian gravity field.

The rings of Uranus are hard to observe directly because they are narrow, they lie close to the planet and the material which constitutes them has a low albedo. However, at 2.2 microns, the planet's albedo is very low because of a strong methane absorption band. Matthews, Neugebauer and Nicholson (1978) obtained contour maps of light reflected by the rings at 2.2 microns. Matthews and his co-workers inferred from these data that the space between rings is essentially empty, with an average optical depth $\leqslant 0.01$. They noticed an azimuthal asymmetry in the ε ring which has been seen on some but not all of the subsequent observations.

Orbital elements, widths and optical depths of the Uranian rings are given in table 1.

2-2 - The rings of Saturn

The rings of Saturn were discovered by Galileo in 1610 but it was only recognized by Huygens in 1655 that they were "a thin flat ring inclined to the ecliptic". For several centuries, the rings of Saturn were described as consisting of three major components : (i) the bright B ring, (ii) the outer A ring, separated from the B ring by the Cassini division (found by Cassini in 1675) and cut in its outer part by a narrower gap named the Encke division (discovered by Keeler in 1888) and (iii) the faint C ring lying inside the B ring. Evidence for the presence of two tenuous rings was reported : the wide outermost E ring (Feibelman, 1967, Kuiper, 1974) and the innermost D ring (Guérin, 1970). Near infrared spectroscopy showed that the ring particle surfaces are covered with water ice (Pilcher et al, 1970, Kuiper, Cruikshank and Fink, 1970). Particle sizes between a few centimeters and few meters were inferred from ground-based radio and radar observations (Cuzzi and Pollack, 1978). The A ring is not azimuthally symmetric : it is fainter in the quadrants following geocentric conjunction than in those preceding it (Camichel, 1958, Ferrin, 1975, Reitsema, Beebe and Smith, 1976, Lumme and Irvine, 1976, 1979, Lumme et al, 1977). Colombo, Goldreich and Harris (1976) and Franklin and Colombo (1978) proposed an explanation for this phenomenon : spiral wakes caused by the gravitational attraction of the larger ring particles.

The flybys of Saturn by spacecraft, in particular Voyager 1 and Voyager 2, which encountered Saturn on 12 November 1980 and 26 August 1981 respectively, revealed that the structure of the rings is unexpectedly complex and varied (Collins et al, 1980, Smith et al, 1981, 1982) : even though the classical divisions between the three major rings A, B and C kept their meaning, tens of thousands of radial structures were discovered and in many places the rings appeared to be non axisymmetric; two additional rings were found : the narrow F ring, just outside the A ring, and the faint and little known G ring.

Ring	a (km)	e ($\times 10^3$)	I (degrees)	W	τ	δa	δe
6	41877.3 ± 16.6	1.01 ± 0.10	0.066 ±0.012	0.4-2	> 0.5		
5	42275.2 ± 16.6	1.85 ± 0.08	0.050 ±0.010	0.8-2	> 1.0		
4	42609.6 ± 16.8	1.15 ± 0.04	0.022 ±0.005	0.7-2	> 0.8		
α	44758.3 ± 16.4	0.78 ± 0.02	0.017 ±0.003	4.9 → 10.1	≈1.4 → 0.7	7.5 ± 0.2	5.8×10^{-5}
β	45701.0 ± 16.5	0.43 ± 0.02	0.006 ±0.002	5.0 → 10.6	≈1.5 → 0.35	7.8 ± 0.3	6.0×10^{-5}
η	47214.9 ± 16.5	(0.03 ± 0.04)	(0.003 ± 0.004)	0.5 - 2	> 0.6		
γ	47666.3 ± 16.4	(0.04 ± 0.02)	0.006 ± 0.002	∼ 3	> 1.5		
δ	48338.7 ± 16.5	0.06 ± 0.02	0.012 ± 0.003	2 - 3	> 1.5		
ε	51188.1 ± 17.0	7.94 ± 0.02	(0.003 ± 0.003)	19.8 → 96.3	?→ 1.2	58.0 ± 0.4	7.4×10^{-4}

Table 1 : Semi-major axes (a), eccentricities (e), inclinations (I), widths (W), ranges of semi-major axes (δa), ranges of eccentricities (δe) of the 9 uranian rings, from Elliot and Nicholson (1984). Arrows indicate measured variations.

Uranus data : mass = $8.669\ 10^{28}$ g
equatorial radius = 26, 200 km
$J_2 = (3.349 \pm 0.005) \times 10^{-3}$
$J_4 = (-3.8 \pm 0.9) \times 10^{-5}$

Feature	Distance from Saturn center (R_s)
D ring inner edge [a]	1.11
C ring inner edge [b]	1.2374
B ring inner edge [b]	1.5244 → 1.5259
B ring outer edge [b]	1.9477
A ring inner edge [b]	2.0230 → 2.0280
A ring outer edge [b]	2.2670
F ring center [b]	2.3267
G ring center [a]	2.8
E ring inner edge [a]	3
E ring outer edge [a]	8

<u>Table 2</u> : Boundaries of Saturn's rings (R_s = 60,330 km).
a : from Stone and Miner (1982);
b : from Esposito <u>et al</u> (1983b).

Name	Diameter (km) [a]	Mean Motion (degrees per day) [b]	a (km) [b]	e [b]	I (degrees) [b]
1980S28	10 × 20	598.08±.05	1.37670	.002±.003	.3 ±.2
1980S27	70 × 50 × 40	587.28±.02	1.39353	.0024±.0006*	.0 ±.15
1980S26	55 × 45 × 35	572.77±.02	1.41700	.004±.002*	.05±.15
1980S3	70 × 60 × 50	518.49±.01	1.51422	.009±.002	.34±.05
1980S1	110 × 100 × 80	518.236±.01	1.51472	.007±.002	.14±.05

<u>Table 3</u> : Data for the five innermost satellites of Saturn
[a] From Smith <u>et al</u>, 1982.
[b] From Synnott <u>et al</u>, 1981.
* From Synnott <u>et al</u>, 1983.

The Voyager spacecraft also discovered a number of small satellites orbiting between the rings and the innermost previously known satellite Mimas (one of them, Janus, had been detected in 1966 by Dollfus, from ground-based observations). Table 2 and 3 list respectively the major ring boundaries and satellite data.

Analysis of Voyager data provided constraints on the properties of individual ring particles such as their sizes and their velocity dispersions. Davis et al (1984) and Weidenschilling et al (1984) have stressed that individual particles have not yet been seen, even by the Voyager spacecraft, and the rings may consist of ever changing aggregates of particles. However, a simple picture in which the particles look more like the classical ice balls which have conventionally been assumed, probably captures much of the relevant physics.

During the Voyager experiment, three types of observations of Saturn's rings were obtained : (i) images, whose resolution reached 10 km/lp; (ii) radio-occultation measurements of the rings, performed when Voyager 1 as seen from the Earth, was behind the rings ; the signals were at wavelengths of 3.6 and 13 cm and they consisted of two components, the attenuated direct signal and a scattered component; the basic resolution was of order 1 km, but in some cases it has been improved by one order of magnitude through the removal of the Fresnel diffraction; (iii) stellar-occultation measurements made by the photopolarimeter and the UV spectrometer onboard Voyager 2 with resolutions of 100 m and 3 km respectively.

In what follows, we describe and discuss the recent data obtained by Voyager concerning the morphology of the Saturn's ring-satellite system and the properties of individual particles.

a - Structures

Some of the structures in Saturn's rings are narrow rings which remind us of the rings of Uranus. The F ring, discovered by Pioneer 11 (Gehrels et al, 1980), is one famous example. Two "shepherd" satellites, 1980S27 and 1980S26, orbit on each side of the ring; both the ring and the satellite orbits are eccentric (From Synnott et al, 1983, the eccentricity of the F ring is .0026 ± .0006; see table 3 for the shepherd eccentricities). On a few narrow angle pictures the ring has a braided appearance which arises because the ring separates into several components which have a wavy shape. The wavelength of these irregular undulations varies from 7,000 to 10,000 km and the amplitude ranges from 15 to 20 km. Bright knots which may be kilometer size particles or temporary accumulations of smaller particles, are observed with a characteristic spacing of 10,000 km (Smith et al, 1982). As reported by Burns, Showalter and Morfill (1984) the F ring appears dramatically different with different techniques of observations : the Voyager 1 cameras showed us three components, each about 20-30 km wide; the more sensitive cameras of Voyager 2 identified faint material spread over a radial extent of about 500 km; the 3.6 cm radio signal detected a 2.4 km wide F ring while the photopolarimeter experiment found a 50 km wide ring.

Besides the F ring, major narrow ringlets are found in the C ring at 1.29 R_s and 1.45 R_s, and in the inner Cassini division at 1.95 R_s, where R_s is the equatorial radius of Saturn, equal to 60,330 km. Like the Uranian rings, these rings have sharp edges, are eccentric and are widest at their greatest radius. Two to them precess under the influence of Saturn's oblateness and the third (at 1.29 Rs) has its kinematics significantly affected by its interaction with Titan (Porco et al, 1984).

The second type of asymmetric ring feature seen in the Voyager images is found at several edges of broad rings. The outer edge of the B ring follows an ellipse which is centered at Saturn and whose short axis is directed towards Mimas (Smith et al, 1982). The outer edge of the A ring has a seven lobed shape (Porco, 1984). The edges of the Encke division have a wavy shape at some places (Cuzzi, 1983). All these distorted edges are sharp as evidenced by the photopolarimeter and radio occultation experiments : microwave edge diffraction patterns at the outer edge of the A ring and at the boundaries of the Encke gap have been reported by Marouf and Tyler (1982).

Waves, found at resonances located inside the A and B rings, are the third type of structure. Most of them are density waves, associated with resonant forcing of particle eccentricities (Cuzzi, Lissauer and Shu, 1981, Lane et al, 1982, Esposito, O'Callaghan and West, 1983, Holberg, Forrester and Lissauer, 1982, Holberg, 1983). The locations and strengths of the most important of these resonances have been calculated by Lissauer and Cuzzi (1982). A few waves are bending waves, associated with resonant forcing of particle inclinations (Shu, Cuzzi and Lissauer, 1983).

The fourth type of structure is the irregular axially symmetric fluctuation of brightness, mostly in the optically thick regions of the B ring.

b - Particle properties

Differences of optical depth between the 3.6 cm and the 13 cm radio wavelengths provide information on the abundance of centimeter sized particles (Tyler et al, 1983). Marouf et al (1983) found that the data were consistent with a power law distribution of sizes of the form $N(R) = N_o R^{-3.3}$ (where N(R)dR is the number of particles per unit area having radii between R and R+dR) between a few centimeters and a few meters. The power law is such that the optical depth is mostly due to the small particles and the mass is dominated by the large ones. The existence of the two cutoffs is supported by independent information (Borderies, Goldreich and Tremaine, 1984a).

Another quantity characteristic of the particles is their velocity dispersion v, defined as the root mean square velocity of a particle with respect to an imaginary one on a circular orbit with the same angular momentum. It is therefore related to the eccentricities and inclinations of the particle orbits and can be translated into a thickness h of the disk by the approximate relation :

$$v = h\,\Omega\,, \tag{1}$$

where Ω is the orbital angular velocity of the particles. Several independent determinations of h have been made, using the occultation data near the edges (Marouf and Tyler, 1982, Lane et al, 1982), damping of bending waves (Lissauer, Shu and Cuzzi, 1984) and the attenuation and scattering of the radio signal (Zebker and Tyler, 1984).

The picture which emerges shows us a very thin disk, extending vertically over a few tens of meters. The thickness of 1.4 km inferred from ground-based observations when the ring is seen edge-on (Brahic and Sicardy, 1981), is probably due to a general warp of the disk (Burns et al, 1979) or to the presence of bending waves (Shu, Cuzzi and Lissauer, 1983).

III - KINEMATICS

The numerous features associated with satellite resonances in Saturn's ring show that in many regions the morphology of these rings is primarily governed by satellite perturbations. The narrowness of Uranus'rings suggests strongly the presence of small, yet undiscovered, satellites which confine them (Goldreich and Tremaine, 1979a). This hypothesis is supported by the fact that most of the uranian rings have non zero eccentricities and inclinations which can plausibly be excited by a nearby satellite (Goldreich and Tremaine, 1981, Borderies, Goldreich and Tremaine, 1984b).

Near an isolated resonance or near a satellite orbit, a set of test particles on a common circular orbit is given a forced eccentricity with a well defined direction of periapse, which translates itself into a distorted streamline. In this section we review the shape of streamlines perturbed by a satellite, and then we describe a model which represents distorted streamlines in the presence of satellite perturbations as well as other forces such as viscous stresses and self-gravity. We approach these questions by first introducing a description of the unperturbed motion which is different from the classical view of Celestial Mechanics.

We consider a test particle in the equatorial plane of a planet and describe its position by cylindrical coordinates r, θ with respect to a system of coordinate axes with its origin at the planet. The motion of the test particle in a gravitational potential φ (r,θ) is governed by the equations :

$$\begin{aligned} \frac{d^2r}{dt^2} - r\left[\frac{d\theta}{dt}\right]^2 &= -\frac{\partial\varphi}{\partial r}\,, \\ \frac{d}{dt}\left[r^2\frac{d\theta}{dt}\right] &= -\frac{\partial\varphi}{\partial\theta}\,. \end{aligned} \tag{2}$$

In the case where φ is only a function of r (for instance if we consider the potential φ_p of an oblate, axisymmetric planet), we can obtain a solution of (2) in the form of a near circular orbit specified by :

$$r = a\,(1 - e\cos\kappa t),$$
$$\theta = \Omega t + \left[\frac{2\Omega e}{\kappa}\right]\sin\kappa t\ , \qquad (3)$$

with :

$$\Omega^2 = \left[\frac{1}{r}\frac{d\varphi}{dr}\right]_{r=a}\ , \qquad (4)$$

$$\kappa^2 = \left[\frac{3}{r}\frac{d\varphi}{dr} + \frac{d^2\varphi}{dr^2}\right]_{r=a}\ . \qquad (5)$$

Relative to an imaginary particle which would move on a circular orbit defined by $r = a$, $\theta = \Omega t$, our test particle describes an ellipse with the epicyclic frequency κ. In the gravitational field of an oblate planet, κ is slightly smaller than the mean motion Ω :

$$\kappa = \Omega - \frac{d\hat{\omega}}{dt}\ , \qquad (6)$$

where $\hat{\omega}$ is the longitude of the periapse.

"Ring people" use the following terminology : the circular motion of the imaginary particle is the mean motion; the unperturbed mean orbits of particles, or streamlines, are circles; the difference between the real motion and the mean motion is the random motion; it is associated to the existence of a free eccentricity e.

3-1 Orbits of test particles perturbed by a satellite

a - Isolated resonance

Let us now turn to the case where a satellite is present, so that :

$$\varphi = \varphi^p + \varphi^s , \tag{7}$$

where φ^s is the satellite potential. Goldreich and Tremaine (1982) show that when the satellite has zero inclination φ^s may be expanded in a double Fourier series as :

$$\varphi^s(r,\theta,t) = R_e \sum_{m=0}^{+\infty} \sum_{k=-\infty}^{+\infty} \varphi^s_{mk}(r) \exp[im(\theta-\Omega_p^{mk}t)] , \tag{8}$$

where φ^s_{mk} is real and Ω_p^{mk} is the pattern speed :

$$\Omega_p^{mk} = \Omega_s + \frac{k}{m}\kappa_s , \tag{9}$$

and the subscript s refers to the satellite. Goldreich and Tremaine (1982) solved equations (2) by a perturbation method, the small parameter being the ratio of the mass M_s of the satellite to the mass M_p of the planet. The first-order variation of r and θ, obtained by solving the linearized equations around the unperturbed solution with zero free eccentricity, is a sum of terms with different Ω_p and m, all of the form

$$r_1 = R_e \left\{ \frac{e^{im[(\Omega-\Omega_p t)+\theta_0]}}{m^2(\Omega-\Omega p)^2 - \kappa^2} \left[\frac{d\varphi^s_{mk}}{dr} + \frac{2\Omega}{m(\Omega-\Omega_p)} \frac{\varphi^s_{mk}}{r} \right] \right\} , \tag{10}$$

$$\theta_1 = R_e \left\{ \frac{i\, e^{im[(\Omega-\Omega_p t)+\theta_0]}}{m^2(\Omega-\Omega p)^2-\kappa^2} \left[\frac{2\Omega}{m(\Omega-\Omega_p)} \frac{1}{r} \frac{d\varphi^s_{mk}}{dr} + \left(\frac{4\Omega^2-\kappa^2}{m^2(\Omega-\Omega_p)^2} + 1 \right) \frac{m\varphi^s_{mk}}{r^2} \right] \right\} \tag{11}$$

plus a homogeneous solution of the form (3). This solution is singular when $\Omega = \Omega_p$ or when $D = \kappa^2 - m^2(\Omega-\Omega_p)^2 = 0$.

In the first case we are at a corotation resonance. A particle at a corotation resonance feels a constant force applied to it by the satellite. The resonance condition is :

$$(m+k)\,\Omega_s - m\,\Omega - k\,\dot{\tilde{\omega}}_s = 0 . \tag{12}$$

Near a corotation resonance (at radius r_c), equation (10) becomes :

$$r_1 \approx \left[\frac{4\varphi^s_{mk}}{3\Omega^2} \right]_{r_c} \frac{\cos m\,(\theta-\Omega_p t)}{(r-r_c)} . \tag{13}$$

The amplitude of the perturbation is of order at least one with respect to the satellite eccentricity e_s since we must have $k \neq m$ (we do not want the test particle to have the same semi-major axis as the satellite) and since $\varphi^s_{mk} = 0\,[(GM_s/a_s)e_s^{|k|}]$.

Lindblad resonances are defined by $D = 0$. They occur where the forcing frequency $m\,(\Omega - \Omega_p)$ is equal, in absolute value, to the natural frequency of oscillation κ of the particles. The resonance condition is

$$(m+k)\,\Omega_s - (m-\varepsilon)\,\Omega - k\,\dot{\tilde{\omega}}_s - \varepsilon\,\dot{\tilde{\omega}} = 0 , \tag{14}$$

where $\varepsilon = +1$ at an inner Lindblad resonance and $\varepsilon = -1$ at an outer Lindblad resonance ; these resonances are labelled as $(m + k) : (m - \varepsilon)$. The linearized radial response of a particle to the satellite perturbation near a Lindblad resonance (at radius r_L) is :

$$r_1 = \left[-\frac{1}{3(m-\varepsilon)\Omega^2} \left(2m\varphi^s_{mk} + \varepsilon r \frac{d\varphi^s_{mk}}{dr} \right) \right]_{r_L} \frac{\cos m\,(\theta-\Omega_p t)}{(r-r_L)} . \tag{15}$$

The solutions (13) and (15) can be summarized in the form :

$$r = a\,\{ 1 - e \cos m\,(\phi+\Delta) \} , \tag{16}$$

where :

$$\phi = \theta - \Omega_p t , \tag{17}$$

is the longitude of the particle with respect to a frame centered at the planet and rotating with the pattern speed in the direction of the motion, $m\Delta = 0$ or π and e is a forced eccentricity which depends on a. Consider a set of non-interacting particles which, in the absence of a satellite, would orbit on a circle of a radius a. If the satellite pertubation were slowly turned on the particles would become organized along a distorted streamline whose shape is given by equation (16). With respect to an inertial reference frame, each particle would travel on an elliptic orbit with its focus at the planet. But, because of the proximity of the resonance, each particle's phase in its orbit would be related to its apsidal line orientation such that equation (16) is satisfied.

The m radial excursions which the particles in the neighborhood of a Lindblad resonance undergo in following the satellite perturbing force, are larger closer to the resonance. It follows that, at some distance Δr_I from the resonance, the periodic orbits cross. The condition at intersection of non-interacting streamlines is $\partial r/\partial a = 0$; it implies :

$$\frac{\Delta r_I}{r} = 0\left[\left(\frac{M_s}{M_p}\, e_s^{|1-m|}\right)^{1/2}\right]. \qquad (18)$$

Δr_I provides a crude measure of the "width" of the resonance.

b) Close encounters

We consider now a test particle and a satellite on nearby trajectories. The two bodies interact gravitation aly mainly during brief encounters lasting on the order of one rotation period, which occur when the particle passes the satellite. For this reason, it is convenient to write the equations of motion in a local reference frame Sxy, whose origin coincides with the satellite, the x axis directed radially outwards and the y axis in the direction of motion. To first-order with respect to x/a_s and y/a_s, the equations of motion read (Spitzer and Schwarzschild, 1953) :

$$\frac{d^2x}{dt^2} + 4\Omega A x - 2\Omega\frac{dy}{dt} = -\frac{\partial\varphi^s}{\partial x},$$
$$\frac{d^2y}{dt^2} + 2\Omega\frac{dx}{dt} = -\frac{\partial\varphi^s}{\partial y}, \qquad (19)$$

where :

$$A = \frac{r}{2}\frac{d\Omega}{dr}, \qquad (20)$$

and $\varphi_s = -GM_s / |\vec{r} - \vec{r}_s|$.

Goldreich and Tremaine (1980) apply the method of variation of constants to solve these equations. The orbit is expressed in the form :

$$x = \alpha - \varepsilon \cos(\kappa t + \delta) ,$$
$$y = 2A\alpha t + \gamma + \frac{2\Omega}{\kappa} \varepsilon \sin(\kappa t + \delta) , \qquad (21)$$

and equations (19) are transformed into :

$$\frac{d\alpha}{dt} = -\frac{1}{2B}\frac{\partial\varphi^s}{\partial\gamma} , \qquad \frac{d\gamma}{dt} = \frac{1}{2B}\frac{\partial\varphi^s}{\partial\alpha} ,$$
$$\frac{dk}{dt} = -\frac{1}{\kappa}\frac{\partial\varphi^s}{\partial h} , \qquad \frac{dh}{dt} = \frac{1}{\kappa}\frac{\partial\varphi^s}{\partial k} , \qquad (22)$$

with :

$$B = \Omega + A , \qquad (23)$$
$$k = \varepsilon \cos\delta \text{ and } h = \varepsilon \sin\delta . \qquad (24)$$

If the satellite and the particles are initially on circular orbits, the first-order solution well after encounter, expressed in polar coordinates relative to a reference frame translated from Sxy so that its origin coïncides with the planet center, is :

$$r = a - ae \cos\left[\frac{2}{3(a_s - a)}(a\phi + \xi)\right] , \qquad (25)$$

with the forced eccentricity :

$$e = f\frac{M_s}{M_p}\left[\frac{a}{a_s - a}\right]^2 , \qquad (26)$$

and the numerical factor :

$$f = \frac{8}{9}\left[K_1\left(\frac{2}{3}\right) + 2K_0\left(\frac{2}{3}\right)\right] \approx 2.24 ; \qquad (27)$$

K_ν denotes the modified Bessel function of order ν. Equation (25) gives the shape of a streamline in the frame rotating with the satellite. The wavelength, $3\pi|a_s - a|$, is due to the difference of angular velocity of the two bodies. The amplitude ae, on the other hand, depends on the strength of the perturbation. Neighbouring streamlines have slightly different wavelengths. For this reason they cross at longitude ϕ_I given by :

$$\phi_I = \frac{2}{3f}\frac{M_p}{M_s}\left[\frac{a_s-a}{a}\right]^4 - \xi\ , \tag{28}$$

where $2\xi/3(a_s-a) = \pi/2$. Showalter and Burns (1982) have computed the modifications of the orbital elements of the particle in the case where the initial satellite and particle orbits are both eccentric. Because of the variable ring-to-satellite distance in this case, the streamlines no longer have a smooth sinusoidal form but exhibit irregular radial variations.

3-2 Streamline representation

Ring particle orbits are distinct from the test particle orbits found above for two reasons : (i) the linear approximation fails very near resonances ; for instance a more careful treatment removes the crossing of periodic orbits at a corotation resonance (Hagihara, 1972, Goldreich and Tremaine,1981) (this is not true for Lindblad resonances); (ii) the ring particles interact through their self-gravity and mutual collisions.

However, more detailed theory predicts and observations confirm that there are considerable similarities between ring particle orbits and test particle orbits . For example, as suggested by the periodic orbit theory, the outer edge of the B ring, which is associated with a 2 : 1 resonance with Mimas, is a Saturn centered ellipse whose short axis points towards the satellite. Furthermore the outer edge of the A ring, which corresponds to a 7 : 6 resonance with 1980S1 seems to have a seven lobed shape. Cuzzi (1983) found that the wavy shape of the edges of the Encke division at several azimuths can be explained by the presence of undiscovered satellites inside the gap. He and Showalter also realized that perturbed streamlines can be observed in the optical depth profile obtained with the photopolarimeter in this region : zones of compression and of rarefaction of the ring material result from the fact that neighbouring streamlines have different wavelengths. The optical depth is maximum along straight lines defined by :

$$y = \frac{3(a-a_s)}{2}\left[\frac{3\pi}{2} + 2\pi N\right] - \xi\ , \tag{29}$$

where N is the number of wavelengths from the satellite. A radial cut of the perturbed streamlines should show an oscillation of the optical depth with the wavelength :

$$\Delta x = \frac{|a_s-a|}{N+3/4}\ , \tag{30}$$

and the amplitude :

$$\Delta\tau = \tau \left[\frac{(1-q^2)^{1/2}}{1-q} - 1 \right], \tag{31}$$

with :

$$q = f \frac{M_s}{M_p} \left| \frac{a}{a_s - a} \right|^3 \left[\frac{3\pi}{2} + 2\pi N \right]. \tag{32}$$

The variation of optical depth exhibited by the stellar occultation profile has the expected characteristic behaviour. Finally, the close encounter model gives the right order of magnitude for the wavelengths and amplitudes of the oscillations which are observed in the F ring. The separation of the ring into several components is more difficult to understand ; its explanation may involve the presence of large particles inside the ring or periodic intersections of 1980S27 with it (Borderies, Goldreich and Tremaine, 1983c).

Hence observations show that streamlines having similar characteristics to test particle orbits exist in rings. We describe them by :

$$r = a - ae(a) \cos m \{\phi + \Delta(a)\}. \tag{33}$$

The idea is that self-gravity and collisions modify the forced eccentricity and introduce a phase lag in the lobe longitudes. In particular, streamline crossing is prevented at a Lindblad resonance by interactions between streamlines. Another justification for choosing the representation given by (33) is that the mean velocity components which can be inferred from (33) and from Kepler's second law satisfy the unperturbed momentum equations to first-order in eccentricity (Borderies, Goldreich and Tremaine, 1983). We define :

$$J(a,\phi) = \left.\frac{\partial r}{\partial a}\right|_{\phi} = 1 - q \cos\{m(\phi + \Delta) + \gamma\}, \tag{34}$$

where :

$$q \cos \gamma = a \frac{de}{da}, \quad q \sin \gamma = mae \frac{d\Delta}{da}, \tag{35}$$

with $q > 0$ and $0 \leq \gamma < 2\pi$. Neighbouring streamlines cross when $q = 1$ and $E = m(\phi + \Delta) = -\gamma$. The geometric interpretation of q and γ is given in Figure 1. We define a perturbed (respectively an unperturbed) streamline by the condition $q \neq 0$ ($q = 0$).

Equation (33) is quite general. It can be applied to an isolated resonance (associated with a perturbing potential varying as $\cos(m\phi)$), to close encounters with a satellite (with $m \approx 2a/3|a_s - a|$) and to elliptical ringlets (with $m = 1$).

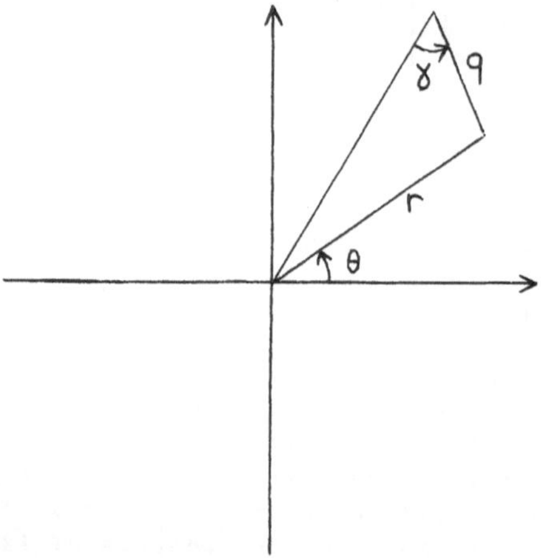

Figure 1 : Geometrical interpretation of q and γ. Each streamline is characterized by a vector with polar coordinates $r = ae/\Delta a$, $\theta = -m\Delta$. In the discrete approximation, q and γ specify differences between the vectors corresponding to two neighbouring streamlines. If neighbouring streamlines differ mainly by their eccentricity, as it is the case near Lindblad resonances, one can see that $\gamma \approx 0$: the streamlines come close together near periapse. If neighbouring streamlines differ mainly by their phase lag Δ, as in the close encounter case, one can see that $\gamma \approx \pi/2$: the streamlines come close together near quadrature.

IV - DYNAMICS

We now have at our disposal a kinematic model (see equation (33)) in which each streamline is characterized by three variables : a, e, and Δ. To study the dynamical evolution of the streamlines, we have to formulate and solve differential equations for the evolution of a, e and Δ (using for instance Gauss equations since Δ is related to $\tilde{\omega}$). In the most general case, the evolution is due to four types of forces : (i) the multipole coefficients of the planet potential, (ii) mutual collisions between the particles, (iii) self-gravity and (iv) satellite potentials. In this section we review the effects of the three latter forces.

4-1 Effects of collisions

The collision frequency is :

$$\omega_c = A\, n\, v\ , \qquad (36)$$

where A is the cross-section of a typical particle, n is the particle density and v is the rms relative speed. The optical depth is :

$$\tau = A\, n\, h\ . \qquad (37)$$

By combining equations (1), (36) and (37), we obtain :

$$\omega_c \approx \Omega\, \tau\ . \qquad (38)$$

A typical particle would have experienced on the order $10^{13}\tau$ collisions in the lifetime of the solar system. Collisions are an important process in rings !

Collisions transport angular momentum and energy and convert energy into heat. To discuss the effects of collisions we shall first describe an elementary approach. Later we develop the fluid-dynamic approximation and the kinetic theory approach which give successively more precise quantitative results.

a - Elementary approach

This approach applies to an unperturbed ring. Its presentation is inspired by Stewart, Lin and Bodenheimer (1984).

Because the ring possesses a symmetry with respect to the rotation axis of the planet, the most important effect of a collision is on excursions in the radial direction. If $\tau > 1$, on average a particle suffers several collisions per orbit and its radial mean free path is given by $l \approx v/\omega_c \approx v/\Omega\tau$. On the other hand, if $\tau < 1$, a particle collides relatively rarely with other particles; its freedom to travel in the radial direction is limited by its motion on an epicyclic ellipse, so that $l \approx ae \approx v/\Omega$. These two cases can be combined in a single formula (Cook and Franklin, 1964) giving l for any τ :

$$l = \frac{v}{\Omega\ (1+\tau^2)^{1/2}} . \tag{39}$$

After N collisions, the radial excursion is given by the random walk formula :

$$\Delta r = N^{1/2}\ l . \tag{40}$$

The time needed for a particle to random walk over a distance Δr is :

$$T = \frac{(\Delta r)^2\ \Omega\ (1+\tau^2)}{v^2\ \tau} . \tag{41}$$

Radial diffusion of particles tends to make the ring spread ; the total angular momentum is conserved, but it is redistributed in the radial direction.

Since collisions produce radial excursions of the particles, in the absence of dissipation, the energy associated with eccentric motion tends to increase. After one elastic collision, the change of energy E per unit of mass is $\Delta E = (\mu^2/2H^2)(\Delta e)^2$, where H is the angular momentum per unit of mass and μ is the product of the gravitational constant G by the planet mass. We evaluate ΔE by replacing Δe by l/a : $\Delta E \approx \Omega^2 l^2$. Since there are ω_c collisions per unit time, the rate at which the energy per unit mass of random motions increases is :

$$\left.\frac{dE}{dt}\right|_c \approx \Omega^2\ \omega_c\ l^2 . \tag{42}$$

In the absence of dissipation the velocity dispersion would increase indefinitely. But collisions have an inelastic nature, which we describe by the coefficient of restitution ε, the fraction by which the relative normal velocity of two colliding particles is reduced after their collision (hence, collisions are perfectly elastic for $\varepsilon = 1$ and completly inelastic for $\varepsilon = 0$). An amount $\Delta E=(1-\varepsilon^2)v^2$ of energy per unit mass is dissipated into heat per collision. The dissipation rate is:

$$\left.\frac{dE}{dt}\right|_d \approx (1-\varepsilon^2)\, v^2\, \omega_c \,. \tag{43}$$

Since we have an energy generating mechanism and a dissipating one, an equilibrium may exist. By using (39) the equilibrium condition has the form (Cook and Franklin, 1964, Goldreich and Tremaine, 1978a) :

$$(1-\varepsilon^2)\ (1+\tau^2) = \text{const.} \tag{44}$$

Equilibrium is possible because ε is a function of v and it is stable because ε is a decreasing function of v (Goldsmith, 1960, Goldreich and Tremaine, 1978a, Borderies, Goldreich and Tremaine, 1984a, Bridges, Hatzes and Lin, 1984). Shukhman (1984) incorporates the effects of tangential friction and particle spin in the $\varepsilon(\tau)$ relation. In a planetary ring, the velocity dispersion adjusts itself to a value such that equation (44) is satisfied.

b - Fluid-dynamics approach

The collective behaviour of the ring particles can be approximated by a fluid-dynamics approach. In this approach, the effects of collisions are modelled as a viscous stress. The force per unit length is the i direction is :

$$F_i = \sigma_{ik}\, n_k \,, \tag{45}$$

with the normal to the streamline being :

$$\hat{n} = n_k\, \hat{e}_k \,, \tag{46}$$

where the stress tensor :

$$\sigma_{ik} = -\,p\,\delta_{ik} + \eta \left[\frac{\partial u_i}{\partial x_k} + \frac{\partial u_k}{\partial x_i} \right] + \delta_{ik}\,(\zeta - \frac{2}{3}\eta)\, \frac{\partial u_l}{\partial x_l} \,. \tag{47}$$

In equation (47), p is the pressure, the u_j are the components of the mean velocity and η and ζ are the dynamic viscosity and the bulk viscosity respectively. The ratio :

$$\nu = \frac{\eta}{\Sigma} \,, \tag{48}$$

where Σ is the surface density, is called the kinematic viscosity. The basic expression for ν is $\nu \approx \omega\, l^2$; using (38) and (39) we obtain (Goldreich and Tremaine, 1978a)^c:

$$\nu = \frac{v^2}{\Omega} \frac{\tau}{(1+\tau^2)} . \tag{49}$$

In the unperturbed case, we find that the ring material inside a streamline at radius a exerts the torque on the material outside :

$$L_H(a) = 3\pi\Sigma\nu\Omega a^2 . \tag{50}$$

If we now consider a narrow ringlet of width Δa centered on radius a, the torque exerted on it (that is to say the derivative with respect to time of its angular momentum) is :

$$T_v = L_H(a-\Delta a/2) - L_H(a+\Delta a/2) . \tag{51}$$

We see that $L_H(a)$ can be interpreted as the rate at which angular momentum crosses a circle of radius a; it is a luminosity of angular momentum. An important result is that since L_H is positive, angular momentum flows outwards : particles in the inner parts of the disk lose angular momentum and move inwards ; particles near the outer boundary gain angular momentum and move outwards ; again, the ring tends to spread. Results obtained by the elementary approach can be recovered by the fluid dynamics treatment : the diffusion time (to compare with equation (41)) is (Goldreich and Tremaine, 1978a) :

$$T = \frac{(\Delta r)^2}{\nu} ; \tag{52}$$

and the rate at which energy is dissipated (to compare with equation (42)) is :

$$\left.\frac{dE}{dt}\right|_d = \frac{9}{4} \nu\, \Omega^2 . \tag{53}$$

In addition to being more rigorous than the elementary approach, the fluid dynamics equations may be applied to the perturbed regions. We find that for high values of q the effects of viscosity are drastically different from those in the unperturbed case. As a matter of fact, we have (Borderies, Goldreich and Tremaine, 1982) :

$$L_H(a) = 4\pi\Sigma\nu\Omega a^2 K(q^2) , \tag{54}$$

where :

$$K(q^2) = \frac{3-4q^2}{4(1-q^2)} . \tag{55}$$

We see that for $q^2 > 3/4$, $L_H(a)<0$, the angular momentum flows inward and the ring tends to contract. This result can be understood as follows : examination of the equations in cylindrical coordinates shows that the tangential viscous force exerted on a unit length of a streamline by the material inside it has the opposite sign of $d\dot{\theta}/da=(2\Omega/a)[q\cos(E+\gamma)-3/4]$; it follows that the angular momentum flows in the direction in which the angular velocity decreases. If $q > 3/4$ the angular momentum flows inwards in some range of azimuth ; for $q^2>3/4$, there is more angular momentum flowing inward than outward ; in other words, $L_H < 0$.

c - Kinetic approach

The evolution of a system of particles, the statistical properties of which are expressed in terms of a phase-space number density f $(\vec{x}, \vec{v}, t)$ ($\vec{x}$ is the position, $\vec{v}$ the velocity), is governed by the Boltzmann equation :

$$\frac{\partial f}{\partial t} + v_\alpha \frac{\partial f}{\partial x_\alpha} - \frac{\partial U}{\partial x_\alpha}\frac{\partial f}{\partial v_\alpha} = \left.\frac{\partial f}{\partial t}\right|_c , \tag{56}$$

where U is the gravitational potential of the bodies which are not included in the collection of particles, and $(\partial f/\partial t)_c$ is the rate of change of f due to collisions. Collisional theories derived from the Boltzmann equations have been developed by Cook and Franklin (1964), Goldreich and Tremaine (1978a), Hämeen - Anttila (1978, 1981, 1982), Borderies, Goldreich and Tremaine (1983d) and Shukhman (1983).

Collisional systems of particles have been studied through numerical simulations by Trulsen (1972), Brahic (1975, 1976, 1977), Brahic and Henon (1977), Hämeen-Anttila and Lukkari (1980).

Comparisons with the fluid-dynamic approach show that the latter is better at higher optical depth but remains qualitatively correct at smaller τ (Goldreich and Tremaine, 1978a, Borderies, Goldreich and Tremaine, 1983d).

We conclude this subsection devoted to collisions by sketching how viscosity materializes itself in rings.

First, collisions affect two microscopic properties :

(i) they determine the velocity dispersion and the thickness of the ring (Goldreich and Tremaine, 1978a, Cuzzi et al, 1979a, 1979b, Henon, 1981, Borderies, Goldreich and Tremaine, 1983d and Stewart, Lin and Bodenheimer, 1984).

(ii) collisions must play a part in the determination of the particle size distribution; the physics by which this is accomplished is not yet understood.

Second, collisions affect the structure of the rings in three ways:

(i) sharp edges apparently require that $L_H = 0$ at finite Σ (Borderies, Goldreich and Tremaine, 1982).

(ii) chaotic structures inside the B ring may be due to viscous instabilities (Lin and Bodenheimer, 1981, Lukkari, 1981, and Ward, 1981).

(iii) viscosity has a damping role which tends to dissipate eccentricities and inclinations, streamline librations, and density and bending waves.

4-2 - Effects of self-gravity

Self-gravity is generally modeled by considering the gravitational force between two streamlines (Goldreich and Tremaine, 1979b, Borderies, Goldreich and Tremaine, 1983a and 1983b, Yoder, 1984). The acceleration exerted on a streamline i by a streamline j is :

$$\vec{\Gamma} = \frac{GM_j \ (1 - e_i \cos f_i)}{\pi a_i \ (a_i - a_j) \ J_{ij}} \hat{e}_N , \quad (57)$$

where $\hat{e}_N$ is the normal to streamline j (oriented from streamline i to streamline j), M_j is the mass of streamline j and $J_{ij}=(r_i-r_j)/(a_i-a_j)$.

The effect of self-gravity is best illustrated by its role in the rigid precession of eccentric ringlets. The fact that several narrow rings are elliptical means that the periapses of the individual particle orbits remain almost aligned : differential precession around the oblate parent planet must be prevented by some force. Goldreich and Tremaine (1979b) and Farinella et al (1980) proposed that particle apsides are locked together by the ring self-gravity. The idea at the heart of their model is that, if the ring is narrower at periapse, the ring gravity accelerates (retards) the apse precession rate of particles at the outer (inner) boundary, making therefore an equilibrium configuration possible. Dissipation associated with particle collisions drives narrow elliptic rings toward this stable equilibrium (Borderies, Goldreich and Tremaine, 1983b). A similar model applies to inclined ringlets (Borderies, Goldreich and Tremaine, 1983a). These models make four predictions :

(i) eccentric rings should be narrower at pericenter ; this is true for all elliptical rings where a gradient of eccentricity has been measured (Esposito et al, 1983a, Porco et al, 1984).

(ii) the surface density of elliptical rings can be computed from their shape. In Saturn's rings these surface density estimates are comparable with independent determinations by other means (Cuzzi, Lissauer and Shu, 1981, Holberg, Forrester and Lissauer, 1982 and Tyler et al, 1983) at other places in the rings.

(iii) the periapse of the inner edge of a narrow elliptical ring should slightly precede the periapse of the outer edge; this has been verified by Nicholson (1983) for the rings α and β of Uranus; the size of the apsidal shift provides a measure of the velocity dispersion.

(iv) eccentric and inclined rings with small differences of

eccentricities Δe and inclinations ΔI between their outer and inner edges must be such that $\Delta e/e \approx \Delta I/I$. Since ΔI has not yet been measured, this relation has not been checked.

To conclude, even if self-gravity helps a ring to keep its elliptical shape, on a longer time-scale, viscosity damps its mean eccentrity (Borderies, Goldreich and Tremaine, 1983b); therefore an excitation mechanism is needed.

4-3 Effects of satellites

By preventing the perturbed streamlines from crossing, self-gravity and viscosity keep the particles response to satellite perturbations coherent.

Near an isolated Lindblad resonance, q is not equal to zero. The loci of constant surface density take the shape of tightly wound spirals; this is a density wave. The non axisymmetric distribution of mass associated with the wave produces a gravitational torque T_s on the satellite. The ring loses (gains) angular momentum at the same rate at which the satellite gains (loses) it. This is accomplished as follows : there is a luminosity of angular momentum equal to $-T_s$; negative (positive) angular momentum flows across the streamlines, transported by the wave ; as the wave damps, this angular momentum is deposited in the ring.

The prediction that density waves are excited near inner Lindblad resonances in Saturn's rings, and that a transfer of angular momentum and of energy takes place between the perturbed ring and the excitating satellite, was made prior to Voyager 1 Saturn encounter by Goldreich and Tremaine (1978b, 1978c, 1979c, 1980, 1981). They applied a two dimensional fluid dynamical model to study the dynamics near a Lindblad resonance (Goldreich and Tremaine, 1978b) and derived T_s to be :

$$T_s = m\pi^2 \left[\Sigma \left(r \frac{dD}{dr} \right)^{-1} \right]_{r_L} \Psi_{lm}^2 , \tag{58}$$

where :

$$\Psi_{lm} = \left[r \frac{d\varphi_{lm}^s}{dr} + 2 \frac{\Omega}{(\Omega-\Omega_p)} \varphi_{lm}^s \right]_{r_L} . \tag{59}$$

and $l = m + k$. The sign of T_s is such that positive angular momentum is always transfered outwards. In the particular case of a m : (m-1) (m>1) inner Lindblad resonance we have :

$$T = \frac{m\pi^2}{3(m-1)} \left[\frac{M_s}{M_p} \right]^2 \Sigma a^2 \left[\alpha^2 \frac{db_{1/2}^{(m)}}{d\alpha} + 2m\alpha b_{1/2}^{(m)} \right]^2 \Omega^2 a^2 \tag{60}$$

where $\alpha = r_L/a_s$ and $b_{1/2}^{(m)}(\alpha)$ is a Laplace coefficient (Brouwer and Clemence, 1961), defined by :

$$b_{1/2}^{(m)}(\alpha) = \frac{2}{\pi}\int_0^{\pi} \frac{\cos mx \, dx}{(1 - 2\alpha\cos x + \alpha^2)^{1/2}} . \qquad (61)$$

The torque is of second order with respect to the ratio of the masses of the satellite and the planet, and is proportional to the unperturbed surface density at the Lindblat resonance. Note the remarkable fact that the torque depends on the strength of the perturbing potential but not on the viscosity which plays only a secondary role.

The agreement between the linear theory of density waves developed by Goldreich and Tremaine, and the observations, is excellent (Cuzzi, Lissauer and Shu, 1981, Lane et al, 1982, Holberg, Forrester and Lissauer, 1982 and Esposito, O'Callaghan and West, 1983); these authors used the observed dispersion relation to compute the surface density at several places of Saturn's rings; a coefficient of viscosity could in principle be extracted from damping measurements if the waves are linear; in practice however, reliable estimates are prevented by the fact that most of the observed density waves are nonlinear.

Density waves are not the only waves which can be found in Saturn's rings. Shu, Cuzzi and Lissauer (1983) showed that near vertical resonances, which occur where the forcing frequency is equal to the natural frequency of oscillations of the particles in the vertical direction, bending waves involving a warp of the ring may develop. Bending waves have been used by these authors as diagnostics of the surface density (Shu, Cuzzi and Lissauer, 1983) and of the viscosity (Lissauer, Shu and Cuzzi, 1984).

Next let us turn to the case of a ring and a nearby satellite. If the ring were made of test particles, no secular torque would be exerted; conditions at closest approach would be different from one encounter to the next one. On the other extreme, if the particle and the satellite both traveled on circular orbits before each encounter, having completely lost the memory of the previous encounter, the transfer of angular momentum between the particle and the satellite would always be in the same direction. Goldreich and Tremaine (1980) showed that interactions between ring particles perturbed by a nearby satellite result in the torque density inside the ring :

$$\frac{dT_r}{dr} = \pm g \left[\frac{M_s}{M_p}\right]^2 \frac{\Sigma\Omega^2 r^7}{(a_s - r)^4} , \qquad (62)$$

where g is a numerical factor given by :

$$g = \frac{2^5}{3^4}\left[K_1\left(\frac{2}{3}\right) + 2\,K_0\left(\frac{2}{3}\right)\right]^2 \approx 2.51\,, \tag{63}$$

and where the signs + and - apply if the satellite is inside and outside the ring. Greenberg (1983) showed that the torque formula (62) is valid for a wide range of damping magnitudes.

Gravitational torques between rings and satellites produce secular variations in the orbital elements.

Realizing that rings are repelled away from satellites, Goldreich and Tremaine (1979a) proposed that small satellites, orbiting on each side of the rings of Uranus may keep them from spreading. The discovery of two small satellites confining the F ring of Saturn gave strong support to their prediction. After this discovery, Lissauer, Shu and Cuzzi (1981) suggested that embedded moonlets within the rings may explain many of the gaps. The Voyager 2 imaging team performed a careful search for such satellites inside two empty gaps of the Cassini division, but they obtained a negative result (Smith et al, 1982).

Satellites are also repelled from rings and there is a short-time scale problem with the evolution of these satellites (Goldreich and Tremaine, 1982, Borderies, Goldreich and Tremaine, 1984a, Lissauer, Peale and Cuzzi, 1984).

Moreover, gravitational torques are able to excite eccentricities of rings and satellites (Goldreich and Tremaine, 1980, 1981) as well as their inclinations (Borderies, Goldreich and Tremaine, 1984b). If they do so, the equilibrium state of the rings follows from a competition between satellite excitation and viscous damping.

Another model of the uranian rings had been proposed by Dermott, Gold and Sinclair (1979). They postulated that each narrow ring contains a small satellite and that the particles travel on nested horseshoe orbits around it. Later on, they recognized that their model could not account for some of the observed features of the uranian rings. However their idea was ingenious and the discovery of the coorbital satellites 1980S1 and 1980S3 provided the first example of bodies on horseshoe orbits. Dermott (1984) suggested that the horseshoe-orbit model might apply to the narrow ring from which the coorbital satellites presumably formed.

V - CONCLUSION

Thanks to space exploration which provided a lot of detailed information, planetary rings became laboratories in which we study the physics of numerous colliding bodies in a gravitational field. Several of the physical processes which determine the various features observed in Saturn's rings are now understood. The concept of perturbed streamlines accross which the luminosity of angular momentum may vanish is an important one, relevant to sharp edges, confinement of narrow rings etc... But our understanding is incomplete : we do not have yet a detailed picture involving effects of viscous instability, size distribution, nonlinear waves etc...; we do not know what the inner edges of the major rings are due to; the puzzle of the short time-scale of evolution of the shepherd satellites is not solved; and the origin of planetary rings is still a matter of speculation. These questions will be answered one day and we may expect that the answers will require some novel physical concepts.

ACKNOWLEDGMENTS

I thank Peter Goldreich and Scott Tremaine for discussions, as well as Andrea Milani who reviewed this paper and made useful comments. I also aknowledge grants from ATP Planétologie 1983 and 1984.

References

BORDERIES, N., GOLDREICH, P. and TREMAINE, S. : 1982, Nature 299, 209.

BORDERIES, N., GOLDREICH, P. and TREMAINE, S. : 1983a, Astron. J. 88, 226.

BORDERIES, N., GOLDREICH, P. and TREMAINE, S. : 1983b, Astron. J. 88, 1560.

BORDERIES, N., GOLDREICH, P. and TREMAINE, S. : 1983c, Icarus 53, 84.

BORDERIES, N., GOLDREICH, P. and TREMAINE, S. : 1983d, Icarus 55, 124.

BORDERIES, N., GOLDREICH, P. and TREMAINE, S. : 1984a, Planetary Rings, R. Greenberg and A. Brahic, Eds., University of Arizona press.

BORDERIES, N., GOLDREICH, P. and TREMAINE, S. : 1984b, Ap. J, in press.

BRAHIC, A. : 1975, Icarus 25, 452.

BRAHIC, A. : 1976, J. Comput. Phys. 22, 171.

BRAHIC, A. : 1977, Astron. Astrophys. 54, 895.

BRAHIC, A. and HENON, M. : 1977, Astron. Astrophys. 59, 1.

BRAHIC, A. and SICARDY, B. : 1981, Nature 289, 447.

BRIDGES, F.G., HATZES, A. and LIN, D.N.C. : 1984, "On the structure, stability and evolution of Saturn's rings : preliminary measurements of the coefficient of restitution of ice-ball collisions", preprint.

BROUWER, D. and CLEMENCE, G.M. : 1961, Methods of Celestial Mechanics, New-York, Academic.

BURNS, J.A., HAMILL, P., CUZZI, J.N. and DURISEN, R.H. : 1979, Astron. J. 84, 1783.

BURNS, J.A., SHOWALTER, M.R. and MORFILL, G. : 1984, Planetary Rings, R. Greenberg and A. Brahic, Eds., University of Arizona press.

CAMICHEL, H. : 1958, Ann. Astrophys. 21, 231.

COLLINS, S.A. and 10 co-authors : 1980, Nature 288, 439.

COLOMBO, G., GOLDREICH, P. and HARRIS, A.W. : 1976, Nature 264, 344.

COLOMBO, G. : 1978, Private communication to P.D. Nicholson et al, 1978.

COOK, A.F. and FRANKLIN, F.A. : 1964, Astron. J. 69, 173.

CUZZI, J.N. and POLLACK, J.B. : 1978, Icarus 33, 233.

CUZZI, J.N., DURISEN, R.H., BURNS, J.A. and HAMILL, P. : 1979a, Icarus 38, 54.

CUZZI, J.N., BURNS, J.A., DURISEN, R.H. and HAMILL, P.M. : 1979b, Nature 281, 202.

CUZZI, J.N., LISSAUER, J.J. and SHU, F.H. : 1981, Nature 292, 703.

CUZZI, J.N. : 1983, Private communication.

DAVIS, D.R., WEIDENSCHILLING, S.J., CHAPMAN, C.R. and GREENBERG, R. : 1984, Science 224, 744.

DERMOTT, S.F., GOLD, T. and SINCLAIR, A.T. : 1979, Astron. J. 84, 1225.

DERMOTT, S.F. : 1984, Planetary Rings, R. Greenberg and A. Brahic Eds., University of Arizona press.

ELLIOT, J.L., DUNHAM, E. and MINK, D. : 1977, Nature 267, 328.

ELLIOT, J.L., DUNHAM, E., WASSERMAN, L.H., MILLIS, R.L. and CHURMS, J. : 1978, Astron. J. 83, 980.

ELLIOT, J.L., FROGEL, J.A., ELIAS, J.H., GLASS, I.S., FRENCH, R.G., MINK, D.J. and LILLER, W. : 1981a, Astron. J. 86, 127.

ELLIOT, J.L., FRENCH, R.G., FROGEL, J.A., ELIAS, J.H., MINK, D. and LILLER, W. : 1981b, Astron. J. 86, 444.

ELLIOT, J.L. and NICHOLSON, P.D.: 1984, Planetary Rings, R. greenberg and A. Brahic, Eds., University of Arizona press.

ESPOSITO, L.W., O'CALLAGHAN, M. and WEST, R.A. : 1983, Icarus 56, 439.

ESPOSITO, L.W., BORDERIES,N., GOLDREICH, P., CUZZI, J.N., HOLDBERG, J.B., LANE, A.L., POMPHREY, R.B., TERRILE, R.J., LISSAUER, J.J., MAROUF, E.A. and TYLER, G.L. : 1983a, Science 222, 57.

ESPOSITO, L.W., O'CALLAGHAN, M., SIMMONS, K.E., HORD, C.W. and WEST, R.A. : 1983b, J.G.R. 88, 8643.

FARINELLA, P., PAOLICCHI, P., FERRINI, F. and NOBILI, A.M. : 1980, The Moon and the Planets 22, 103.

FEILBELMAN, W.A. : 1967, Nature 214, 793.

FERRIN, I.R. : 1975, Astrophys. Space Sci. 33, 453.

FRANKLIN, F.A. and COLOMBO, G. : 1978, Icarus 33, 279.

FRENCH, R.G., ELLIOT, J.L. and ALLEN, D.A. : 1982, Nature 298, 827.

GEHRELS, T., and 22 co-authors : 1980, Science 207, 434.

GOLDREICH, P. and TREMAINE, S. : 1978a, Icarus 34, 227.

GOLDREICH, P. and TREMAINE, S. : 1978b, Icarus 34, 240.

GOLDREICH, P. and TREMAINE, S. : 1978c, Ap. J. 222, 850.

GOLDREICH, P. and TREMAINE, S. : 1979a, Nature 277, 97.

GOLDREICH, P. and TREMAINE, S. : 1979b, Astron. J. 84, 1638.

GOLDREICH, P. and TREMAINE, S. : 1979c, Ap. J. 233, 857.

GOLDREICH, P. and TREMAINE, S. : 1980, Ap. J. 241, 425.

GOLDREICH, P. and TREMAINE, S. : 1981, Ap. J. 243, 1062.

GOLDREICH, P. and TREMAINE, S. : 1982, Ann. Rev. Astron. Astrophys. 20, 249.

GOLDSMITH, W. : 1960, Impact (London : Arnold).

GREENBERG, R. : 1983, Icarus 53, 207.

GUERIN, P. : 1970, Sky and Telescope 40, 88.

HAGIHARA, Y. : 1972, Celestial Mechanics II, 286-312.

HÄMEEN-ANTTILA, K.A. : 1978, Astrophys. Space Sci. 58, 477.

HÄMEEN-ANTTILA, K.A. and LUKKARI, J. : 1980, Astrophys. Space Sci. 71, 475.

HÄMEEN-ANTTILA, K.A. : 1981, Moon and Planets 25, 477.

HÄMEEN-ANTTILA, K.A. : 1982, Moon and Planets 26, 171.

HENON, M. : 1981, Nature 293, 33.

HOLBERG, J.B., FORRESTER, W.T. and LISSAUER, J.J. : 1982, Nature 297, 115.

HOLBERG, J.B. : 1982, Astron. J. 87, 1416.

KUIPER, G.P., CRUIKSHANK, D.P. and FINK, U. : 1970, Sky and Telescope 39, 80.

KUIPER, G.P. : 1974, Cel. Mech. 9, 321.

LANE, A.L., HORD, C.W., WEST, R.A., ESPOSITO, L.W., COFFEEN, D.L., SATO, M., SIMMONS, K.E., POMPHREY, R.B. and MORRIS, R.B. : 1982, Science 215, 537.

LIN, D.N.C. and BODENHEIMER, P. : 1981, Astrophys. J. Lett. 248, L83.

LISSAUER, J.J., SHU, F.H. and CUZZI, J.N. : 1981, Nature 292, 707.

LISSAUER, J.J. and CUZZI, J.N. : 1982, Astron. J. 87, 1051.

LISSAUER, J.J., SHU, F.H. and CUZZI, J.N. : 1984, Planetary Rings, Proceedings IAU Symp. No. 75, 1982, CNES, Toulouse, France.

LISSAUER, J.J. PEALE, S.J. and CUZZI, J.N. : 1984, Icarus 58, 159.

LUKKARI, J. : 1981, Nature 292, 433.

LUMME, K. and IRVINE, W.M. : 1976, Ap. J. Lett. 204, L55.

LUMME, K., ESPOSITO, L.W. IRVINE, W.M. and BAUM, W.A. : 1977, Ap. J. Lett. 216, L123.

LUMME, K. and IRVINE, W.M. : 1979, Astron. Astrophys. 71, 123.

MAROUF, E.A. and TYLER, G.L. : 1982, Science 217, 243.

MAROUF, E.A., TYLER, G.L., ZEBKER, H.A., SIMPSON, R.A. and ESHELEMAN, V.R. : 1983, Icarus 54, 189.

MATTHEWS, K., NEUGEBAUER, G. and NICHOLSON, P.D. : 1978, Bull. Am. Astron. Soc. 10, 581.

NICHOLSON, P.D., PERSSON, S.E., MATTHEWS, K., GOLDREICH, P. and NEUGEBAUER, G. : 1978, Astron. J. 83, 1240.

NICHOLSON, P.D., MATTHEWS, K. and GOLDREICH, P. : 1981, Astron. J. 86, 596.

NICHOLSON, P.D., MATTHEWS, K. and GOLDREICH, P. : 1982, Astron. J. 87, 433.

NICHOLSON, P.D. : 1983, Private communication.

PILCHER, C.P., CHAPMAN, C.R., LEBOFSKY, L.A. and KIEFFER, H.H. : 1970, Science 167, 1372.

PORCO, C., NICHOLSON, P.D., BORDERIES, N., DANIELSON, G.E., GOLDREICH, P., HOLBERG, J.B., and LANE, A.L. : 1984, Icarus, in press.

PORCO, C. : 1984, Ph. D. Thesis, California Institute of Technology, Pasadena, CA.

REITSEMA, H.J., BEEBE, R.F. and SMITH, B.A. : 1976, Astron. J. 81, 209.

SHOWALTER, M.R. and BURNS, J.A. : 1982, Icarus 52, 526.

SHU, F.H., CUZZI, J.N. and LISSAUER, J.J. : 1983, Icarus 53, 185.

SHUKHMAN, I.G. : 1983, "On Collisional Dynamics of Particles in Saturn's Rings", preprint.

SMITH, B.A. and 26 co-authors : 1981, Science 212, 163.

SMITH, B.A. and 28 co-authors : 1982, Science 215, 504.

SPITZER, L. and SCHWARZSCHILD, M. : 1953, Ap. J. 118, 106.

STEWART, G.R., LIN, D.N.C. and BODENHEIMER, P. : 1984, Planetary Rings, R. Greenberg and A. Brahic, Eds., University of Arizona press.

STONE, E.C. and MINER, E.D. : 1982, Science 215, 498.

SYNNOTT, S.P., PETERS, C.F., SMITH, B.A. and MORABITO, L.A. : 1981, Science 212, 191.

SYNNOTT, S.P., TERRILE, R.J., JACOBSON, R.A. and SMITH, B.A. : 1983, Icarus 53, 156.

TRULSEN, J. : 1972, Astrophys. Space Sci. 17, 241.

TYLER, G.L., MAROUF, E.A., SIMPSON, R.A., ZEBKER, H.A. and ESHLEMAN, V.R. : 1983, Icarus 54, 160.

WARD, W.R. : 1981, Geophys. Res. Lett. 8, 641.

WEIDENSCHILLING, S.J., CHAPMAN, C.R., DAVIS, D.R. and GREENBERG, R. : 1984, Planetary Rings, R. Greenberg and A. Brahic, Eds., University of Arizona press.

YODER, C.F. : 1984, Planetary Rings, Proceedings IAU Sy.p. No. 75, 1982, CNES, Toulouse, France.

ZEBKER, H.A. and TYLER, G.L. : 1984, Science 223, 396.

FORMATION OF THE KIRKWOOD GAPS IN THE ASTEROID BELT

A. LEMAITRE
Research Assistant
DEPARTMENT OF MATHEMATICS
Rempart de la Vierge, 8
5000 Namur
BELGIUM

ABSTRACT. A possible mechanism to explain the depletion of the Kirkwood gaps in the asteroid belt would be the slow dissipation of the solar nebula at the origin of the Solar System. The effects of this dissipation on a uniform distribution of asteroids are explored by means of the adiabatic invariant theory for the 2/1, 3/1 and 5/2 resonance cases. The framework is the restricted, circular and planar three body problem.

1 THE MODEL OF RESONANCE

To modelize the motion around the Sun of an asteroid perturbed by Jupiter, we use the planar circular restricted three body problem ; the Hamiltonian H is given by :

$$H = -\frac{\mu^2}{2L^2} - \frac{m_J}{m_S} P + n' \Lambda' \qquad (1)$$

$$\text{where } P = \mu \left(\frac{1}{\Delta} - \frac{\vec{r}.\vec{r}'}{r'^3} \right) \qquad (2)$$

$\mu = G\, m_S$ G is the universal gravitational constant

$\vec{r}$ and $\vec{r}'$ are the positions of the asteroid and of Jupiter with respect to the Sun

Δ is the distance between Jupiter and the asteroid.

The canonical variables are λ, the asteroid's mean longitude, p=-g, the opposite of its argument of pericenter and λ', Jupiter's mean longitude; the conjugate momenta are $L = \sqrt{\mu a}$, $P = L\,(1 - \sqrt{1-e^2})$ and Λ', a and e being the semi major axis and the eccentricity of the

Celestial Mechanics **34** (1984) 329–341. 0008–8714/84.15

asteroid's orbit.
First we expand P in a series of increasing powers of e (Abu-El-Ata and Chapront 1975) : the result can be written as :

$$P = \frac{\mu}{a'} \sum_{l,m} e^{l} \sum_{k} P^{k}_{l,m} \cos (k\lambda' - (k-m)\lambda + mp) \tag{3}$$

where the $P^{k}_{l,m}$ are functions of $\frac{a}{a'}$, combinations of Laplace's coefficients (a' is Jupiter's semi mjor axis).
Note that the Hamiltonian system possesses two first integrals : H and ($\Lambda' + L - P$). It can be shown that the Jacobi integral is :
$H - n'(\Lambda' + L - P)$
The second step is to average the perturbation over the short periodic terms in the neighbourhood of a resonance ($jn = (j+i)n'$).Following the notation of Poincaré (1902) and Schubart (1966) we introduce two new angles :

$$\sigma = \frac{j+i}{i} \lambda' - \frac{j}{i} \lambda + p \tag{4a}$$

$$\nu = - \frac{j+i}{i} \lambda' + \frac{j}{i} \lambda + p' \tag{4b}$$

(p' is the opposite of Jupiter's argument of pericenter; it is obviously constant in this context).
The conjugate momenta are :

$$S = P \tag{5a}$$

$$N = \frac{i}{j} L + P \tag{5b}$$

λ' is now conjugated to $\Lambda = \Lambda' + \frac{j+i}{j} L$

It leads to a new expression for H :

$$H = n'(\Lambda + (S-N) \frac{j+i}{i}) - \frac{i^2\mu^2}{2j^2(N-S)^2} - \frac{m_J}{m_S} P \tag{6}$$

$$P = \frac{\mu}{a'} \sum_{l,m} e^{l} P^{k^*}_{l,m} \cos m\sigma$$

where k^* is defined by : $k^* = m \frac{j+i}{i}$

$$e^2 = \frac{2Si}{j(N-S)} \left(1 - \frac{iS}{2j(N-S)^2} \right) \tag{7}$$

Note that ν and λ' are ignorable and so N and Λ are constants.
The third step is a truncation of H with respect to S (i-e to e)

$$K = \alpha S + \beta S^2 + \varepsilon (\sqrt{2S})^{i} \cos i\sigma \tag{8}$$

where we choose the constant term so to get H=0 when S=0.

$$\alpha = \frac{j+i}{i} n' - \frac{i^2}{j^2} n^* \qquad n^* = \frac{\mu^2}{N^3}$$

$$\beta = - \frac{3i^2}{2j^2N^4}$$

$$\varepsilon = - \frac{m_J}{m_S} \frac{\mu}{a'} P^{j+i}_{i,i} \left(\frac{i}{jN} \right)^{i/2}$$

The fourth and last step is a scaling of th momentum S and of the time t, different for each case of resonance : for example, for the (j+1)/j resonance case :

$$R = \left| \frac{2\beta}{\varepsilon} \right|^{2/3} S$$

$$r = \begin{cases} (\text{sign } \beta)\ \sigma & \text{if } \beta\varepsilon < 0 \\ (\text{sign } \beta)\ \sigma + \pi & \text{if } \beta\varepsilon > 0 \end{cases}$$

$$\tau = \left| \frac{\beta\varepsilon^2}{4} \right|^{1/3} t$$

The hamiltonian function (8) is replaced by (9) :

$$K = R^2 - 3(\delta+1)\ R - 2\ \sqrt{2R} \cos r \qquad (9)$$

$$\text{where } \delta = - \text{sign}(\beta\varepsilon)\ \frac{\alpha}{3}\ \left| \frac{4}{\beta\varepsilon^2} \right|^{1/3} - 1 \qquad (10)$$

becomes the only independant parameter of the problem.
Similar scalings and Hamiltonian functions for the (j+2)/j and (j+3)/j resonance cases have been calculated. (Lemaitre 1983).

2 THE PHASE SPACE

We can easily deduce the phase space : each value of δ gives a plane and in that plane, each curve corresponds to a value of K.
In the (j+1)/j case, for $\delta < 0$, there is no homoclinic orbit, and only one stable equilibrium. For $\delta > 0$, there are two homoclinic orbits and three equilibria (one unstable and two stable).
We call 'resonance zone' the area enclosed between the two homoclinic orbits,'internal zone' or 'I zone' the area enclosed by the smallest homoclinic orbit and 'external zone' or ' E zone' the area outside the largest homoclinic orbit (Henrard and Lemaitre 1983 a and b).
We can characterize each trajectory by a value of δ and a value of the area index, which is defined as the area enclosed by the trajectory, if it lies in the I or in the E zone, and as the area enclosed by the trajectory plus the area enclosed by the smallest homoclinic orbit, if it lies in the resonant zone.So we get a one-to-one correspondance between trajectories and points in the area index diagram (figure 1).

By comparable scalings we can give new Hamiltonian functions and area index diagrams for the (j+2)/j and (j+3)/j resonance cases (lemaitre 1983).

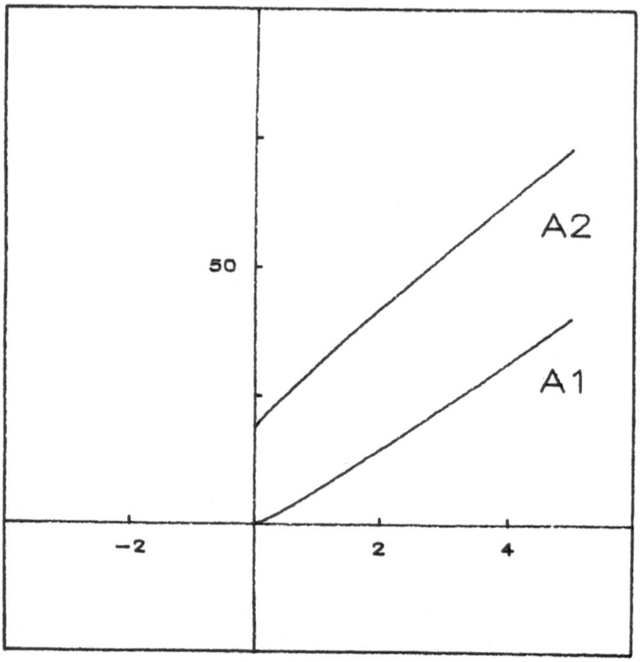

Figure 1 : representation of the three characteristic zones (external, resonant and internal) on the (δ,area index) diagram.

3 EVOLUTION THROUGH THE RESONANCE

Now let us assume that the parameter δ slowly varies with the time :

$|\dot{\delta}| \leq \eta$ and $|\ddot{\delta}| \leq \eta^2$ where η is some small quantity.

The solutions associated to the Hamiltonian (9) are no longer the closed curves described before, but they are closed to them ; more specifically the solutions with varying δ stay for a moderate amount of time close within η to a solution of the system with constant δ (which we call the guiding trajectory).
In the long run, the guiding trajectory evolves. To identify it from phase space to phase space, we use the adiabatic invariant theory, which states that the area enclosed by the trajectory does not change by more than η for times smaller than $1/\eta$. (see for example Henrard 1982).

It may happen that at some point in the evolution the guiding trajectory becomes an homoclinic orbit. In that case the adiabatic invariant theory breaks down, and under some assumptions the evolution of the system can be described in terms of probability of capture : a trajectory situated in one of the three zones (E, I or R) must jump into one of the two others; the choice depends on the phase of the system at the time of transition. The area enclosed by the trajectory undergoes a discontinuity and then stays constant again.
Let us assume now that δ decreases with the time; we can calculate the probability of capture by another zone for each type of trajectory and the conclusions are the following, for the j+1/j resonance case :
a trajectory initially in the I zone is going to jump into the E zone with probability 1;
a trajectory initially in the R zone with a large initial enclosed area is going to jump in the E zone with probability 1;
a trajectory initially in the R zone with a small initial enclosed area becomes external without jump (when $\delta<0$);
a trajectory initially in the E zone stays in the E zone .
The conclusion is that, for variations of δ large enough, all the guiding trajectories become external; in particular, the resonance zone is depleted.
The same mechanism can be applied for the (j+2)/j and the (j+3)/j resonance cases; the only difference is that a trajectory initially in the resonant zone must always jump into it; it never glides into it, whatever its initial enclosed area is.
The figures 2,3 and 4 give the evolution of a uniform distribution of asteroids (truncated at e = 0.25), applying different variations of δ to show the progressive depletion of the resonance zone.

4 APPLICATION TO THE ASTEROID BELT

Now let us compare the distributions with the present observed distribution of astreroids : we use the file TRIAD where we select asteroids with $e<0.35$ and $i<15°$; for the 2/1 case, $0.6 < a < 0.66$, for the 3/1 case, $0.47 < a < 0.491$,; for the 5/2 case, $0.5355 < a < 0.5509$.The unit is Jupiter's semi major axis.
It roughly corresponds to variations of δ of the order of : -3.3 for the 2/1 case, -30. for the 3/1 and -35000. for the 5/2 (figure 5).

5 MECHANISM OF VARIATION OF δ

The question to answer now is the reason for δ to decrease : we are going to present a possible mechanism but it is obvious that there could be many others.
We consider the prehistory of the Solar System ; at some time in the evolution, we assume that there was an enormous flat disk of dust and gas rotating about a proto-Sun, in which Jupiter and the planets were forming. This cloud slowly dissipated, a part of its mass was captured by the Sun, the rest was ejected out of the System.

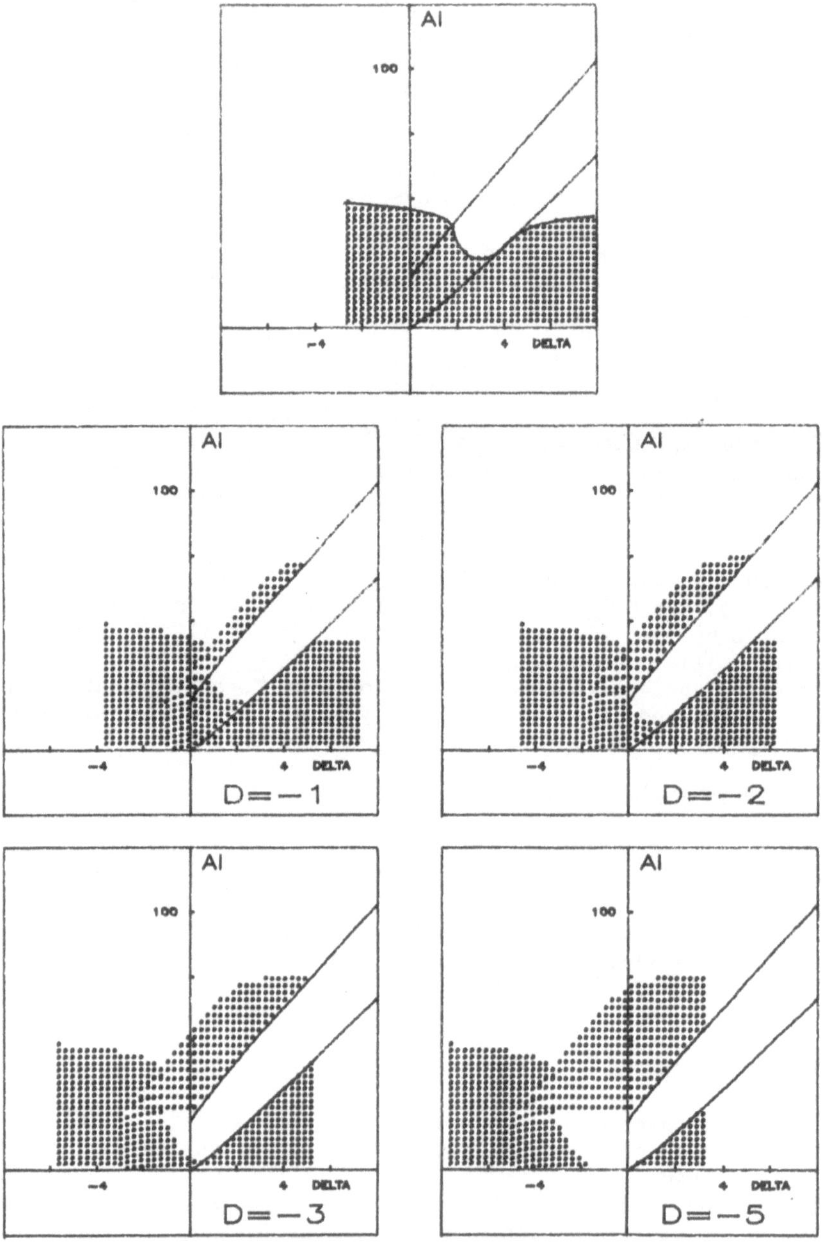

Figure 2 : evolution of a uniform distribution of asteroids (truncated at e= 0.25) for several variations of δ in the (j+1)/j case.
AI : Area Index D = Δδ

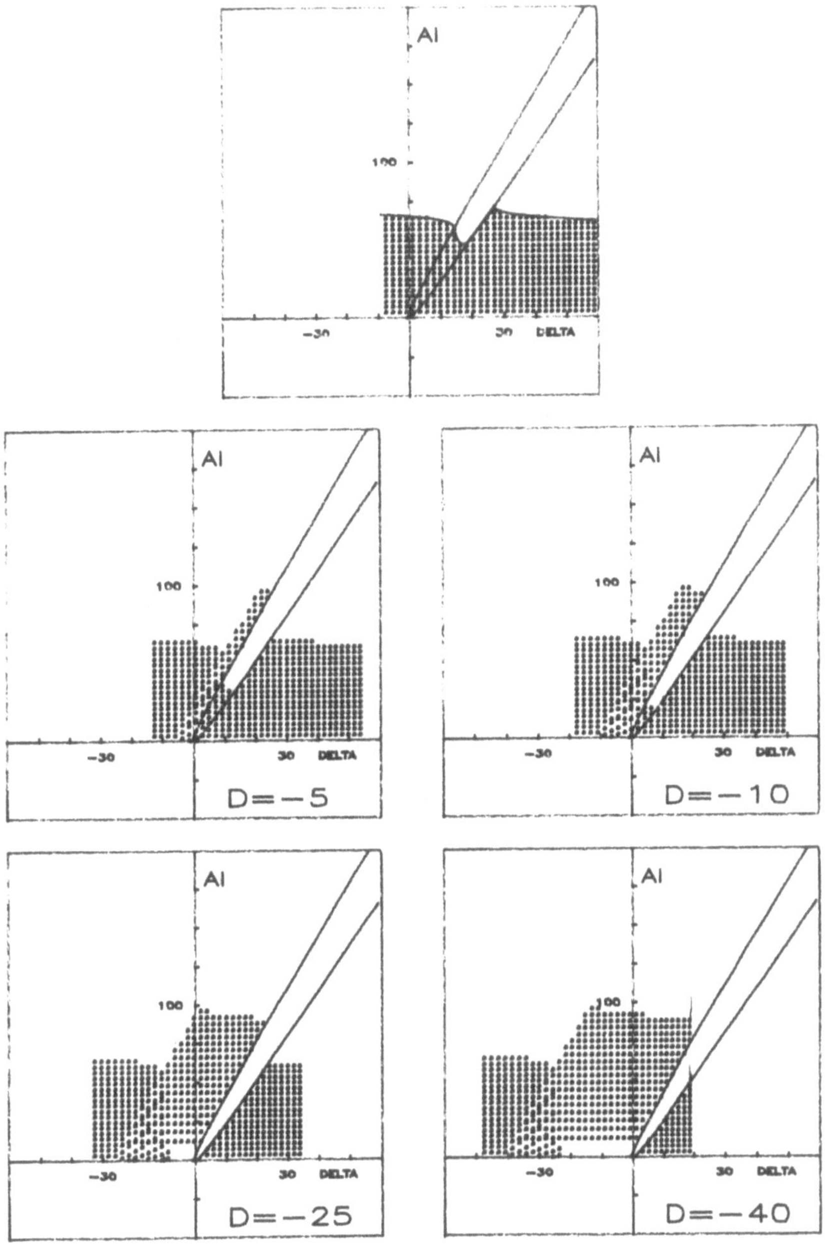

Figure 3 : evolution of a uniform distribution of asteroids (truncated at e= 0.25) for several variations of δ in the (j+2)/j case.
AI : Area Index   D = Δδ

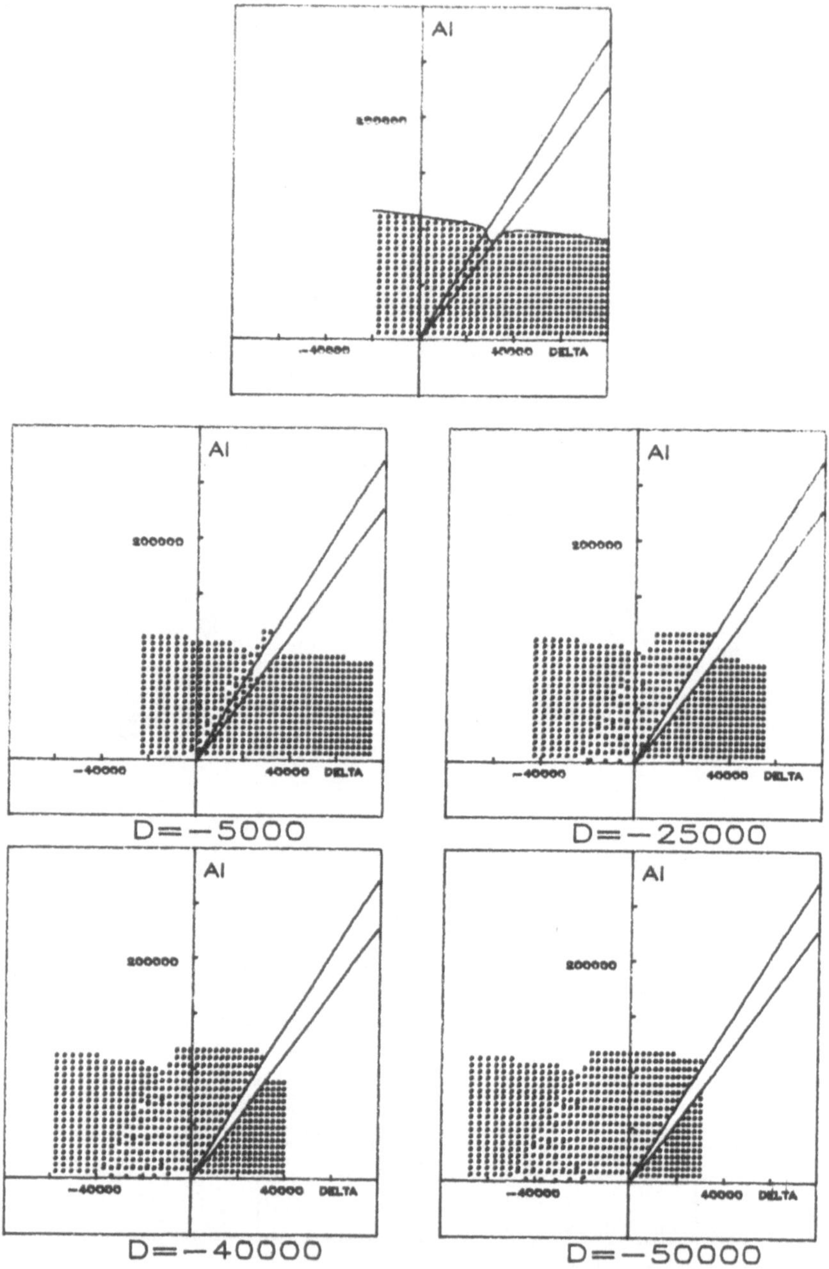

Figure 4 : evolution of a uniform distribution of asteroids (truncated at e= 0.25) for several variations of δ in the (j+3)/j case.
AI : Area Index $D = \Delta\delta$

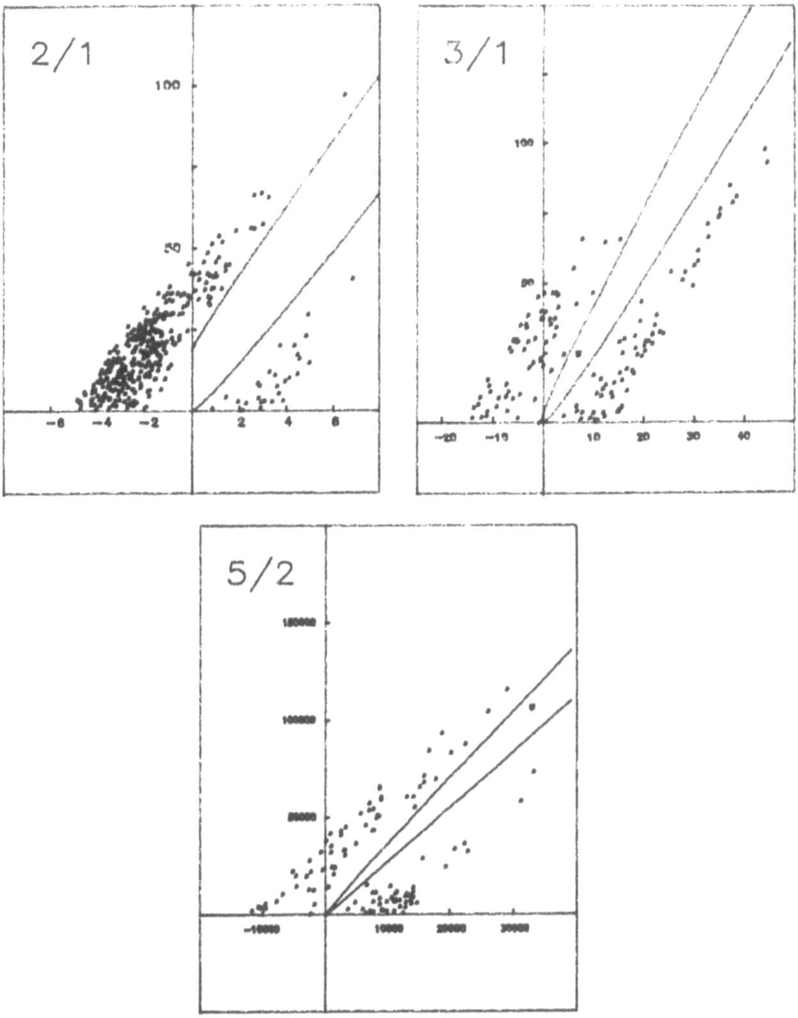

Figure 5 : Observed distributions of the asteroids for the 2/1, 3/1 and 5/2 cases in the (δ,area index) diagram.

Because of this disk, the total mass acting on Jupiter (disk + Sun) is larger than the total mass acting on the asteroid. This difference is responsible for a variation of δ through the parameter α (more precisely through the difference of the mean motions n' and n^*).
In a first model we assume that the surface density in the disk is given by the following law :

$$\sigma(R) = \sigma_0 \left(\frac{a_J}{R}\right)^{3/2} \tag{11}$$

where $\sigma_0 = \sigma(a_J)$.
(a_J being the present value of Jupiter's semi major axis a_J = 5.2026 AU)

We can deduce a corresponding potential using the formula of Lass and Blitzer (1983) :

$$V(R) = V_0 \left(\frac{a_J}{R}\right)^{1/2} + \text{neglected constant terms}$$

$$\text{where } V_0 = -8\, G\, \sigma_0\, K^2\left(\frac{\sqrt{2}}{2}\right) \tag{12}$$

K(.) is the elliptic function of the first type.
We develop this potential in a series of increasing powers of e, we average it over the short periodic terms, we truncate it and we add it to the Hamiltonian (3), so to get :

$$H = \alpha S + \beta S^2 + \varepsilon (\sqrt{2S})^i \cos i\sigma + V_0\, \eta\, S + V_0\, \xi\, S^2 \tag{13}$$

$$\text{where } \eta = (\pi G a_J)^{1/2} \frac{i}{8jN^2} \left(8 - \frac{i}{j}\right)$$

$$\xi = (\pi G a_J)^{1/2} \frac{i}{256jN^3} \left(\frac{i^2}{j^2} - 64\frac{i}{j} + 256\right)$$

This potential also acts on Jupiter's motion ; adding it to the two body problem it leads to a new expression of a', the semi major axis of Jupiter's orbit, which is a solution of :

$$V_0\, a_J^{1/2}\, a'^{3/2} - 2GM\, a' + 2h^2 = 0 \tag{14}$$

where M is the Sun's mass and h the angular momentum (constant)
So we calculate the variation of δ, denoted by $\Delta\delta$, and defined as :

$$\Delta\delta = \delta\Big|_{\substack{M = MF \\ V_0 = 0}} - \delta\Big|_{\substack{M = MI \\ V_0}}$$

where MI is the initial value of M, before dissipation of the disk,

MF is the final value of M, after dissipation, when $V_0 = 0$.

Exactly in the same way, we develop a second model where the density function is :

$$\sigma(R) = \sigma_0 \frac{a_J}{R} \operatorname{Arcos}\left(\frac{R}{R_0}\right) \quad \text{if } R < R_0$$
$$= 0 \quad \text{if } R \geq R_0 \qquad (15)$$

where $\sigma_0 = \sigma(a_J) / \operatorname{Arcos}\left(\frac{a_J}{R_0}\right)$

R_0 is the radius of the disk

This leads to the potential function :

$$V(R) = - V_0 \ln \frac{R}{R_0} + \text{neglected constant terms} \qquad (16)$$

where $V_0 = - \pi^2 G \sigma_0 a_J$

and after extension, averaging and truncation :

$$H = \alpha S + \beta S^2 + \varepsilon (\sqrt{2S})^i \cos i\sigma + V_0 \eta S + V_0 \xi S^2 \qquad (17)$$

where $\eta = \frac{1}{2N}\left(4 - \frac{i}{j}\right)$

$$\xi = \frac{-1}{8N^2}\left(\frac{i^2}{j^2} + 4\frac{i}{j} - 8\right)$$

and a' is solution of the equation (18):

$$V_0 a'^2 - GM a' + h^2 = 0 \qquad (18)$$

The results of the two models are given in Table 1; we see that a density $\sigma(a_J)$ of about 2000 gr/cm² will fit with the actual distribution for the 2/1 resonance case, for all the data of the Table. Similar calculations with the same data for the 3/1 or the 5/2 cases confirm this result (see Table 2).

Let us note that massive nebulae (Cameron 1978) with MF = $M_\odot$ or 'minimum mass nebulae' (Weidenschilling 1977) with MF < $M_\odot$ are both possible in this context ; the only important parameter of our models is the density in the region of the present position of Jupiter.

Table 1 : Values of $\Delta\delta$ in the 2/1 case

σ (a_J)	500	1000	1500	2000	2500
Model 1 MI : 0.7 MF : 1.3	-0.986	-1.956	-2.914	-3.863	-4.803
Model 1 MI : 1.0 MF : 1.1	-0.735	-1.483	-2.184	-2.900	-3.610
Model 2 MI : 0.7 MF : 1.3 MD : 1.0	-0.924	-1.821	-2.686	-3.549	-4.382
Model 2 MI : 0.7 MF : 1.3 MD : 0.1	-0.950	-1.969	-3.060	-4.283	-5.782
Model 2 MI : 1.0 MF : 1.1 MD : 1.0	-0.580	-1.153	-1.721	-2.283	-2.841
Model 2 MI : 1.0 MF : 1.1 MD : 0.1	-0.602	-1.249	-1.960	-2.774	-3.793

Table 2 : Values of $\Delta\delta$ for the other resonance cases
(the mass unit is M_o)

example 1 : first model MI = 1.0, MF = 1.1, σ (a_J) = 2500 gr/cm²
$\Delta\delta$ = -3.610 for the 2/1,
-45.810 for the 3/1
-51420.3 for the 5/2

example 2 : second model, MI = 0.7, MF = 1.3, MD = 1.0,
$\sigma(a_J)$ = 2000 gr/cm²
$\Delta\delta$ = -3.549 for the 2/1 ,
-36.643 for the 3/1
-27874.8 for the 5/2

REFERENCES

ABU-EL-ATA N. and CHAPRONT J. (1975) : Astron. et Astrophys. , **30** ,57-68

CAMERON A.G.W. (1978) : 'Physics of the primitive Solar accretion disk' , The Moon and the Planets , **18** , 5-40

HENRARD J. (1982) : 'Capture into resonance : an extension of the use of adiabatic invariants' , Celest.Mech. , **27** , 3-22

HENRARD J. and LEMAITRE A. (1983a) : 'A Second Fundamental Model of Resonance' , Celest.Mech. **30** ,197-218

HENRARD J. and LEMAITRE A. (1983b) : 'A Mechanism of formation for the Kirkwood gaps' , ICARUS , **55** ,482-494

LASS H. and BLITZER L. (1983) : 'The gravitational potential in uniform disks and rings' , Celest.Mech. , **30** , 225-228

LEMAITRE A. (1983) : 'High order resonances in the restricted three body problem', Celest.Mech. **32** , 109-126

POINCARE H. (1902) : 'Sur les planetes du type d'Hécube' , Bulletin Astronomique , T XIX

SCHUBART J. (1966) : 'Special cases of the restricted three body problem' Proc.Symposium IAU **25** , 187-193

WEIDENSCHILLING S.J. (1977) : 'The distribution of mass in the planetary system and solar nebula' , Astrophys. and Space Science , **51** ,153-158

RESONANT STRUCTURE OF THE OUTER ASTEROID BELT

Andrea Milani and Anna M. Nobili
Dipartimento di Matematica
Universita' di Pisa
I-56100 Pisa Italy

ABSTRACT. An analysis of ordered and chaotic regions of motion in the outer asteroid belt has shown that once the eccentricity of Jupiter is introduced the chaotic regions of the circular model are quite easily depleted. This suggests that also objects in neighbouring regions must be strongly perturbed. Therefore it is not surprising that many outer belt asteroids have been reported in the literature as resonant or anyway dynamically protected. By using the planar elliptic restricted 3-body model we have investigated the motion of outer belt asteroids which had not been suspected to librate. We find 3 cases of ϖ libration and 11 cases of e,ϖ coupling that can be explained within the theory of secular resonances. It is thus established that in the outer belt only resonant and dynamically protected asteroids can have lifetimes of the same order as the age of the Solar System.

1. JUPITER'S PERTURBATIVE EFFECTS ON OUTER BELT ASTEROIDS

If the numbered outer belt asteroids (say with semimajor axis greater than 3.2 A.U.) of the TRIAD file (Bender,1979) are plotted in the a,e plane (a,e are the osculating semimajor axis and eccentricity) we are struck with the evident depletion and extremely uneven distribution (see Fig.1). As far as the depletion problem is concerned, the hypothesis that it is mainly due to the pure gravitational perturbation of Jupiter has been investigated in the last 10 years with numerical experiments. Lecar and Franklin (1973) used a planar elliptic restricted 3-body model with Jupiter's eccentricity equal to 0.06. They simulated an initial uniform distribution of asteroids for 2400y. Their conclusion was that "the region between the 2/1 and 3/2 resonances (i.e. between

Celestial Mechanics **34** (1984) 343–355. 0008–8714/84.15

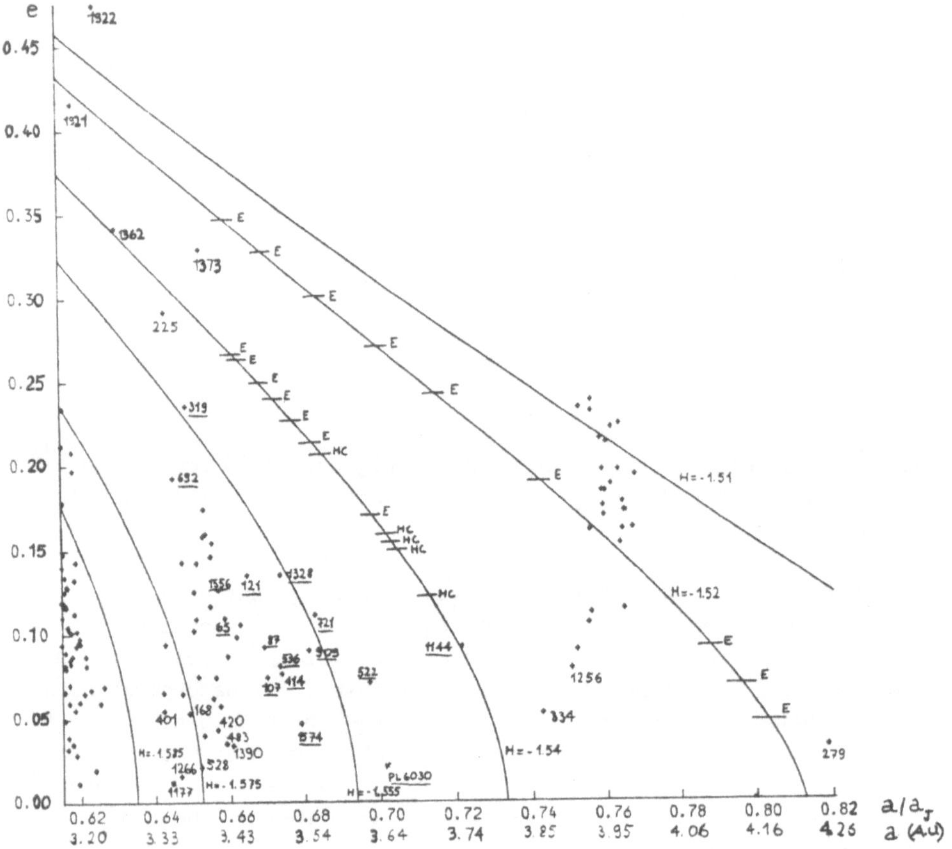

Figure 1.Numbered asteroids (plus PL-6030) are plotted in the a, e plane. The curves are level lines of the Jacobi function H. E means that we have found ejection; HC means that the orbit is highly chaotic (as defined in our paper 1983b). The asteroids whose number is drawn, plus the Hilda group ($a \simeq 3.95$ A.U.) are dynamically protected. The asteroids whose number is underlined have been analysed by ourselves within a planar elliptic 3-body model.

about 3.2 and 4 A.U.) remains well populated". Froeschlé and Scholl (1979) used a 3-dimensional elliptic restricted 3-body model and investigated the region 3.6 A.U. $<$ a $<$ 3.9 A.U. for a longer time interval of 10^5 y. They succeeded in depopulating this region to a larger extent than Lecar and Franklin, but still not enough in comparison with the actual distribution of asteroids. They also showed that, were Saturn included the results remained essentially the same. In fact, if we estimate Jupiter's and Saturn's perturbation on outer belt asteroids with the ε's introduced by Walker et al.(1980) the perturbation of Saturn turns out to be about 20 times weaker than that of Jupiter. Perturbations from inner planets are even weaker.

We have recently investigated the effect of Jupiter's eccentricity on outer belt asteroids with numerical experiments based on different methods (Milani and Nobili, 1983a and 1983b). We first established regions of ordered and chaotic motion in the framework of the restricted circular 3-body problem on different level manifolds of the Jacobi integral (see e.g. Henón and Helies, 1964). Then, the eccentricity of Jupiter was introduced (taking into account its oscillation with about 54000 y period because of Saturn's perturbation) and escapers were searched for in the planar elliptic restricted 3-body model. Instead of starting with a random distribution of asteroids we chose objects in the chaotic regions of the restricted circular problem and with an initial angle $\varpi_0 - \varpi_J$ between asteroid's and Jupiter's pericenter close to 180°. In this way, because of the chaotic behaviour, a very close approach to Jupiter is ensured within few synodic periods. In fact we found ejections, after 10^2-10^3 years only, in the actually depleted region (see Fig.1) thus supporting the gravitational hypothesis more strongly than previous numerical experiments.

The set of initial conditions giving rise to very rapid ejections, like the ones we have found, is very narrow indeed. However, our experiments show that in the chaotic regions the asteroid is actually allowed to go everywhere (i.e. also in the vicinity of Jupiter). This confirms the conjecture that the chaotic behaviour and the absence of a dynamical protection mechanism in these regions ensure that asteroids belonging to them will sooner or later, in $\sim 10^9$ years, pass through the very narrow set of initial conditions giving rise to rapid ejection. The need for non-gravitational mechanisms as major responsible of the lack of asteroids in the outer belt is thus ruled out. According to our experiments, if the eccentricity of Jupiter is set equal to its maximum value of 0.061, ejections occur within a few thousands years at most. Since Jupiter's eccentricity oscillates between 0.030 and 0.061 in about 54000y because of Saturn's perturbation (Cohen et al.,1973) we were forced to conclude that the outer belt was largely depleted over quite short timescales of hundreds of thousands of years.

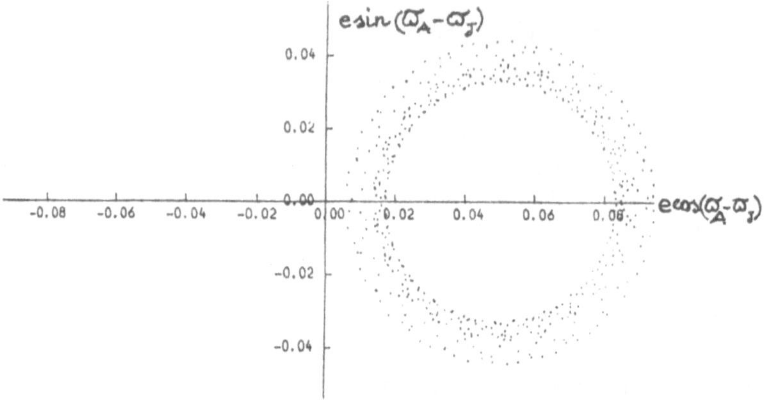

Figure 2. 1144 Oda for 250 synodic periods (1 for each point) in the planar elliptic restricted model. The angle $\varpi_A - \varpi_J$ between the asteroid's and Jupiter's pericenter librates around 0 within about $\pm 70°$.

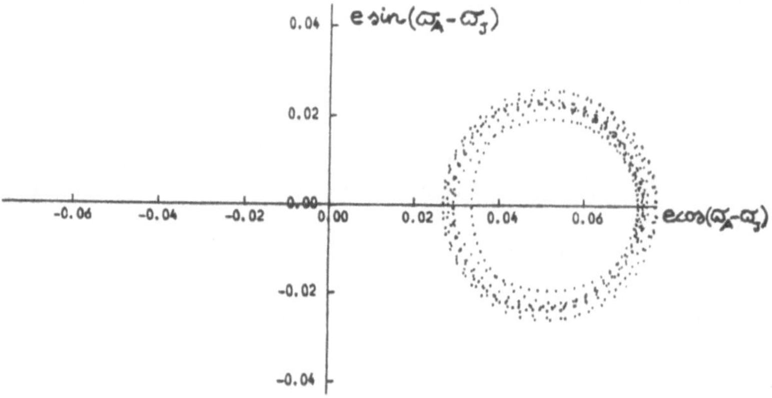

Figure 3. 522 Helga for 250 synodic periods (planar model). The angle $\varpi_A - \varpi_J$ librates around 0 within about $\pm 30°$.

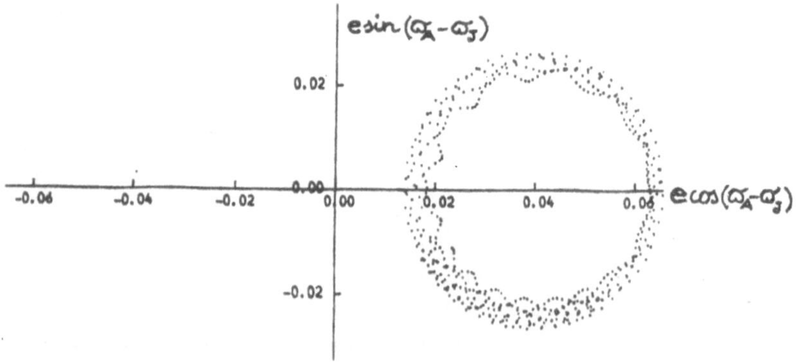

Figure 4. PL-6030 for 250 synodic periods (planar model). The angle $\varpi_A - \varpi_J$ librates arond 0 within about $\pm 40°$.

2. RESONANCE MECHANISMS AND DYNAMICAL PROTECTION

If the eccentricity of Jupiter can have such a destabilizing effect on outer belt asteroids, even the ones that have not been ejected must be strongly perturbed. This suggests the hypothesis that they must be in some peculiar dynamical configuration such that Jupiter's perturbation is minimized. In fact, several kinds of libration and protection mechanisms are presently known to prevent close approaches to Jupiter of observed asteroids in the outer belt. We can mention:

1.) The Trojans. They are in a 1/1 resonance in mean motion with Jupiter and are trapped in two libration islands around the triangular equilibrium points L_4 and L_5 . Because of this libration they never approach Jupiter more than 2.5 A.U. (Cheboratev et al., 1974).

2.) The Hilda group (Schubart, 1968). They have semimajor axis around 3.95 A.U. and are protected from close encounters to Jupiter through a 3/2 resonance in mean motion with it. As it can be appreciated in Fig.2 of our previous paper (1983b), where the orbit of 153 Hilda is also shown, the small amplitude of the libration island where the asteroid is trapped in ensures that it never gets close to Jupiter. In Fig.4 of the same paper we see the surrounding chaotic region where an asteroid wandering therein actually gets as close to Jupiter as 0.5 A.U. It is not surprising that such asteroids are ejected as soon as Jupiter's eccentricity is taken into account (see Fig.1). The critical argument of the libration is $\sigma = 2\lambda_A - 3\lambda_J + \varpi_A$ (λ_J is the mean longitude of Jupiter; λ_A, ϖ_A are mean longitude and longitude of pericenter respectively of the asteroid). It gives the difference between mean longitude of Jupiter and asteroid's perihelion at conjuction. A libration of σ around $0°$ ensures that the configuration with Jupiter at pericenter and asteroid at apocenter (i.e. the closest possible approach) can never occur. According to Schubart (1968), all but two of the Hilda asteroids show a libration of σ around $0°$. None of them can go closer to Jupiter than 1.4 A.U. (Marsden, 1970). In the two cases in which σ does not librate (334 Chicago and 1256 Normannia) a σ,e coupling protection mechanism is operating (Schubart, 1968; Sinclair, 1969): σ circulates, but whenever it is close to $180°$ and a very close approach to Jupiter might occur the eccentricity is at its minimum.

3.) 1362 Griqua, 1921 Pala and 1922 Zulu. They have all been confirmed to be in a 2/1 mean motion resonance with Jupiter (Franklin et al., 1975). The critical argument $\sigma = \lambda_A - 2\lambda_J + \varpi_A$ librates around $0°$ so that, in spite of their very high eccentricity (see Fig.1), they never get closer to Jupiter than about 2 A.U. At lower eccentricity 978 Aidamina, 1125 China and 1101 Clematis have been mentioned to be 2/1 librators around 0°, but only 1125 China has been confirmed by Franklin et al. (1975). They analyse also asteroids of the Palomar-Leiden (PL) Survey (van Houten et al., 1970) which might be trapped in the 2/1 resonance, and find several candidates. Unfortunately their orbits are quite poorly determined (they are of quality class 3 or 4). It is worth noting that, at low eccentricity Franklin et al. find a different protection mechanism: in two cases (1280 Baillauda and 1390 Abastumani) the angle $\varpi_A - \varpi_J$ between the asteroid's and Jupiter's pericenter oscillates around 0°. This is definitely a protection mechanism, since the closest possible approach (i.e. asteroid at apocenter and Jupiter at pericenter) can occur only if the relative position of the two orbits is such that $\varpi_A - \varpi_J$ is close to 180°. Franklin et al.(1975) find also 8 low eccentricity asteroids (1 is a PL object) for which $\varpi_A - \varpi_J$ is, alternatively, either librating around 0° or circulating; when it circulates, the critical argument σ of the mean motion resonance librates around 180°, otherwise it circulates. Although a σ libration around 180° is quite surprising, we must remember that these asteroids have very low eccentricities. The existence of coupled librations near the 2/1 resonance has been studied theoretically by Greenberg and Franklin (1975).

4.) 279 Thule. Takenouchi (1962) has studied this asteroid which is trapped in a 4/3 resonance in mean motion with Jupiter. The critical argument $\sigma = 3\lambda_A - 4\lambda_J + \varpi_A$ librates around 0° with an amplitude of 80° so that, although it could in principle approach Jupiter within 0.3 A.U. (the maximum eccentricity is 0.14), in reality it never gets closer to it than 1.1 A.U. (Marsden, 1970).

5.) 1373 Cincinnati. Kozai (1962) and later Marsden (1970) and Froeschlé and Scholl (1979), have shown that the argument of pericenter of this asteroid librates around 90°. Since it is also highly inclined ($i \simeq 39°$), this means that whenever it is at aphelion, and therefore might be very close to Jupiter because of the high eccentricity, it is about 39° below the ecliptic. As far as we know this is the only example of libration, but there are examples of e,ω coupling: as established by Kozai (1962), when ω circulates, e tends to be a maximum and i a minimum for $\omega = 90°$ or 270° , and viceversa for $\omega = 0°$ or 180° . In the outer belt this happens for 1006 Lagrangea and 225 Henrietta, although the range of variation of both e and i is not very large (Marsden, 1970); and for 721 Tabora and 522 Helga (Froeschlé and Scholl, 1979). The perturbation theory explaining the e,ω coupling can be found in Williams (1969).

The results we have mentioned from 1.) to 5.) clearly show that most of the outer belt asteroids are dynamically protected, in different ways, from very close approaches to Jupiter. As we see in Fig.1, there are still several asteroids between about 3.43 and 3.8 A.U. whose motion has not been studied, with the exception of 721 Tabora and 522 Helga.

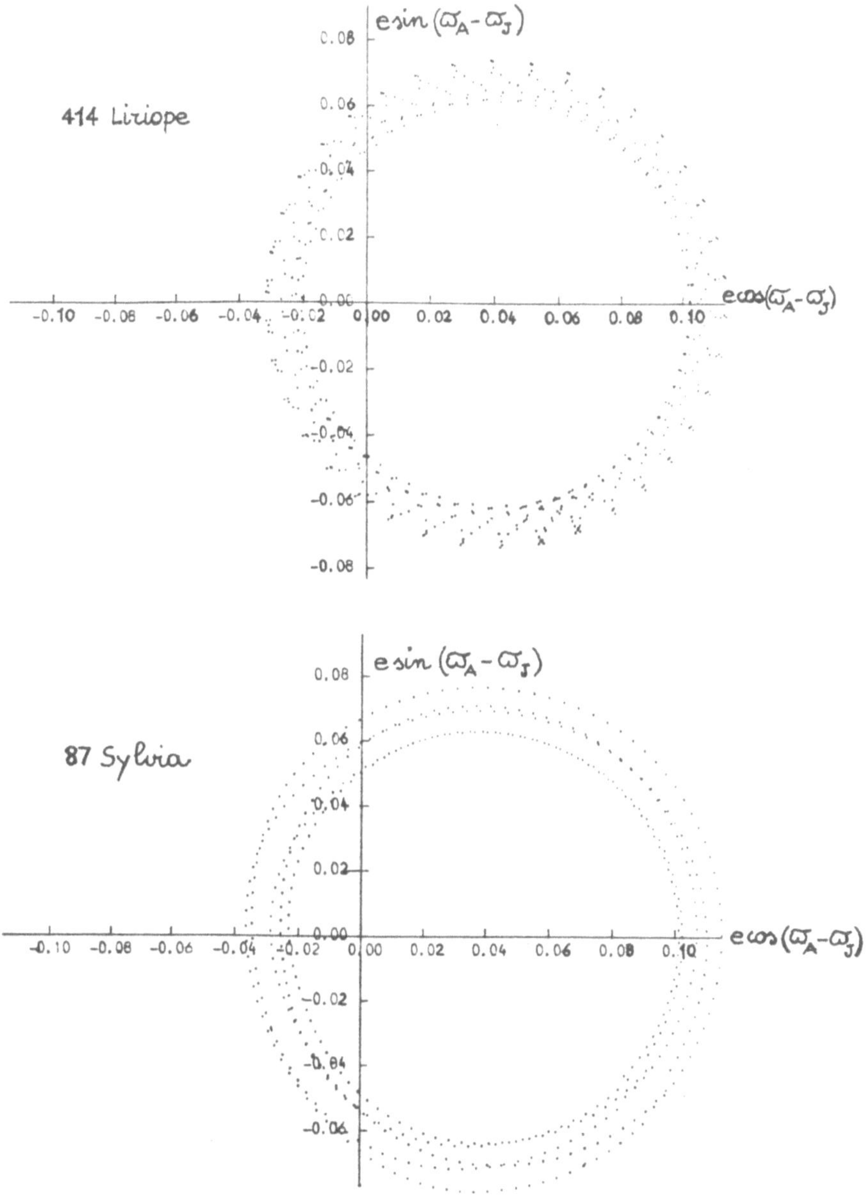

Figure 5. 414 Liriope and 87 Sylvia (planar model). The e, ϖ coupling mechanism is apparent: whenever $\varpi_A - \varpi_J$ is close to 180°, and a very close approach might occur, e is at its minimum. The proper eccentricity is in agreement with the theoretical value in the case of 414 Liriope ($i=9^\circ.542$) but not in the case of 87 Sylvia ($i=10^\circ.879$).

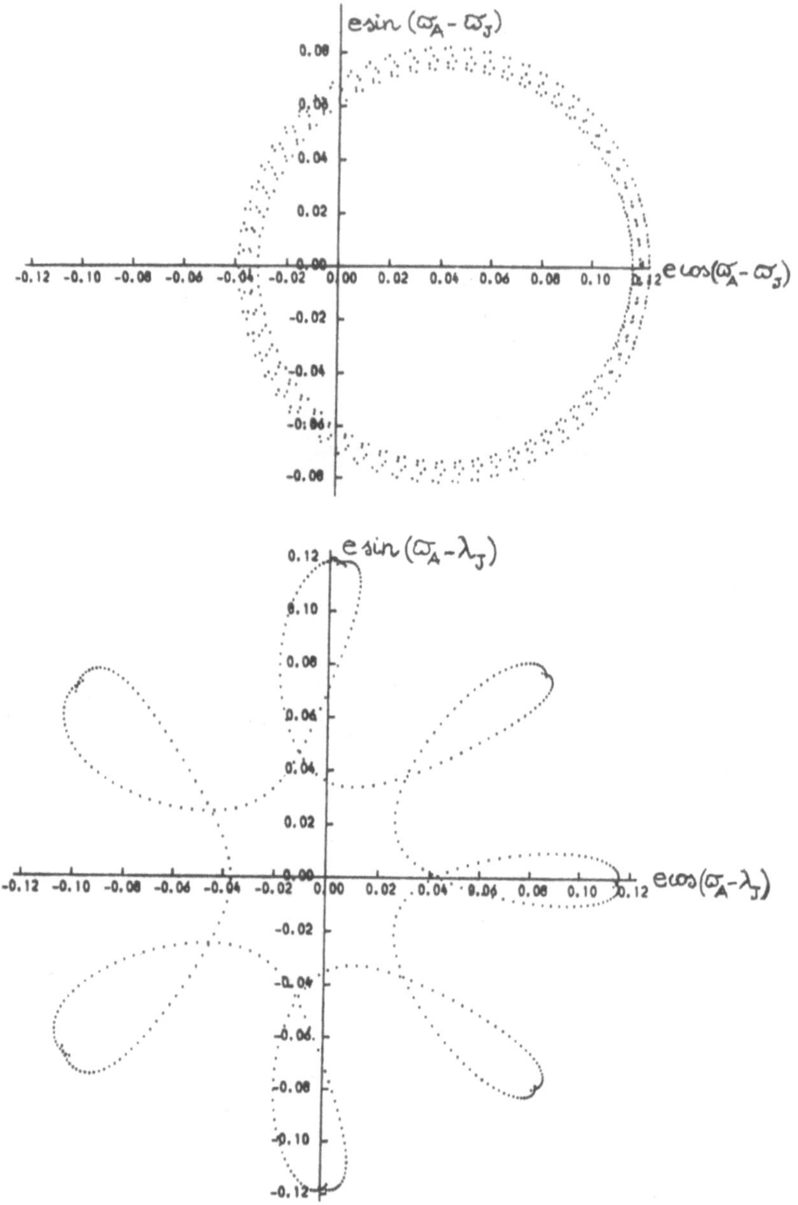

Figure 6. 721 Tabora in the planar elliptic restricted model. The e, ϖ coupling is apparent in the upper diagram. In the lower one we can see that when Tabora's aphelion is close to a conjunction with Jupiter e is at its minimum; when e is large a temporary 16/9 resonance in mean motion prevents close approaches to Jupiter.

1144 Oda and Pl-6030 (a quality class 1 object with a=3.651 A.U.) have also been analysed by Froeschle and Scholl (1979), and they conclude that there is no clear e,ω coupling. Nevertheless, it is reasonable to think that like other outer belt asteroids they are strongly perturbed by Jupiter. Moreover, in the region between 3.54 and 3.8 A.U. there are so few asteroids to make us think that Jupiter's perturbation must have been very effective in almost depleting it (the collisional hypothesis cannot be invoked anyway because of the very few objects therein). Therefore, any asteroid that survived must have been dynamically protected.

We have studied these asteroids within the planar elliptic 3-body model, with the eccentricity of Jupiter equal to its present value. The pericenter of Jupiter was fixed, since it completes one revolution in a period of about 300000 y (Cohen et al., 1973), which is by far longer than the circulation or libration period of the asteroid's pericenter. In every experiment, the asteroid's initial semimajor axis and eccentricity are the osculating values from the TRIAD file. Also the angle between the asteroid's and Jupiter's perihelion is given by the osculating value; the integrations start at pericenter conjunction. We have found the following results. The angle $\varpi_A - \varpi_J$ of 1144 Oda oscillates around 0° in about 2000y. The annulus shown in Fig.2 is centered on the eccentricity forced by Jupiter, which in this region is almost the same as Jupiter's present eccentricity (Brouwer and Clemence, 1961 pp.524-529). The radius is the so called proper eccentricity of the asteroid. It is in agreement with the theoretical value given by Williams (1979).

As in the two cases found by Franklin et al. (1975), the ϖ libration mechanism ensures that the configuration of closest approach to Jupiter cannot occur. In fact Oda approaches Jupiter within 1.14 A.U. at most. Were it not librating, its aphelion distance could reach 4.48 A.U., i.e. only 0.72 A.U. from Jupiter's mean orbit. In the same almost void region near Oda two more asteroids, 522 Helga and PL-6030 show a similar ϖ libration mechanism (see Fig.3 and 4). A comparison with the theoretical value of the proper eccentricity gives a good agreement for PL-6030 (van Houten et al., 1970), but not for 522 Helga (Williams, 1979); the theoretical value of its proper eccentricity is larger than the value found with our planar model by about 0.02. 522 Helga is not highly inclined ($i=4^\circ.42$), but we remind that according to Froeschlé and Scholl (1979) it is protected through the e,ω coupling mechanism which exploits the third dimension. It is possible that both mechanisms (ϖ libration and e,ω coupling) are operating together, and a 3-dimensional analysis of both of them would probably account for the disagreement we have found with the planar model. For values of the semimajor axis smaller than that of 522 Helga, where the asteroids are also quite sparse, we find 11 asteroids that clearly show the e,ϖ

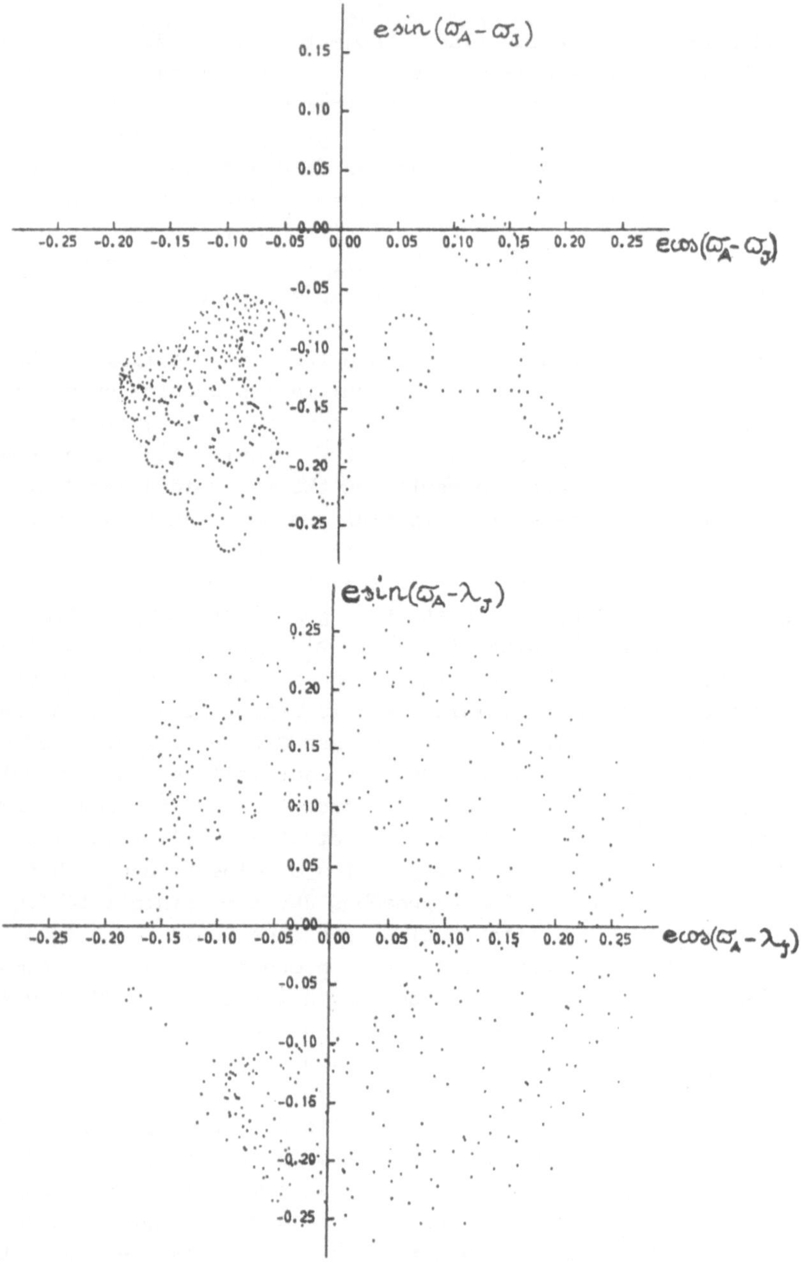

Figure 7. 692 Hippodamia (planar model). The angle $\varpi_A - \varpi_J$ reaches 150° and then starts to come back. The closest it is to 180°, the smallest is the amplitude of the2/1 libration (lower diagram). When $\varpi_A - \varpi_J$ is close to 0° the amplitude ofthe 2/1 libration is about as large as 360°.

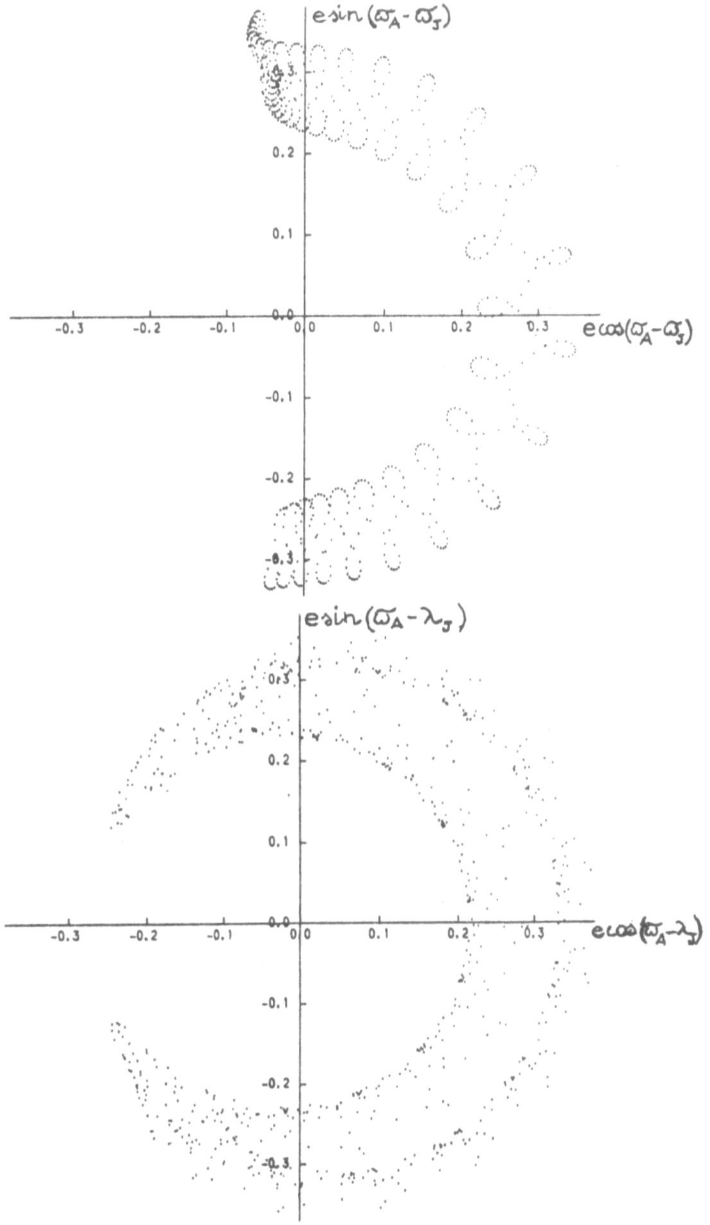

Figuré 8. 319 Leona (planar Model) for 700 synodic periods. The angle $\varpi_A-\varpi_J$ starts to come back to 0° once it has reached about 110° . The closest it is to this value, the smallest is the amplitude of the 2/1 libration (lower diagram). The latter becomes very large when $\varpi_A-\varpi_J$ close to 0° and the closest approach cannot occur.

coupling mechanism: 1556 Wingolfia, 65 Cybele, 121 Hermione, 87 Sylvia, 107 Camilla, 1328Devota, 536 Merapi, 414 Liriope, 1574 Meyer, 909 Ulla and 721 Tabora. Fig.5 shows two of them. The angle $\varpi_A - \varpi_J$ circulates, but when it is close to 180° the eccentricity is minimum. The theoretical and experimental value of the proper eccentricity are in agreement for asteroids inclined less than about 10°, with the exception of 121 Hermione, for which we find disagreement although it is inclined by only 7°.56. A 3-dimensional analysis of its motion would probably provide an answer to this question. Both the ϖ libration and the e,ϖ coupling are explained in the theory of secular resonances.

The case of 721 Tabora is interesting because it shows, besides the e,ϖ coupling mechanism, a 16/9 resonance in mean motion with Jupiter (Fig.6). With also the e,ω coupling found by Froeschlé and Scholl (1979) this asteroid has the surprising characteristic of being protected through 3 different mechanisms.

In the cases of 692 Hippodamia and 319 Leona (two quite highly eccentric asteroids close to the 2/1 resonance) we have found the coexistence of two different mechanisms, even in the planar model. Hippodamia (Fig.7) is a good 2/1 librator as long as $\varpi_A - \varpi_J$ is greater than 90°, but the 2/1 libration reaches a very high amplitude (becoming almost a circulation) when $\varpi_A - \varpi_J$ is close to 0° and therefore there is no need to be protected through another mechanism. In the case of Leona (Fig.8), the farther is $\varpi_A - \varpi_J$ from 0°, the smaller is the amplitude of the 2/1 libration an viceversa, so that when one mechanism is weak the other is strong. During our integration of 700 synodic periods $\varpi_A - \varpi_J$ never becomes larger than 100° and starts to decrease once it has reached this value.

3. CONCLUSIONS

Our numerical experiments in the chaotic regions of the outer asteroid belt (Milani and Nobili, 1983a and b) clearly show the destabilizing effects of Jupiter's eccentricity. This suggests that also objects belonging to the ordered regions of the restricted circular model must be in reality strongly perturbed. If this is so, the observed asteroids must have found their way of minimizing Jupiter's perturbations through some peculiar dynamical configuration. In this paper we have quoted all the outer belt asteroids that have been reported in the literature as being dynamically protected. We have also investigated (in the framework of the planar elliptic restricted 3-body model) most of the outer belt asteroids about which, as far as we can tell, not much was previously known. We have found 3 cases of ϖ libration and 11 cases of e,ϖ coupling. In Fig.1 we show all the protected asteroids in the outer belt. It is apparent that they are by far the great majority. Their number could become larger if an even more systematic analysis will be performed in the future. We definitely conclude that in the outer belt only dynamically protected asteroids managed to survive. The ones we now observe are those which have been "naturally selected" because of their "favourable" dynamical configuration.

REFERENCES

Bender, D. F.: 1979, in Asteroids (T. Geherls ed.), Univ. of Arizona press, Tucson, p.1011.

Brouwer, D. and Clemence, G. M.: 1961, Methods of Celestial Mechanics, Academic press, NY-London.

Cohen, C. J., Hubbard, E. C. and Oesterwinter, C.: 1973, Astr. Pap. Am. Ephemeris 22, 1.

Cheboratev, G. A., Belyaev, N. A.and Eremenko, R. P.: 1974, in The Stability of the Solar System and of Small Stellar Systems, Kozai ed., Reidel Pu. Co., Dordrecht, p.63.

Franklin, F. A., Marsden, B. G., Williams, J. G. and Bardwell,C.M.: 1975, Astron.J. 80, 729.

Froeschlé, C. and Scholl, H.: 1979, Astron. Astrophys. 72, 246.

Greenberg, R. and Franklin, F. A.: 1975, M.N.R.A.S. 173, 1.

Henón, M. and Helies, C.: 1964, Astron. J. 69, 73.

Kozai, Y.: 1962, Astron. J. 67, 591.

Lecar, M. and Franklin, F. A.: 1973, Icarus 20, 422.

Marsden, B. G.: 1970, Astron. J. 75, 206.

Milani, A. and Nobili, A. M.: 1983a, in Asteroid, Comets, Meteors (Lagerkvist and Rickman eds.), Univ. of Uppsala Press, Uppsala, p.127.

Milani, A. and Nobili, A. M.: 1983b,'The Depletion of the Outer Asteroid Belt', submitted to Astronomy and Astrophysics.

Sinclair, A. T.: 1969, M.N.R.A.S. 142, 289.

Schubart, J.: 1968, Astron. J. 73, 99.

Takenouchi,T.: 1962, Ann. Tokyo Obs. 7, 191.

Van Houten,C. J., Van Houten-Groeneveld, I., Herget,P. and Gehrels, T. : 1970, Astron.Astrophys. Suppl. 2, 339.

Walker, I. W., Emslie, A. G. and Roy, A. E.: 1980, Cel. Mech. 22, 371.

Williams, J. G.: 1969, Ph.D thesis, University of California at Los Angeles.

Williams, J. G.: 1979, in Asteroids, T: Geherls ed., Univ. of Arizona Press, Tucson, p.1040.

THE DYNAMICS OF BODIES WITH VARIABLE MASSES

H. Lichtenegger
Institut für Astronomie
Karl-Franzens-Universität
Universitätsplatz 5
A-8010 Graz
Austria

ABSTRACT. The Lie-series provide a convenient and simple method to solve systems of differential equations, especially in problems dealing with variable masses. A system of n-bodies moving inside a cloud and collecting mass is considered. The equations of motion are derived whereby the interchange of momentum is treated as a perturbation. It is shown that the solutions, represented by Lie-series, can be expressed by binomial expansions plus a perturbation which can be solved by interation. In addition, the motion of massless bodies experiencing frictional forces is briefly discussed.

1. INTRODUCTION

We consider the motion of n-bodies inside a nebula and assume that the bodies acquire matter in a continuous way, a situation that may have occured in the early solar system when the planets were formed. The motion will not only be determined by the variable gravitational potential but also by the increase of mass due to the accretion of matter from the surrounding space and by the interchange of momentum. In addition, the interplanetary cloud also exerts a gravitational attraction on the bodies. The equations describing the motion are non autonomous since the masses depend explicitly on time. In the following we deduce the solutions by means of Lie-series and represent them in a way that allows rapid numerical computation.

2. GENERAL REMARKS ON LIE-SERIES

2.1. Introduction

We define a Lie-series by

$$\sum_{\nu=0}^{\infty} \frac{t^\nu}{\nu!} D^\nu f(z) = e^{tD} f(z) \tag{1}$$

Celestial Mechanics **34** (1984) 357–368. 0008–8714/84.15

and the operator D by

$$D = \theta_1(z)\frac{\partial}{\partial z_1} + \theta_2(z)\frac{\partial}{\partial z_2} + \ldots + \theta_n(z)\frac{\partial}{\partial z_n} \tag{2}$$

with holomorphic functions f(z) and $\theta_i(z)$ of the complex variables $z_1,\ldots z_n$, and t denotes another complex variable, independent of z. It can be shown (e.g. Gröbner, 1967) that the series (1) converges absolutely within

$$|t| < \frac{\rho}{(n+1)N}\left(1 - \frac{y}{\rho}\right)^{n+1} \tag{3}$$

where

$$0 \leq y < \rho$$
$$\rho_i \geq |z_i| \tag{3a}$$
$$\rho = \min(\rho_i)$$

and

$$N = \max\{|\theta_i(z)|\} \quad , \quad z_1 = z_2 = \ldots = z_n = \rho \tag{3b}$$

Let us consider a non autonomous system of n differential equations of first order

$$\frac{dX_j}{dt} = \theta_j(X,t) \qquad j = 1,2,\ldots,n \tag{4}$$

with initial conditions

$$(X_j)_{t=t_0} = x_j \qquad j = 1,2,\ldots,n \tag{4a}$$

If we define

$$X_o = t$$
$$(X_o)_{t=t_0} = x_o = t_o \tag{5}$$

then it is easily seen by means of the theorem of commutation (see Gröbner-Knapp, 1967)

$$f(e^{tD}z) = e^{tD}f(z) \tag{6}$$

that the Lie-series

$$X_j = e^{(t-t_o)D}\, x_j \qquad j = 0,1,\ldots,n \tag{7}$$

represent the solutions of system (4), which must be completed via (5)by $\frac{dX_o}{dt} = 1$. The operator (2) now becomes

$$D = \frac{\partial}{\partial x_o} + \theta_1\,(x,x_o)\,\frac{\partial}{\partial x_1} + \ldots + \theta_n\,(x,x_o)\,\frac{\partial}{\partial x_n} \tag{7a}$$

We note that the (numerical) values (4a) may be inserted into the solution (7) only after the required differentiations have been performed. The series (7) converges within the circle (3) and can be continued analytically along any path P in the t-plane provided that there appear no singularities on P.

2.2. Perturbation theory

We assume that the operator D belonging to a system of differential equations of the form (4) can be decomposed into two parts

$$D = D_1 + D_2 \tag{8}$$

with

$$D_1 > D_2 \tag{8a}$$

We denote the solution of the unperturbed problem, formed with D_1, by

$$s(x,t) = e^{tD_1}\, x \qquad x = x_1,\ldots,x_n \tag{9}$$

and the total solution by

$$r\,(x,t) = e^{t(D_1+D_2)}\, x = e^{tD}\, x \qquad x = x_1,\ldots,x_n \tag{10}$$

Since in general D_1 and D_2 do not commute we have

$$e^{t(D_1+D_2)}x = \sum_{\nu=0}^{\infty} \frac{t^\nu}{\nu!} D_1^\nu\, x + \sum_{\nu=1}^{\infty} \frac{t^\nu}{\nu!} D_1^{\nu-1}\, D_2\, x + \sum_{\nu=2}^{\infty} \frac{t^\nu}{\nu!} D_1^{\nu-2}\, D_2\, D\, x + \ldots$$

$$\ldots + \sum_{\nu=\alpha}^{\infty} \frac{t^\nu}{\nu!} D_1^{\nu-\alpha}\, D_2\, D^{\alpha-1}\, x + \ldots \tag{11}$$

With the aid of the formula

$$\frac{t^\nu}{\nu!} = \int_0^t \frac{(t-\tau)^{\alpha-1}}{(\alpha-1)\,!}\,\frac{\tau^{\nu-\alpha}}{(\nu-\alpha)\,!}\, d\tau \qquad \nu \geq \alpha > o \tag{12}$$

and the theorem of commutation we finally find (see Gröbner, 1967)

$$r(x,t) = e^{tD} x = s(x,t) + \int_0^t [D_2 \; r(x,t-\tau)]_{x=s} \, d\tau \tag{13}$$

The subscript x=s indicates that the variables $x_1,\ldots,x_n$ must be substituted by $s(x,\tau)$ after the differentiations have been performed. Equ.(13) can easily be solved by iteration. If we set

$$r_o(x,t) = s(x,t)$$

$$r_{\beta+1}\;(x,t) = s(x,t) + \int_0^t [D_2 \; r_\beta \;(x,t-\tau)]_{x=s} \, d\tau \quad \beta=0,1,2,\ldots \tag{14}$$

we obtain

$$r_1(x,t) - r_o(x,t) = \int_0^t [D_2 \sum_{\nu=0}^{\infty} \frac{(t-\tau)^\nu}{\nu!} \; D_1^\nu \; x]_{x=s}$$

$$= \sum_{\alpha,\nu=0}^{\infty} \frac{t^{\alpha+\nu+1}}{(\alpha+\nu+1)!} D_1^\alpha \; D_2 \; D_1^\nu \; x \tag{15}$$

By successive continuation of this procedure we can represent the solution (10) by

$$r(x,t) = s(x,t) + \sum_{\beta=0}^{\infty} [r_{\beta+1}\;(x,t) - r_\beta(x,t)] \tag{16}$$

3. THE EQUATION OF MOTION

Let us envisage a system of n-bodies moving inside a cloud, each of them acquiring matter. Then the behaviour of the system will mainly be governed by the variation of mass and the Lagrangean equations of motion in its general form are

$$\frac{d}{dt}\left(\frac{\partial T}{\partial \dot{q}_j}\right) - \frac{\partial T}{\partial q_j} = Q_j \qquad j = 1,2,\ldots,s \tag{17}$$

where s is the number of degrees of freedom, q_j and $\dot{q}_j$ are the generalized coordinates and velocities and T is given by

$$T = \sum_{i=1}^{n} \frac{1}{2} m_i(t) \left(\sum_j \frac{\partial \vec{r}_i}{\partial q_j} \dot{q}_j + \frac{\partial \vec{r}_i}{\partial t}\right)^2 \tag{17a}$$

Q_j denotes the components of the sum of the external forces

$$Q_j = \sum_i \vec{F}_i \cdot \frac{\partial \vec{r}_i}{\partial q_j} \tag{17b}$$

Although the gravitational potential V now depends through the masses m_i on the time t we can derive the mutual gravitational forces $\vec{F}_i^G$ from the potential V

$$\vec{F}_i^G = - \vec{\nabla}_i V \tag{18}$$

$$Q_j = - \frac{\partial V}{\partial q_j}$$

since

$$\oint \vec{F} \cdot d\vec{s} = 0 \tag{18a}$$

still holds ($d\vec{s}$ is the line-element).

Further, the cloud not only supplies mass to the bodies but also interchanges momentum with them. This is taken into account by

$$Q_j^{int} = \dot{m}_i \dot{\vec{r}}_i{}' \cdot \frac{\partial \vec{r}_i}{\partial q_j'} \tag{19}$$

where $\dot{\vec{r}}'$ denotes the velocity of the mass before it falls on the body. Finally, if we assume that the mass density ρ of the cloud only depends on the distance $|\vec{r}-\vec{r}_o|$ from the center of the cloud, the generalized force exerted by the cloud can be written as

$$Q_j^c = \frac{4\pi}{3} G \rho \left(|\vec{r}-\vec{r}_o|\right) m_i |\vec{r}_{io}| \frac{\partial \vec{r}_i}{\partial q_j} \tag{20}$$

where $|\vec{r}_{io}| = |\vec{r}_i - \vec{r}_o|$ is the distance between the center of the cloud and the i-th body; G is the gravitational constant.

With the expression L = T - V equ. (17) becomes

$$\frac{d}{dt}\left(\frac{\partial L}{\partial \dot{q}_j}\right) - \frac{\partial L}{\partial q_j} = Q_j^{int} + Q_j^c \qquad j = 1,2,\ldots,s \tag{21}$$

In order to simplify equ. (21) we put Q_j^c into the gradient $\frac{\partial V}{\partial q_j}$. Then the potential V reads

$$V(\vec{r})\Big|_{\vec{r}=\vec{r}_k} = - \sum_{\substack{l=o \\ l\neq k}}^{n} \frac{G m_l m_k}{|\vec{r} - \vec{r}_l|}\Big|_{r=r_k} \tag{22}$$

m_o is the mass of the cloud within a sphere S of radius $|\vec{r}-\vec{r}_o|$ at $\vec{r}=\vec{r}_k$

$$m_o\,(|\vec{r}-\vec{r}_o|) = \int_S \rho(|\vec{r}-\vec{r}_o|)\, d\upsilon \tag{22a}$$

Here $d\upsilon$ is the volume element.

4. THE SOLUTION WITH LIE-SERIES

We attach the origin of our frame of reference to the center of mass of the system which at any time coincides with the center of mass of the cloud ($\vec{r}_o = 0$) and assume that the cloud remains always at rest with respect to the frame. Hence it follows $\dot{\vec{r}}_i' = 0$. If we reduce the s equations (21) of second order to 2s equations of first order, the corresponding Lie-operator reads, using cartesian coordinates ($\vec{r}=(x,y,z)$)

$$D = \sum_{k=1}^{n} \{\dot{\vec{r}}_k \cdot \frac{\partial}{\partial \vec{r}_k} - [\frac{1}{m_k}\frac{\partial V}{\partial \vec{r}_k} + \frac{\dot{m}_k}{m_k}\dot{\vec{r}}_k]\frac{\partial}{\partial \dot{\vec{r}}_k}\} + \frac{\partial}{\partial t} \tag{23}$$

with V given by (22) and the notation

$$\frac{\partial}{\partial \vec{r}_k} = \frac{\partial}{\partial \vec{r}}\Bigg|_{\vec{r}=\vec{r}_k}, \quad \dot{\vec{r}}_k \cdot \frac{\partial}{\partial \vec{r}_k} = \dot{x}_k \frac{\partial}{\partial x_k} + \dot{y}_k \frac{\partial}{\partial y_k} + \dot{z}_k \frac{\partial}{\partial z_k} \tag{23a}$$

If the gravitational attraction of the cloud were neglected (i.e. $m_o=0$ in equ. (22)) it is readily seen that the angular momentum $\vec{I}$ of the system is conserved, since

$$D\vec{I} = D\,(\sum_k m_k\, \vec{r}_k \times \dot{\vec{r}}_k) = 0 \tag{24}$$

Now we split the operator (23) into two parts

$$D = D_1 + D_2 \tag{25}$$

and assume that the influence of the interchange of momentum is small in comparison with the influence of the gravitational attraction of the bodies. Then

$$D_1 > D_2 \tag{25a}$$

where

$$D_1 = \sum_{k=1}^{n} \{\dot{\vec{r}}_k \cdot \frac{\partial}{\partial \vec{r}_k} - \frac{1}{m_k} \frac{\partial V}{\partial \vec{r}_k} \frac{\partial}{\partial \dot{\vec{r}}_k}\} + \frac{\partial}{\partial t} \tag{25b}$$

and

$$D_2 = - \sum_{k=1}^{n} \frac{\dot{m}_k}{m_k} \dot{\vec{r}}_k \frac{\partial}{\partial \dot{\vec{r}}_k} \tag{25c}$$

Thus the terms of the Lie-series built with D_1 are

$$D_1 \vec{r}_k = \dot{\vec{r}}_k$$

$$D_1^2 \vec{r}_k = \sum_{\substack{l=o\\ l\neq k}}^{n} m_l \vec{r}_{lk} r_{lk}^{-3} = \sum_{\substack{l=o\\ l\neq k}}^{n} m_l \vec{U}_{lk} \tag{26}$$

$$D_1^3 \vec{r}_k = \sum_{\substack{l=o\\ l\neq k}}^{n} [m_l (r_{lk}^{-3} D_1 \vec{r}_{lk} + \vec{r}_{lk} D_1 r_{lk}^{-3}) + \vec{r}_{lk} r_{lk}^{-3} D_1 m_l]$$

$$= \sum_{\substack{l=o\\ l\neq k}}^{n} [m_l D_1 \vec{U}_{lk} + \vec{U}_{lk} D_1 m_l]$$

.
.
.

where we have introduced the vector $\vec{U}_{lk} = \vec{r}_{lk} r_{lk}^{-3}$
and with

$$D_1 r_{lk}^{-3} = - r_{lk}^{-2} a_{11} r_{lk}^{-3} \sigma_{lk} \tag{26a}$$

$$D_1^2 r_{lk}^{-3} = - r_{lk}^{-2} [a_{21} \sigma_{lk} D_1 r_{lk}^{-3} + a_{22} r_{lk}^{-3} D_1 \sigma_{lk}]$$

.
.
.

and

$$D_1\, \sigma_{1k} = \dot{\vec{r}}_{1k}\, D_1\, \vec{r}_{1k} + \vec{r}_{1k}\, D_1\, \dot{\vec{r}}_{1k}$$

$$D_1^2\, \sigma_{1k} = \dot{\vec{r}}_{1k}\, D_1^2\, \vec{r}_{1k} + 2\, D_1\, \vec{r}_{1k}\, D_1\, \dot{\vec{r}}_{1k} + \vec{r}_{1k}\, D_1^2\, \dot{\vec{r}}_{1k} \tag{26b}$$

$$\vdots$$

Here we have set $G = 1$, $\vec{r}_{1k} = \vec{r}_1 - \vec{r}_k$ and $\sigma_{1k} = \vec{r}_{1k} \cdot \dot{\vec{r}}_{1k}$.

Equ. (26) - (26b) are just binomial expansions, and the coefficients a_{ij} in (26a) are given by (Hanslmeier and Dvorak, 1984)

$$\begin{aligned} a_{ii} &= 3 && i = 1,2,\ldots \\ a_{i1} &= a_{i-1,1} + 2 && i = 2,3,\ldots \\ a_{ij} &= a_{i-1,j} + a_{i-1,j-1} && j = 2,3,\ldots \quad j<i \end{aligned} \tag{27}$$

In equ. (26) $D_1\, m_1 = \dot{m}_1$ $(1 = 1,2,\ldots,n)$ whereas $D_1\, m_o$ depends on the special choice of $\rho(r)$.

Now the solution

$$\vec{r}_k\,(t) = e^{tD}\, \vec{r}_k \Big|_{\vec{r}_k = \vec{r}_k^{(o)}} \qquad k = 1,2,\ldots,n \tag{28}$$

with the corresponding operator (23), can be written as

$$\vec{r}_k(t) = \vec{r}_k^{(o)} + t\, \dot{\vec{r}}_k^{(o)} +$$

$$+ \sum_{\nu=2}^{\infty} \frac{t^\nu}{\nu!} \sum_{\mu=0}^{\nu-2} \Big[\sum_{\substack{1=o \\ 1\neq k}}^{n} \binom{\nu-2}{\mu} D_1^\mu\, m_1\, D_1^{\nu-\mu-2}\, \vec{U}_{1k}\Big]_{\vec{r}_{1k} = \vec{r}_{1k}^{(o)}} +$$

$$+ \sum_{\beta=o}^{\infty} [\vec{r}_{k,\beta+1}\,(t) - \vec{r}_{k,\beta}\,(t)] \qquad k = 1,2,\ldots,n \tag{29}$$

where we have made use of equ. (16) and $\vec{r}_k^{(o)} = \vec{r}_k(t)$ at $t = 0$.

$\vec{r}_{1k} = \vec{r}^{(o)}_{1k}$ indicates that $\vec{r}^{(o)}_{1k}$ must be substituted into $\vec{U}_{1k}$ after the differentiations have been performed. The differences $\vec{r}_{k,\beta+1}(t) - \vec{r}_{k,\beta}(t)$ are given by equ. (15) and successive continuation. Further

$$D_1^{\nu}\, \vec{U}_{1k} = \sum_{\alpha=o}^{\nu} \binom{\nu}{\alpha} D_1^{\alpha}\, \vec{r}_{1k}\, D_1^{\nu-\alpha}\, r_{1k}^{-3} \tag{30}$$

There arises no difficulty in determining the velocity $\dot{\vec{r}}_k(t)$. Since $D\,\vec{r}_k = \dot{\vec{r}}_k$ it follows

$$\dot{\vec{r}}_k\,(t) = \dot{\vec{r}}^{(o)}_k + \sum_{\nu=1}^{\infty} \frac{t^{\nu}}{\nu!} \sum_{\mu=o}^{\nu-1} \Big[\sum_{\substack{1=o\\ 1\neq k}}^{n} \binom{\nu-1}{\mu} D_1^{\mu}\, m_1\, D_1^{\nu-\mu-1}\, \vec{U}_{1k}\Big]_{\vec{r}_{1k}=\vec{r}^{(o)}_{1k}} +$$

$$+ \sum_{\beta=o}^{\infty} [\dot{\vec{r}}_{k,\beta+1}\,(t) - \dot{\vec{r}}_{k,\beta}(t)] \qquad k=1,2,\dots,n \tag{31}$$

The solutions (29) and (31) can be simplified, since for practical purposes they will be solved by analytical continuation. If the steps of integration are not chosen too large, the dependence of the density ρ on the distance r may be neglected within one step of integration, thus $D\, m_o = 0$.

Further, if $\frac{\dot{m}_k}{m_k}$ varies only slowly with time, we may also set $D\,(\frac{\dot{m}_k}{m_k}) = 0$ within one step of integration. If an exponential increase of mass is considered, then at any time $\frac{\dot{m}_k}{m_k}$ = constant, which simplifies the operator D_2.

Let us finally consider a number of n' massless bodies also moving inside the cloud. By definition they do not change mass and are therefore only exposed to the variable gravitational potential V due to the massive bodies and, in addition, they may experience a frictional force which we assume to be proportional to the velocity

$$\vec{F}^{f}_k = \kappa_k\,(\rho)\, \dot{\vec{r}}_k \qquad k = n+1,\ n+2,\dots,n+n' \tag{32}$$

The coefficient κ is a function of the density since the cloud is inhomogenous with respect to the distance. Now we decompose the operator again into two parts D_1 and D_2 which are given by

$$D_1 = \sum_{k=n+1}^{n'} \{\dot{\vec{r}}_k \frac{\partial}{\partial \vec{r}_k} + \sum_{1=o}^{n} \frac{m_1\, \vec{r}_{1k}}{r_{1k}^3} \frac{\partial}{\partial \dot{\vec{r}}_k}\} + \frac{\partial}{\partial t} \tag{33}$$

and

$$D_2 = - \sum_{k=n+1}^{n'} \kappa_k(\rho) \dot{\vec{r}}_k \frac{\partial}{\partial \dot{\vec{r}}_k} \tag{33a}$$

If we assume $D_1 > D_2$ the solution can be written similar to (29) and (31) with the only difference that k runs from n + 1 to n' and D_1 and D_2 are given by (33) and (33a) respectively.

Again we can argue as before that within one step of integration the change of the density with the distance from the origin can be neglected. This automatically implies that the coefficient also remains constant during one step of integration, thus $D \kappa = 0$. That again simplifies the operator (33a). After each step m_o and κ must be adjusted to their actual values.

5. EXAMPLE

Let us consider a two-body-system (i.e. n = 2) with masses m_1 and m_2, respectively, moving inside a spherical cloud and collecting matter by an exponential law

$$m(t) = m_1(t) + m_2(t) = m(o)\, e^{\delta t}$$

and

$$m(t) + m_c(t) = \text{const.}$$

The mass m_c of the cloud is calculated by ($d\upsilon$ is the volume element)

$$m_c = \int_0^\infty \rho(r)\, d\upsilon$$

with

$$\rho(r) = \rho_o\, e^{-\alpha r}$$

As a numerical example we compute the motion of m_2 around m_1, using the following initial conditions (the gravitational constant is set to unity):

$$m = m_1 + m_2 = 1\, m_\odot \quad , \quad m_c = 1\, m_\odot$$

$$v = 0 \quad , \quad e = 0 \quad , \quad a = 6 \text{ AU}$$

v is the true anomaly, e the excentricity and a is the semi major axis. The constants ρ_o and δ are chosen to be

$\rho_o = 0.001$

$\delta = 6.931471 \times 10^{-5}$

The value of δ implies that the mass of the cloud will be absorbed by the body within 10000 years.

We use a step length of 15 days and $\nu = 5$ in equ.(29). After 1000 years we get the following values

$m = 1.071773$ $\qquad m_c = 0.928226$

$a = 5.200968$ $\qquad e = 0.000918$

The largest numerical values of the terms appearing in equ. (29) during the integration time are found to be:

	x	y
$\nu = 2$	0.1553×10^{-2}	0.1471×10^{-3}
$\nu = 3$	0.1270×10^{-4}	0.1207×10^{-5}
$\nu = 4$	0.7746×10^{-7}	0.7374×10^{-8}

Thus the error in the coordinates and velocities will be less than 10^{-6}. The computer time for one step of integration is about 0.00018 seconds.

6. DISCUSSION

The method described above allows a clear and simple treatment of the dynamics of bodies with variable mass. The presented solutions (equ.(29) and (31)) are given by recurrence formulae and thus are easy to handle on the computer. As indicated by equ.(3) the circle of convergence is determined by the function θ_i of the operator (2). By inspection of equ. (25b) it is seen that the behaviour of convergence of the series obtained with D_1 ("unperturbed solution") critically depends on the distances $|\vec{r}_{1k}|$ of the bodies and on the rate of the change of mass. The former as the latter can be provided for by reducing the step size as well as by taking into account a greater number of terms. The convergence of the iterative procedure will essentially depend on the operator D_2, i.e. on $\frac{\dot{m}_k}{m_k}$. As long as $|\sum_{1 \neq k} m_1 \vec{r}_{1k} \vec{r}_{1k}^{\,-3}| \gg |\frac{\dot{m}_k}{m_k}|$ only a very few terms in equ. (15) and in its successive continuations will be of significant importance. Finally we note that the simplicity of the method allows the inclusion of additional perturbations with little amount.

REFERENCES

Gröbner, W.: 1967, Die Lie-Reihen und ihre Anwendungen VEB Deutscher Verlag der Wissenschaften, Berlin

Gröbner, W., Knapp, H.: 1967a, Contribution to the Method of Lie-Series Bibliograph.Inst., Mannheim

Hanslmeier, A., Dvorak, R.: 1984, Numerical Integration with Lie-Series Astron.Astrophys. 132, Nr. 1

NUMERICAL EXPERIMENTS ON PLANETARY ORBITS IN DOUBLE STARS

R. Dvorak
Institut für Astronomie
Universität Wien
Türkenschanzstraße 17
A-1180 Wien, Austria

ABSTRACT. This is a numerical study of orbits in the elliptic restricted three-body problem concerning the dependence of the critical orbits on the eccentricity of the primaries. They are defined as being the separatrix between stable and unstable single periodic orbits. As our results are adapted to the existence of planetary orbits in double stars we concentrated first on the P-orbits (defined to surround both primaries). Due to the complexity of the elliptic problem there is no analytical approach possible. Using the results of some 300 integrated orbits for 10^3 to 3.10^3 periods of the primaries we established lower and upper bounds for the critical orbits for different values of the eccentricity.

1. INTRODUCTION

This work is a first attempt to treat the problem of stable single periodic(quasiperiodic) orbits in the elliptic restricted three body problem adaptable to the existence of planets in double stars. There exist approaches to treat this problem in the circular restricted problem by J.Hadjidemetriou (1976) and V.Szebehely (1980). As we know from the systematic survey of Hénon and Guyot (1970) on the stability of periodic orbits in the circular restricted problem as it is used by Dvorak (1983), three different types of possible planetary orbits can be distinguished in this model. In this numerical experiment we concentrate on the planet-type orbits (which surround both primaries) and study numerically the stability behaviour of periodic (quasiperiodic) orbits (= PO) according to the eccentricity of the primaries.

2. GENERAL PROBLEMS OF PLANETARY ORBITS IN DOUBLE STARS

Using the results of Hénon and Guyot (1970; = HG) it is possible to discuss this topic in the circular restricted problem. In the mentioned study a systematic survey of all critical orbits, for $0 \leq \mu \leq 1$, is presented; these are defined as the separatrices between single stable

Celestial Mechanics **34** (1984) 369–378. 0008–8714/84.15

periodic orbits and unstable periodic orbits. This was done using the Jacobian constant and the method of surfaces of section. The stability study was undertaken in the sense that the perpendicular intersections of an orbit with the x axis after one revolution (in the rotating frame) can be regarded as an area preserving mapping, which is then used as stability indicator. The families f,g,h,i,l and m are studied in the range $0 \leq \mu \leq 1$. It turns out that there is a great variety of such critical orbits according to the family and the value of μ. On the basis of these results and the wellknown stability of the Lagrange points L_4 and L_5 three different types of stable planetary orbits in double stars can be defined (Dvorak, 1983):

- The librator type orbits (L-orbits), which are stable orbits around L_4 or L_5.
- The satellite type orbits (S-orbits), which surround only one primary and include the classes f,g,h and i.
- The planet type orbits (P-orbits), which surround both primaries and include the classes l and m.

We limit ourselves in this study to the P-orbits and especially to the Strömgren class l. Due to HG there appear for all classes small strips of unstable POs in the region of stable POs, which disappear for μ=0 and μ=0.5. This interesting phenomenon can be seen in detail in fig.1 (after HG).

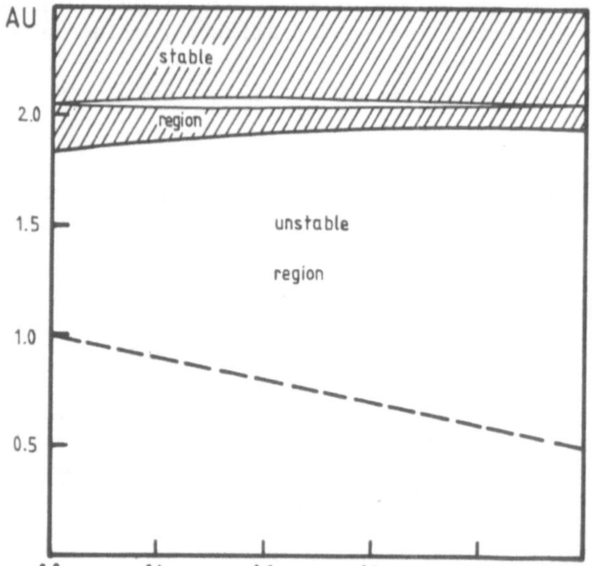

Fig.1 Critical orbits of the family "l"
(The dotted line corresponds to the positions of one of the primaries)
(after: M. Hénon and M. Guyot, 1970)

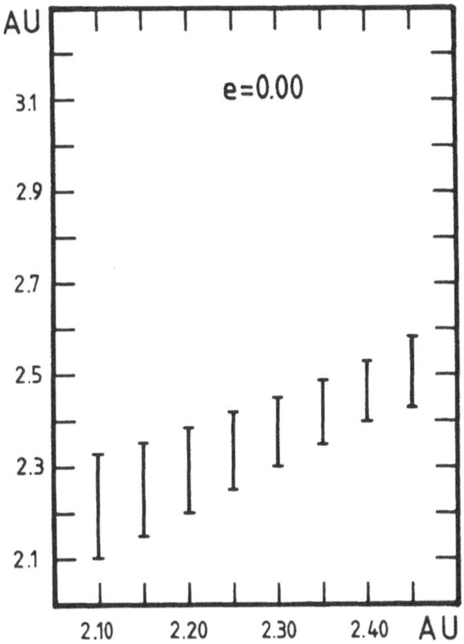

Fig. 2a

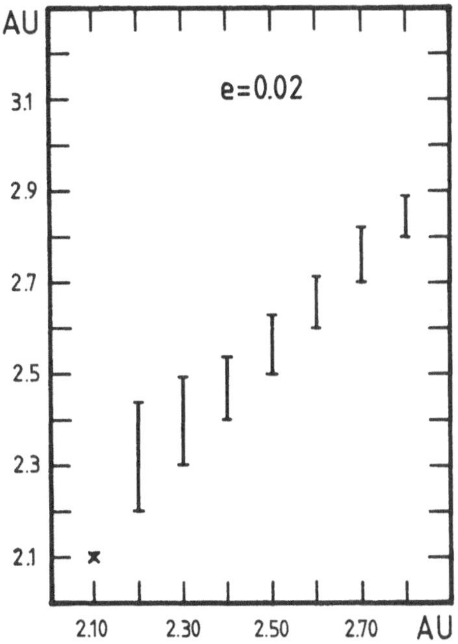

Fig. 2b

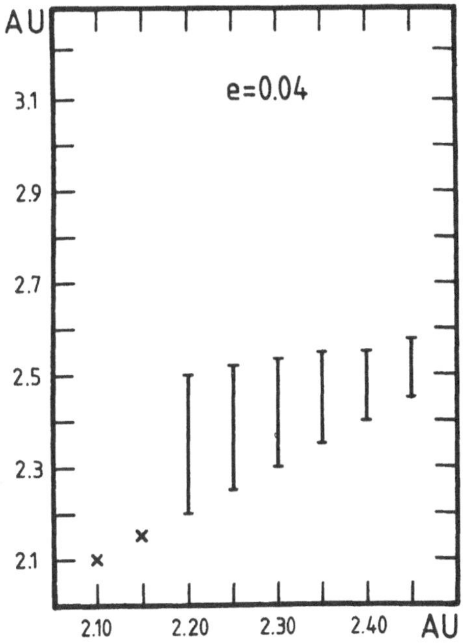

Fig. 2c

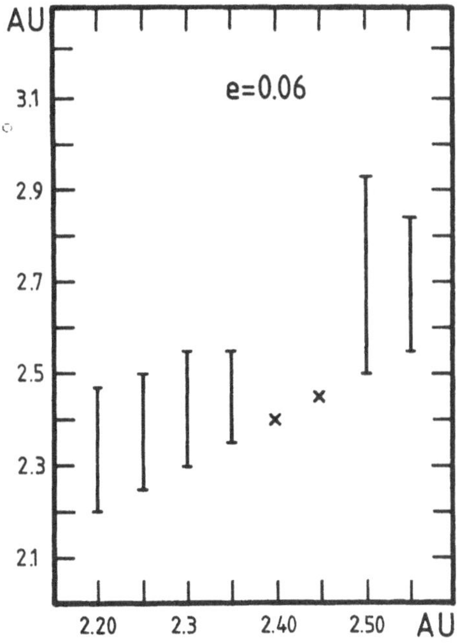

Fig. 2d

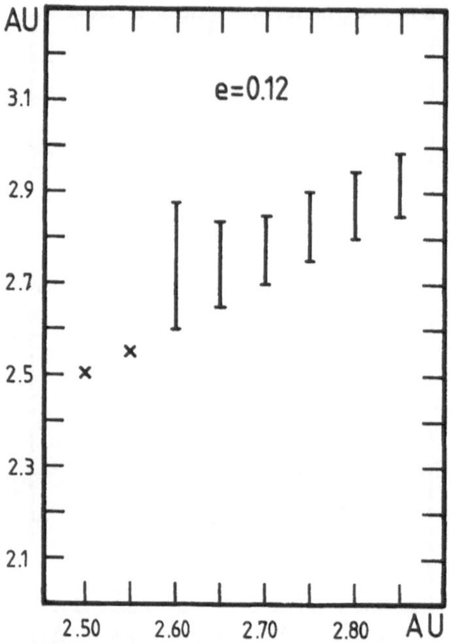

Fig. 2g

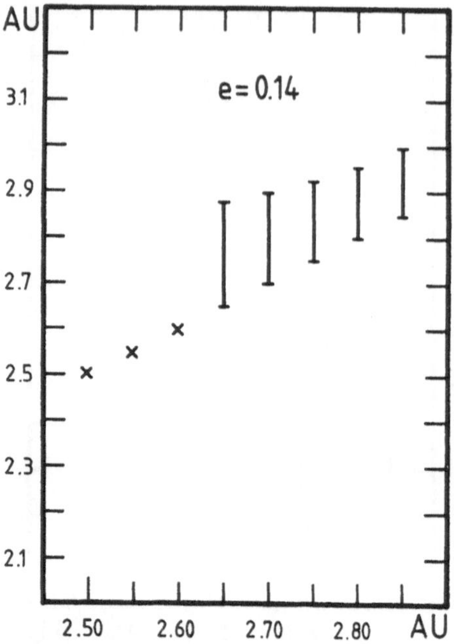

Fig. 2h

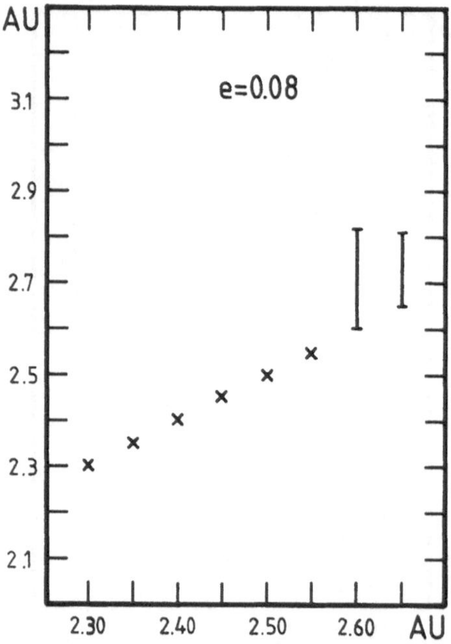

Fig. 2e

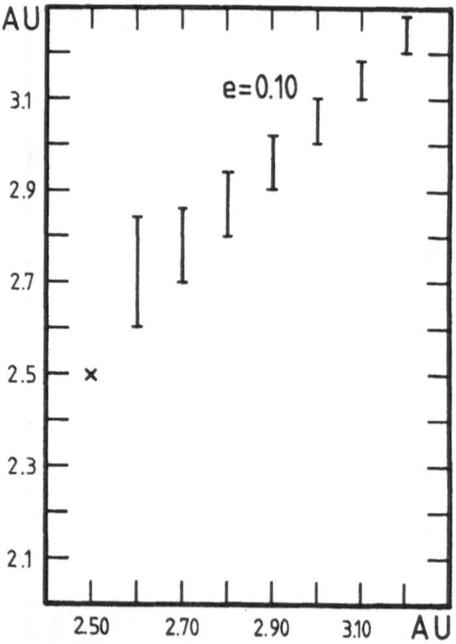

Fig. 2f

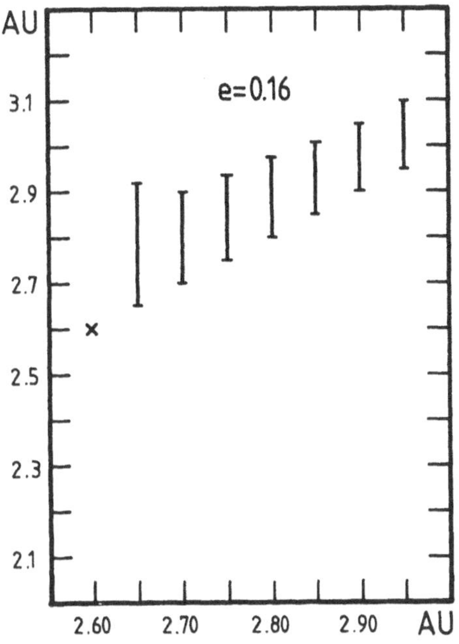

Fig.2j

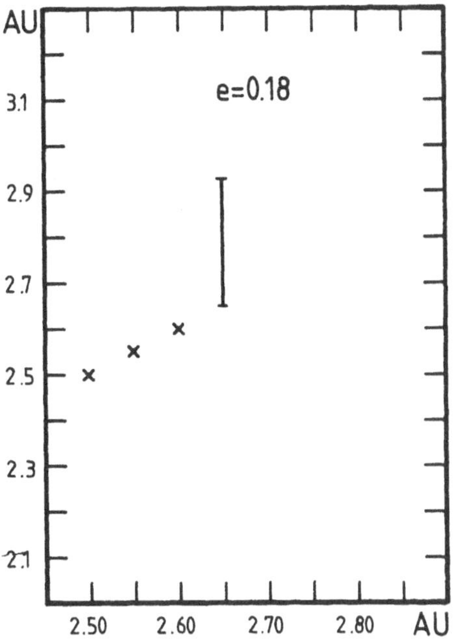

Fig. 2k

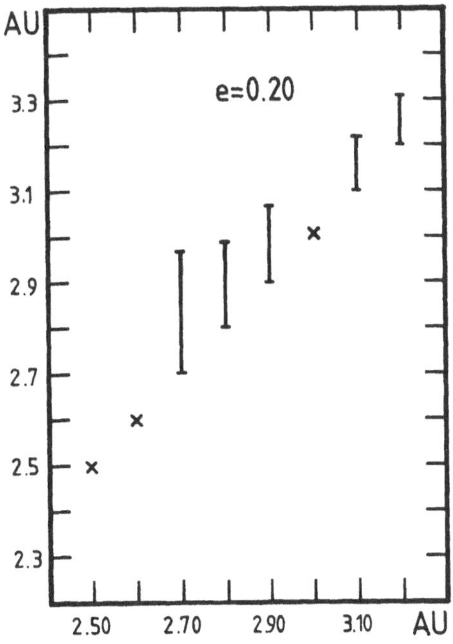

Fig. 2l

Fig. 2a - 2l: The variation of the semimajor axes for selected orbits in experiment "A" for different values of the eccentricity e.
The x-axes correspond to the initial value of the semi-major axes, the y-axes show the variations of the semi-major axes during the integration time of some 10^3 orbits of the primaries.
The unstable orbits are marked by crosses.

3. NUMERICAL EXPERIMENTS FOR μ=0.5 IN THE ELLIPTIC PROBLEM

Because there is no Jacobi-like integral in the elliptic problem the only way of determining the critical orbits (= CPO) is a numerical one. In the first step we limited ourselves to the case of μ=0.5 (for different eccentricities), in order to gain some preliminary results. It seems rather well established that there exist quasi POs around stable POs, whereas unstable POs are surrounded only by chaotic (ergodic) orbits due to the KAM (Hénon & Heiles, Arnold & Avez and C.Froeschlé). In our numerical study we therefore looked for such quasiperiodic orbits which stay stable for a certain time of integration (which was always greater than 1000 times the period of the primaries). Our conclusion was that this orbit can be regarded as "stable" in a certain sense, which is not very different from the Laplace definition of stability in the Planetary system (the semimajor axis stay in a more or less small range and show no secular trend).

A wide variety of orbits was integrated using the Lie-integrator (Hanslmeier & Dvorak, Lichtenegger, Delva) which allowed us to take a relatively large step length compared to the other commonly used methods of integration (30 days for μ=0.5 and a period of 1 year which correspond to the mean distance of the primaries of 1 AU).

3.1. Experiment A: Increasing eccentricity with constant semi-major axis

In experiment A the orbits were calculated with $0 \leq e \leq 0.20$ for different samples of orbits (a to 1, in Table 1, e = 0.0 to e = 0.20; Δe = 0.02), for μ = 0.5 with a constant semi-major axis of the relative orbit a = 1.0 AU of the primaries corresponding to a period of 1 year.

In every sample some twenty orbits over the time period of 10^3 to 3×10^3 years (which correspond consequently to 10^3 to 3×10^3 periods of the primaries) are calculated numerically to determine the CPO defining the separatrix between stable and unstable POs. The initial values are chosen in such a way, that at the beginning they correspond to circular orbits around the barycentre of the primaries.

The first sample (fig. 2a) is given for reasons of comparison and shows the numerical results for the well-known behaviour of the orbits of the circular restricted problem. There we found a limit of $1.90 < a_{CPO} < 1.95$ corresponding quite well to the theoretical one given by HG with a_{CPO} = 1.90796. Corresponding in figs 2a to 2l to the sample a to l respectively, the variations of the semi-major axis of the orbit of the test particles are drawn along with their integrated orbit. It turns out that the variation width is something which is quite close to 0.2 AU for all the considered orbits and a_{init} (which is the initial semi-major axis) is always the lower bound. These initial conditions are chosen in such a way that they represent circular orbits in barycentric coordinates. Additionally the integration always started with the primaries in their Apoastron (test integrations for the primaries in their Periastron showed quite similar results.

As a result of fig. 2a-2l the dependence of the CPO on the eccentricity of the primaries is visible. It is evident that according to the initial conditions chosen (a=1 AU, P=1 year and $0 \leq e \leq 0.20$) we receive a broader and broader unstable region around the primaries: consequently the CPO is further away from the barycentre. No doubt this is due to the closer approaches of the test particles (=m_3 with zero mass) to the primaries moving in this fictious double star system. So we have a systematic increase of the CPO with an increasing eccentricity.

Unfortunately, we cannot yet give satisfactory upper and lower boundaries for the CPOs. Additionally it should not be forgotten that we calculated only quasi-periodic orbits under the assumption that they lie on a torus around a stable PO. Table 1 summarizes the results of experiment A and shows the CPOs of all the 11 samples a to 1.

TABLE I: LOWER AND UPPER BOUNDS FOR THE CRITICAL ORBIT (EXPERIMENT A)

sample	eccentricity	lower bound (AU)	upper bound (AU)
a	0.00	1.90	1.95
b	0.02	2.10	2.15
c	0.04	2.15	2.20
d	0.06	2.15	2.20
e	0.08	2.50	2.55
f	0.10	2.60	2.65
g	0.12	2.55	2.60
h	0.14	2.60	2.65
j	0.16	2.60	2.65
k	0.18	2.60	2.65
1	0.20	2.75	2.80

We need to make many more additional integrations covering a wider range of eccentricity up to 0.5 with a smaller step size for the increasing eccentricity (something like 0.005 would be desirable).

An interesting phenomeon should be remarked: in fig.3 one can see the appearance of unstable POs inside a large region of stable PO and completely outside the CPO. This will be discussed in more detail in the appendix.

3.2. Experiment B: Increasing eccentricity with decreasing semi-major axis

In the second experiment B we let the initial position of the primaries remain unchanged compared with the case $e = 0.0$ (the x values for the primaries being 0.5 and -0.5 AU and $y = 0$) and decreased only the initial velocities $\dot{y}$ ($\dot{x}=0$) from $\dot{y} = 0.5$ (for $e=0$) to $\dot{y} = 0.49$, 0.48, 0.47, 0.46 and 0.45 which correspond approximately to $e = 0.04$, 0.08, 0.12, 0.16 and 0.20. The overall picture is presented in fig.4. As in the former fig.3 an increase of the CPOs can be seen which cannot now be explained by closer approaches of the test particles to the primaries. This effect is only due to the eccentric orbit of the primaries! There seems to be a maximum value for 0.04, but no final conclusion should be drawn from this numerical study, which is only at a preliminary step.

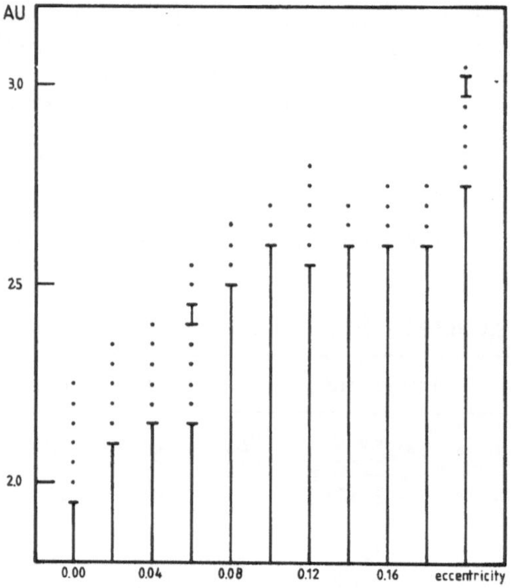

Fig. 3 Stable and unstable quasi periodic orbits in experiment "A"

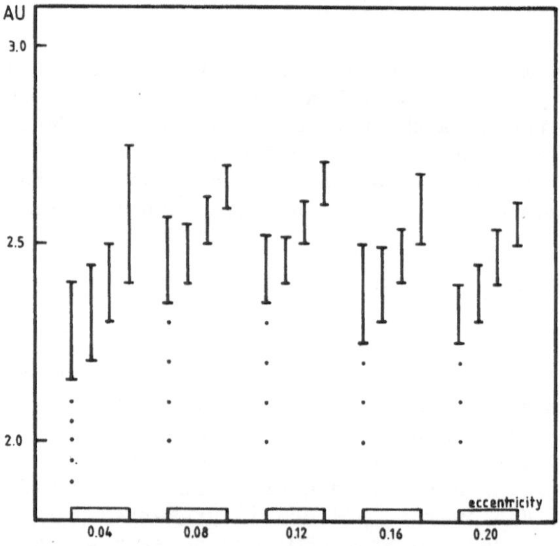

Fig. 4 The variation of the semimajor axes for selected orbits ("experiment B")

TABLE II: LOWER AND UPPER BOUNDS FOR THE CRITICAL ORBIT (EXPERIMENT B)

sample	eccentricity	lower bound (AU)	upper bound (AU)
a	0.00	1.90	1.95
b	∿0.04	2.10	2.15
c	∿0.08	2.30	2.35
d	∿0.12	2.30	2.35
e	∿0.16	2.20	2.25
f	∿0.20	2.20	2.25

We also can see in fig.4 that the variation of the semi-major axis for one and the same orbit for different values of eccentricity of the primaries is about 0.2 AU and lies slightly over the "unperturbed" case (e=0.0).

A word should be said concerning the role of the eccentricity of the test particles. It seems that according to our experiments there is a variation of $0.0 \leq e \lessapprox 0.1$ which does not change the stability characteristics of an orbit in our samples. For values of e higher than 0.1 the orbit becomes unstable in general. This is not a surprising fact, because e=0.1 corresponds to a variation of the distance to the barycentre of 20%, and lead consequently to closer approaches of the test particles to the primaries.

The results presented here show for the first time the dependence fo the CPOs on the eccentricities of the primaries in the elliptic problem. It should not be interpreted more then this as it is simply a first attempt to discuss this complicated topic. As this is a numerical approach many additional numerical integrations should be done to find more rigorous limits for the CPOs depending on e. More needs to be done; for instance, a finer network is required for the eccentricities of the primaries and also a finer network of the initial conditions of the integrated orbits (even for different initial conditions for the primaries). Another step forward would be to extend the range of eccentricity. This asks for much more time for the numerical integrations on high speed computers. The work is still in progress and will hopefully provide more rigorous results in the near future.

APPENDIX: CHAOTIC LAKES

In fig.3 there appear unstable strips in the stability zone outside the CPO in two different samples, for e = 0.06 and e = 0.20. To establish these results, numerical integrations of orbits with initial conditions very close to one another covering the whole strip of instability are undertaken (representing only a small region in phase space). Unfortunately no final conclusion can yet be drawn from these results. In trying to understand their existence the following can be said: In the circular case we have already such strips of instability for special values of μ, which disappear for $\mu = 0.5$ (fig.1). Therefore we could not expect such a behaviour for equal masses; this phenomenon is doubtless due to the eccentricity of the primaries. The unstable strips are found only in figs 2d and 2l; consequently this behaviour is only dependent on the initial conditions. We therefore suppose that there exist in phase space small unstable regions in larger zones of stability, which would resemble in the surface of section something like chaotic lakes in the mainland of POs and quasi POs. Is it dangerous to draw such a conclusion? As a detailed study of these results is till in progress which will hopefully clarify the situation, it is suggested that the "Chaotic Lakes" are regarded with great care.

ACKNOWLEDGEMENTS

I thank G.Lustig, A.Jupp and J.Hadjidemetriou for the help in preparing the manuscript respectively for critical comments.

REFERENCES

Arnol'd,V.J., and Avez,A., 1967, Problèmes Ergodiques de la Mécanique classique, Gauthier-Villar (ed.)
Delva,M., 1984, Integration of the elliptic restricted three body problem with LIE-series, this conference
Dvorak,R., 1983, Planetenbahnen in Doppelsternsystemen, Österr.Akad. d.Wiss., math.-nat.Klasse, Bd. 191, p.423
Froeschlé,C., 1971, Astrophys.Space Sci. 37, 87
Hadjidemetriou,J., 1976, Astrophys.Space Sci. 40, 201
Hanslmeier,A. and Dvorak,R., 1984, Astron.Astrophys. 132, 203
Hénon,M. and Guyot,M., 1970, in G.E.O.Giacaglia (ed.) in Periodic orbits, Stability and Resonances, 349
Hénon,M. and Heiles,C., 1964, Astron.J. 69, 73
Lichtenegger,H., 1984, Die Lie-Reihen als Lösungen des Mehrkörperproblems mit variablen Massen, Thesis, University of Graz
Szebehely,V., 1980, Cel.Mechanics 22, p.7

PERIODIC ORBITS

John D. Hadjidemetriou
University of Thessaloniki
GR-54006 Thessaloniki
Greece

ABSTRACT. Recent results on periodic orbits are presented. Planetary systems can be studied by the model of the general 3-body problem and also some satellite systems and asteroid orbits can be studied by the model of the restricted 3-body problem. Triple stellar systems and planetary systems with two Suns are close to periodic systems. Finally, the motion of stars in various types of galaxies can be studied by finding families of periodic orbits in several galactic models.

1. INTRODUCTION

The study of the periodic orbits and their stability is a powerful tool in understanding the behavior of a nonintegrable system. The aim of this paper is to review the various types of periodic orbits in connection with the study of problems in Celestial Mechanics and Stellar Dynamics. We shall not attempt to extend this review to the study of periodic orbits of nonintegrable systems in general, or to the corresponding nonlinear mapping (fixed points). Reviews on this matter can be found by the interested reader in the article by Helleman (1980) and in several articles contained in the Conference Proceedings of AIP, edited by Jorna (1978).

We shall state first analytic results concerning the existence of periodic orbits and their stability and then we shall review numerical results, stressing mainly those numerical results which represent closely actual systems in our solar system and in stellar systems, and thus enable us to understand better these systems.

2. ANALYTIC RESULTS

2.1. Existence of Periodic Orbits.

The method which is used to prove the existence of periodic orbits is to start from a restricted version of the problem, where periodic orbits are known to exist and extend them by varying a parameter. In order for

Celestial Mechanics **34** (1984) 379–393. 0008–8714/84.15

this method to be applicable, the Hamiltonian of the system must be expressed in the form

$$H = H_o + \varepsilon H_1 \qquad (1)$$

where H_o is the Hamiltonian of the restricted version of the problem, corresponding to $\varepsilon=0$, and εH_1 the perturbation, ε being the small parameter. This method is called the continuation method (see Siegel and Moser, 1971 and Arenstorf, 1966).

We shall present below the most important cases where the continuation method was used to prove the existence of periodic orbits. In each case we shall define the unperturbed Hamiltonian H_o, corresponding to the restricted version of the problem and the meaning of the small parameter ε of the perturbation. We shall first review the older results and we shall then come to the more recent results. We mention at this point that the proof of the existence of a periodic orbit is valid for a "sufficiently small" value of the parameter ε. In practice this value is usually quite large, but this can be found by numerical computations only.

(i) Periodic orbits of the restricted circular 3-body problem.
Consider the classical planar restricted circular 3-body problem (see Szebehely, 1967) where the primaries P_1 and P_2 revolve in circular orbits around their center of mass. For convenience let us call the primary P_1 the Sun and the primary P_2 Jupiter, with masses $1-\mu$ and μ, respectively. If $\mu=0$ the massless body P describes a Keplerian orbit around the Sun, under its gravitational attraction, which we shall study with respect to the well known rotating frame of the restricted problem. We shall consider the simplest cases where the unperturbed orbit of P is a circle around P_1 or an ellipse, with respect to the inertial frame. Let us consider these two cases separately when the value of μ becomes greater than zero. It is evident that in this case H_o is the Hamiltonian for the motion of P around the Sun, in the rotating frame and the small parameter ε is the mass μ of Jupiter.

Circular orbit of P: A circular orbit of P in the inertial frame is evidently periodic in the uniformly rotating frame also, for any radius R. This is true both for a direct orbit (revolving in the same sense as the rotating frame) and a retrograde orbit (revolving in the opposite sense). It is proved (Poincaré, 1899, Birkhof, 1915) that a direct circular orbit is continued to a simple nearly circular periodic orbit of the restricted 3-body problem, for a sufficiently small μ, provided the condition

$$\omega/\omega_0 \neq (n+1)/n \qquad (2)$$

is satisfied, where ω, ω_0 are the frequencies of the circular motion of P (in the inertial frame) and of the rotating frame. If the orbit of P is retrograde, then it is continued for any radius R. These are called the periodic orbits of the 1st kind.

Elliptic Orbits of P: Let a be the semimajor axis of the elliptic motion of P around P_1, for $\mu=0$. In order for this motion to be periodic in the rotating frame, the relation $\omega/\omega_0=p/q$ must be satisfied, where $\omega=a^{-3/2}$ is the mean angular velocity of the elliptic motion and p,q are integers. This is a multiple periodic orbit with multiplicity r=p-q. When $\mu>0$ this orbit is continued to a nearly elliptic periodic orbit (Arenstorf 1963, Barrar 1965. See also Milani 1983)

In both the above cases the mass of Jupiter was considered as the small parameter. This, however, is not valid when the small body P comes to a close approach to the small primary P_2, because the perturbation μH_1 contains the term μ/r_2, where r_2 is the distance of P from P_2. This type of orbits, which tend to collision orbits for $\mu\to 0$ (but do not correspond to the two body problem for $\mu=0$!) are called 2nd species solutions and have important applications, for example in Earth-Moon trajectories. An existence proof of such type of periodic orbits is given by Perko (1974, 1977, 1981) and also by Henrard (1980). Extended work, both analytic and numerical has been made by Breakwell and Perko (1966, 1974), Bruno, (1972, 1981), Broucke (1968), Hénon (1968, 1969), Guillaume (1975a,b), Arenstorf (1962), Hitzl and Hénon (1977a,b) and Hitzl (1977a,b). The connection of 2nd species solutions with periodic orbits of the first and second kind has been studied by Perko (1981). Also, Perko, (1983) proved the existence of symmetric periodic orbits of the restricted problem by analytic continuation of symmetric periodic solutions of Hill's problem, for small μ. Belbruno (1981a,b) proved the existence of periodic orbits in the restricted 3-dimensional problem by continuation of collision orbits.

(ii) Periodic orbits of the planetary type.

Consider N+1 bodies, P_0, $P_1,\ldots P_N$ moving in the plane under their mutual gravitational attraction, with masses m_0, $m_1,\ldots m_N$, respectively. The mass of P_0, called the Sun, is much larger than the masses of the other bodies, called planets. If $m_1=m_2=\ldots=m_N=0$, each planet describes a Keplerian orbit around the Sun. In particular we assume that the orbits of the planets are circular, with commensurable periods. In this case the motion of the whole system is periodic in a rotating frame such that its x-axis is defined by the line P_0P_1. We want to study what happens when the masses $m_1,\ldots m_N$ are no longer zero. The Hamiltonian H_0 in (1) is here the one corresponding to N uncoupled two-body problems and ε is, say, m_1 (the ratios m_i/m_1 are considered fixed). It can be proved that apart from some resonant cases the unperturbed periodic motion of the N+1 bodies is continued to nonzero masses of the planets (periodic orbits of the 1st kind). For a proof see Griffin (1920), also Stellmacher (1976,77).

We can also consider elliptic motion of the planets around the Sun, for zero masses of the planets. By properly selecting the semimajor axes so that their periods are commensurable we can make the whole system periodic in a rotating frame. This motion can be also continued to nonzero masses of the planets (periodic orbits of the second kind). A proof is given by de Sitter (1909) for three small bodies, in the study of the three inner satellites of Jupiter. Four different types of periodic orbits are proved to exist, all with the same resonance of P_1,P_2,P_3 to P_0, equal to 1:2:4 and different phases. One of them, which was found to be the only stable motion, corresponds to the actual configuration. Also, Message (1980) proved the existence of periodic motion of two planets revolving around the Sun in elliptic motion.

The above results are for planar orbits only. Consider now two planets, with negligible masses, to move in Keplerian orbits around the Sun in commensurable periods and zero eccentricity, in different planes. By introducing the masses of the planets it is proved (Message, 1982) that the motion is continued to a periodic orbit which is called periodic orbit of the 3rd kind.

(iii) Periodic orbits of the general 3-body problem and the N-body problem: Continuation by increasing one mass.

It can be proved that periodic orbits of the general planar 3-body problem, in a rotating frame, can be obtained as a continuation of periodic orbits of the restricted circular 3-body problem by increasing the mass of the third body (Hadjidemetriou, 1975a). This result was generalized by Meyer (1981a,b) who proved the existence of periodic orbits of the general N-body problem starting from the following restricted problem: N-1 bodies move on a relative equilibrium configuration and N-th particle, with negligible mass, describes a periodic motion. This solution is proved to be continued to a periodic motion when the mass of the N-th particle is increased.

In both the above cases, H_0 in (1) is the Hamiltonian of the restricted problem and ε is the small mass.

The continuation method has been used by several authors to prove the existence of periodic orbits of the general 3-body problem: Kameyer (1978) proved the continuation of the periodic orbits of the elliptic restricted 3-body problem to the general problem. Katopodis (1979) and Michalodimitrakis (1979) proved the continuation of the periodic orbits of the restricted three dimensional 3-body problem to the general problem. Marchal (1979) proved the existence of periodic orbits with long periods in the neighborhood of arbitrary given initial eccentricities and inclinations when the distance of the outer body is much larger than the dimensions of the inner binary. Kameyer (1983) proved the existence of symmetric periodic orbits in the rectilinear 3-body problem with the middle mass much larger than the outer mass.

(iv) Periodic orbits of the general N-body problem: Continuation by increasing N-2 masses together.

Symmetric periodic orbits of the general planar N-body problem, in a rotating frame, are proved to exist, as a continuation of N-2 commensurable periodic orbits of massless bodies, in the restricted circular 3-body problem (Hadjidemetriou 1977). This problem is related to the existence of planetary systems with two Suns. The continuation by increasing the masses can be carried out numerically to quite large masses. Hadjidemetriou (1976b) gives an example of a periodic orbit of five bodies with comparable masses.

(v) Periodic orbits of the general N-body problem: Continuation from a central configuration.

Consider a central configuration of the N-body problem, i.e., N bodies moving uniformly along concentric circular orbits on the inertial frame

(A study of central configurations has been made by Moulton (1910), Smale (1970), Arenstorf (1982) and Palmore (1976, 1982)). Arenstorf (1968, 1978) proved that a periodic motion of N+1 bodies can be obtained if we replace one of the N bodies by a close binary system, with the same total mass, whose members describe circular or elliptic orbits around their common center of mass. H_0 in this case, in the Hamiltonian (1), corresponds to the central configuration and the small parameter is the dimensions of the close binary.

(vi) Periodic oribts of the general N-body problem: Continuation of periodic orbits coming from infinity.

Let us consider a central configuration of N bodies and an N+1 particle with negligible mass describing a circular orbit around the center of mass of the N bodies, at a great distance. The small parameter is here the inverse of the dimensions of the orbit of the N+1-body. Moulton (1912) proved the existence of such type of periodic orbits for N=2 (restricted circular 3-body problem). Meyer (1981b) generalized this result for any value of N. Also Meyer (1981a) proved that the continuation to the general N+1-body problem is also possible, by varying both the dimensions and the mass of the N+1 body.

The periodic orbits of the general N-body problem mentioned in section (iii) for circular motion of the primaries, and the periodic orbits mentioned in sections (v) and (vi) are presented by Meyer (1981a) in a unified way, by representing them by a Hamiltonian of the form (1). The meaning of the small parameter ε is, of course, different in the above three cases.

2.2. Analytic construction of periodic orbits

In the previous section we proved the existence of various types of periodic orbits. In this section we shall see how we can construct analytically a periodic orbit, in the form of a series expansion. Several authors have worked on this problem, using different methods. Richardson (1980) obtained a third order analytical solution for halo type periodic orbits around the collinear points of the circular restricted 3-body problem. Hitzl and Levinson (1980), using Hamilton's law of varying action obtained approximate power series solutions for the restricted 3-body problem. Presler and Broucke (1981a,b) have developed a theory to obtain the solution of a dynamical system with two degrees of freedom in the form of a series, by the Lindstedt method. The solution is carried out to a high order of the small parameter by a systematic computerization of the lengthy literal algebraic developments. Exact resonance and near resonance cases are studied and several examples are given. Davoust (1983a,b) uses the Lindstedt method to obtain periodic orbits in a three dimensional harmonic oscillator perturbed by cubic terms. Stellmacher (1982) gives an algorithm to construct a nearly circular periodic orbit for the motion of a satellite around an oblade planet. Wiesel (1981) gave a formal solution for the motion near a periodic solution by extending the Floquet problem. Also Kameyer (1980) gave a linearized mapping near a periodic orbit of the plane three body problem. Several properties are presented.

Helleman and Bountis (1978) have devised a rapidly convergent variational method to construct periodic solutions in the form of Fourier series, analytic in time, with arbitrary long period for a two degree of freedom nonintegrable dynamical system (the well known Hénon-Heiles system). Instead of specifying the initial conditions, the periodic orbit was determined by specifying the period and the frequencies of the oscillation along the x and the y axis, so that their ratio is a rational number. In this way the small divisor problem is avoided and convergence is secured. As this method can be applied for a periodic orbit of an arbitrary period, any (non periodic) motion can be approximated by a periodic solution of very long period. In this way the areas which are sensitive to the initial conditions are detected and the chaotic regions are obtained.

3. FAMILIES OF PERIODIC ORBITS - BIFURCATIONS

3.1. Continuation of families

In the previous section 2(a) we studied single periodic orbits. Their existence was proved as a continuation of periodic orbits of a "restricted" problem (corresponding to $\varepsilon=0$ in the Hamiltonian (1)). These latter periodic orbits, however, (for $\varepsilon=0$) belong to monoparametric families along which a parameter varies (for example the period). Most of these orbits can be continued when $\varepsilon=0$, with the exception of some particular periodic orbits. At these latter orbits the continuation theorem is not applicable and the unperturbed family may not be continued at that point.

We shall show the above by studying, in parallel, two systems with two degrees of freedom, one of interest to solar system dynamics, which we shall call case I, and the other to Galactic Dynamics, called case II. In this way the similarities in the above two areas of dynamics will become evident. For both cases the unperturbed Hamiltonian H_o is

$$H_o = \frac{1}{2}\left(p^2 + \frac{J^2}{r^2}\right) - \omega J + V_o(r), \qquad (3)$$

which describes motion in a uniformly rotating frame, with constant angular velocity ω, under the central force potential $V_o(r)$. The variables p and J are the momenta and r is the distance from the origin. In case I the potential is

$$V_o(r) = -\frac{k}{r}, \qquad k>0 \qquad (4)$$

and in case II it is (Contopoulos 1983a,b)

$$V_o(r) = \{1+(1+r^2)^{1/2}\}^{-1}, \qquad (5)$$

which represents the axisymmetric background of a barred galaxy. The potential (4) represents the attraction from the Sun. The angular velocity ω in (3) can be considered to be defined by the circular motion of Jupiter, for case I, or the rotation of the bar of the galaxy, for case II.

In both cases I and II circular orbits exist, around the origin 0 (for $\varepsilon=0$), for any radius R, which are considered as symmetric periodic orbits in the rotating frame. Since a circular orbit can be direct or retrograde (compared to the rotation of the rotating frame) we have, by varying R, two distinct families of circular periodic orbits, corresponding to $J>0$ (direct) and $J<0$ (retrograge). Note that for $\varepsilon=0$ J is an integral of the motion. Along each family the radius R varies. Since to each R there corresponds a period T of the circular orbit with respect to the inertial frame, or a frequency $\omega_1=2\pi/T$, we shall use the ratio ω_1/ω as the parameter along the family.

Case I: We can assume, as an example, that the moving body is an asteroid. The small parameter ε is the mass μ of Jupiter, and the perturbed Hamiltonian, when the effect of Jupiter is added, is that of the circular restricted 3-body problem. It is known that the continuation theorem is not applicable when $\omega_1/\omega=(n+1)/n$, $n=1,2,3,\ldots$, for direct orbits. At these critical resonance cases the family of direct circular orbits (periodic orbits of the 1st kind) is not continued as a whole, when $\mu>0$, but behaves as shown in Fig. 1. The dotted line represents another family of periodic orbits, for $\mu=0$, which bifurcates from the circular

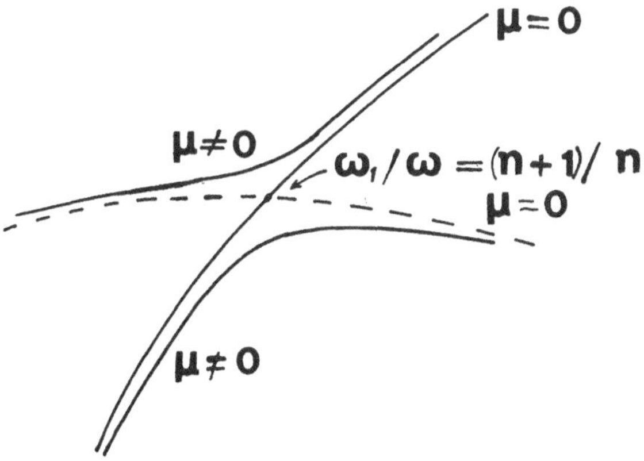

Fig. 1. Continuation of the circular family ($\mu=0$, full line) at the resonant point $\omega_1/\omega=(n+1)/n$. The continued family is deflected and follows the resonant elliptic family ($\mu=0$, dotted line).

family at the resonant orbit $(n+1)/n$. All members of the dotted family are elliptic orbits with the same semimajor axis and different eccentricities, such that their resonance with Jupiter is equal to $(n+1)/n$. The continuation shown in Fig. 1 has been proved by Poincaré (1902), Message (1966), Guillaume (1969) and Schmidt (1972). See also Milani (1983). From the above it is clear that the family of direct circular orbits for $\mu=0$ breaks into an infinite number of families when $\mu>0$, which are separated by the resonant oribts 2/1, 3/2, 4/3,... , (Hadjidemetriou and Ichtiaroglou 1983,84, Colombo et.al. 1968).

All the above are for direct circular orbits. The retrograde asteroid orbits are all continued to $\mu>0$ and consequently the unperturbed family is continued as one family.

Case II: The perturbing potential of the barred galaxy is taken (Contopoulos, 1983a,b) to be

$$V_1 = r^{1/2}(16-r)\cos 2\vartheta \tag{6}$$

It is then proved that along the family of direct circular orbits all resonant cases of the form (n+1)/n with n even are not continued when $\varepsilon\neq 0$. In their vicinity the continued family has the shape shown in Fig. 1. However, the resonant cases (n+1)/n with n odd are continued when $\varepsilon\neq 0$ and the continued family has the form shown in Fig. 2. Note that another family for $\varepsilon=0$ appears, bifurcating from the point A corresponding to the resonant orbit of the circular family, and both these families

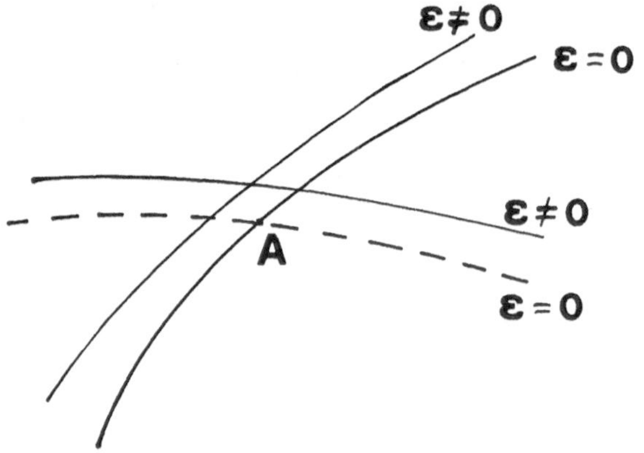

Fig. 2. Continuation of the circular family ($\varepsilon=0$, full line) at a resonant point A, $\omega_1/\omega=p/q\neq(n+1)/n$. The family is continued at that point for $\varepsilon\neq 0$. The resonant family ($\varepsilon=0$, dotted line) is also continued. (In the planetary problem there are two such resonant, elliptic families).

are continued to $\varepsilon\neq 0$. An analytic study for the continuation of the family at the resonant point (n+1)/n has been made by Contopoulos (1983a) for an integrable case (the definition of the resonance is different in this latter paper, the resonance (n+1)/n used here corresponding to n/1).

As far as stability is concerned it is proved (Hadjidemetriou 1982, Hadjidemetriou and Ichtiaroglou 1983, 84) that in the asteroid case all the circular direct orbits are stable, when $\mu>0$ with the exception of the resonant orbits with resonance (2n+1)/(2n-1) with Jupiter. In these latter cases a small unstable region appears. On the other hand all resonant elliptic branches (i.e., the continuation of the elliptic family in Fig. 1) are all structurally unstable, in the sense that there always exists a Hamiltonian perturbation which generates instability. But for $\mu>0$ there do exist stable branches at the resonances 2/1, 3/2, 4/3. This ensures that stable regions in phase space exist where trapping of asteroids could take place near those resonances, but also chaotic regions near these resonances are present, corresponding to the Kirkwood gaps.

In the galactic problem an unstable region exists near the resonant orbits (n+1)/n for n odd on the continued family (Contopoulos 1983a). Note that this type of resonant orbits are not continued in the asteroid case.

We also remark the following with respect to the continuation of elliptic periodic orbits in the asteroid problem, when μ is increased. Let the resonance be of the form (n+1)/n and a the corresponding semi-major axis which is fixed. The eccentricity varies and can be considered as the parameter of the elliptic family. For a symmetric periodic orbit all three bodies, Sun, Jupiter and Asteroid, are on the x-axis at $t=0$. In that case $x_0 = a(1+e)$ and for $a>0{,}5$ an eccentricity e exists so that $x_0=1$ (for $\mu=0$). This is a collision orbit with Jupiter and for $\mu>0$ is continued as a second species solution (see section 2a(i)). The unperturbed family, for $\mu=0$, breaks at this point into two parts, on each side of the collision orbit, when $\mu>0$.

3.2. Bifurcations

In the previous section we discussed the continuation of an unperturbed family of periodic orbits, by increasing a parameter. We shall now start from the continued family and study how new families, with multiple periodic orbits, bifurcate from it. The asteroid problem will serve as an example (see Hadjidemetriou and Ichtiaroglou 1984) but similar results hold for the galactic problem (Contopoulos 1983a,b, Contopoulos and Michailidis 1981, Heggie 1983, Magnenat 1982, Aa, 1983).

Consider a circular direct orbit, for $\mu=0$, with any radius. When $\mu>0$, this orbit is continued as a stable periodic orbit, with the exception of the resonant orbits of the form (2n+1)/(2n-1) (remember that the resonant orbits (n+1)/n are not continued at all). At the resonant orbits (2n+1)/(2n-1), i.e., 3/1, 5/3,..., small unstable regions appear. From both ends of this region there bifurcates a family of doubly symmetric periodic orbits, one of them being stable and the other unstable. From all other resonant orbits of the form p/q ($p\neq q-1$, $p\neq q-2$) there bifurcate two families of multiple periodic orbits, from the same point, with multiplicity equal to $r=p-q$, one stable and the other ustable. All these orbits are nearly elliptic orbits in the inertial frame. Further bifurcations also appear, with higher multiplicity.

A study of the sequence of period doubling pitchfork bifurcations is made by Contopoulos and Pinotsis (1984) in the restricted problem and it is proved that it has the universal ratio 8.72. Such type of bifurcations have been studied by Benettin et.al. (1980), Green et.al. (1981) Bountis (1981) and Contopoulos (1983c) for different dynamical systems.

4. NUMERICAL WORK ON PERIODIC ORBITS

A vast amount of numerical work has been obtained on periodic orbits and especially on the restricted 3-body problem. We are not attempting to review all this material. The interested reader is referred to the book by Szebehely. We shall review here only the recent work on periodic orbits, mainly of the general 3-body problem, and also periodic motions which closely represent actual configurations in the solar system.

4.1. Planetary and satellite orbits.

Extended numerical work has been done recently for the study of periodic orbits of the planetary type in a rotating frame. Hadjidemetriou (1976a) computed numerically families of periodic planetary direct orbits with two planets, for equal masses of the planets (equal to .001 with total mass equal to 1) and studied their stability. It was found that there is an infinite (denumerable) number of planetary families, each one containing periodic orbits of the first kind (nearly circular orbits) and periodic orbits of the second kind (nearly elliptic orbits). Each family contains a branch of elliptic orbits along which the resonance is almost constant equal to (n+1)/n. These resonant orbits are proved by Hadjidemetriou (1982) to be potentially unstable, i.e.,there always exists a Hamiltonian perturbation which generates instability. However, stable resonant elliptic orbits at these resonances exist, with quite large eccentricities. It was also proved that all resonant circular oribts with resonance 3/1, 5/3,..., are unstable.

The above results were extended by Delibaltas (1976) for the masses of Jupiter and Saturn. It was also found that the stability of the resonant elliptic orbits at the resonances 2/1, 3/2, 4/3,..., are affected by the mass ratio m_1/m_2 of the two planets.

The numerical results for two planets have been extended by Hadjidemetriou and Michalodimitrakis (1981) to the case of three "planets". In fact they computed several families of periodic orbits of the general 4-body problem with the aim to study the Galilean satellites of Jupiter. For this reason, the masses used were the mass of Jupiter and the three inner Galilean satellites. On each family there exists a part with nearly circular orbits of the satellites and a part with nearly elliptic orbits. Four different branches were computed, with a constant resonance 1:2:4 (as in the actual case), which differ in phase. One branch, corresponding to the actual case, is stable (the other three are unstable). These numerical results are in agreement with results on stability obtained by de Sitter in 1909!

A particular case of the planetary problem is the so called restricted planetary N-body problem where all planets but one have nonzero masses and the N-th planet has negligible mass. This is a generalization of the restricted N-body problem studied by Meyer (1981b) mentioned in section 2.(2.1)(vi). In Meyer's problem the N-1 bodies move on an equilibrium solution while in the present case they describe a periodic orbit in general. Hadjidemetriou (1980) studied the periodic orbits of a massless body in a rotating frame, under the gravitation attraction of the Sun, Jupiter and Saturn, with their actual masses, describing a periodic orbit in the plane. Several periodic orbits of the massless body were computed and their stability was studied. These orbits are close to the Earth, Mars and several asteroids inside and outside the orbit of Jupiter or Saturn.

The actual motion of Jupiter and Saturn around the Sun can be approximated by a periodic motion. Hadjidemetriou (1976a) computed a periodic orbit with nearly circular orbits of Jupiter and Saturn, with their actual masses, with a resonance equal to 5/2. Also, Hadjidemetriou (1980) computed two distinct families of elliptic orbits for Jupiter and Saturn, with almost constant resonance 5/2 and varying eccentricities. Some of these orbits are close to the actual configuration. Two similar families, for zero mass of Saturn, have been obtained by Kwok and Nacozy (1982). Stable configurations are present in both families. Taylor (1981) has computed several families of symmetric horseshoe periodic orbits of the restricted 3 body problem for the Sun-Jupiter mass ratio.

4.2. Stellar systems.

Periodic orbits of the general 3-body problem with comparable masses have been computed quite recently. The first orbits computed were isolated periodic orbits in the inertial frame. Schubart (1956) computed a rectilinear periodic motion of three bodies. Szebehely (1972) and Standish (1972) have computed periodic orbits with comparable masses in the plane. The first families of periodic orbits in a rotating frame were computed by Broucke (1975), by Hadjidemetriou (1975b), and Hénon (1974). It was found that stable orbits exist, with comparable masses and distance between the three bodies and this means that triple stellar systems with comparable distances between the three stars can exist in nature. Many more families of periodic orbits of the general 3-body problem have been computed recently. For the planar problem we mention the papers by Bozis and Christides (1975), Broucke (1979), Broucke et.al. (1980, 1981), Delibaltas (1976, 1983), Markellos (1977a,b, 1979). The properties of the characteristic surfaces of the symmetric periodic orbits of the planar three-body problem have been studied by Bozis and Hadjidemetriou (1976). Also families of vertical critical orbits in the general three-body problem were computed by Ichtiaroglou (1980a), Ichtiaroglou et.al. (1980c) and Michalodimitrakis (1978). Periodic orbits of the three-dimensional 3-body problem were computed by Michalodimitrakis (1979, 1980), Ichtiaroglou (1980), Ichtiaroglou and Michalomitrakis (1980), Markellos (1980, 1981) and Katopodis et. al. (1980).

4.3. Restricted 3-body problem.

We are not going to review here periodic orbits of the planar restricted 3-body problem. We shall mention briefly recent numerical work on the three-dimensional restricted 3-body problem. Vertical critical periodic orbits of the planar restricted 3-body problem have been computed by Ichtiaroglou et. al. (1980a,b) and the planar restricted N-body problem by Michalodimitrakis (1978a). These orbits generate three-dimensional periodic orbits. Also Michalodimitrakis (1978c, 1979), Ichtiaroglou et. al.(1980a), Robin and Markellos (1980) and Robin (1981), have computed families of three dimensional periodic orbits of the restricted 3-body problem. All the above mentioned periodic orbits are symmetric periodic orbits with respect to the rotating x-axis. Asymmetric periodic orbits of the restricted 3-body problem have been computed by Taylor (1983a,b) using the method given by Message (1970). Also Howell (1984) and Howell and Breakwell (1984) have computed three dimensional Halo type orbits in the restricted 3-body problem.

REFERENCES

AA, Els Van der: 1983, Celes. Mech. 31, 163.
Arenstorf, R. F.: 1966, J. Reine Angew. Math. 221, 113.
Arenstorf, R. F.: 1962, A.I.A.A. 1, 238.
Arenstorf, R. F.: 1968, J. Diff. Eqs. 4, 202.
Arenstorf, R. F.: 1978, Celes. Mech. 17, 331.
Arenstorf, R. F.: 1982, Celes. Mech. 28, 9.
Barrar, R. B.: 1965, Astron, J., 70, 3.
Belbruno, E.A.: 1981a, Celes. Mech. 25, 195.
Belbruno, E.A.: 1981b, Celes. Mech. 25, 397.
Benettin, G., Cercignani, G., Galgani, L. and Giorgilli, A.: 1980, Lett. Nuovo Cim. 28, 1.
Birkhoff, G.D.: 1915, Rend. Circ. Mat. Palermo 39, 1.
Bountis, T.: 1981, Physica 3D, 577.
Bozis, G. and Hadjidemetriou, J.D.: 1976, Celes. Mech. 13, 127
Breakwell, J. V. and Perko, L. M.: 1966, in "Progress in Astronautics and Aeronautics", Vol. 17, Academic Press.
Breakwell, J. V.: 1974, Celes. Mech. 9, 437.
Broucke, R.: 1975, Celes. Mech. 12, 439.
Broucke, R.: 1979, Astron. Astrophys. 73, 303.
Broucke, R. and Walker, D.E.: 1980, Celes. Mech. 21, 73.
Broucke, R., Anderson, J.D., Blitzer, L., Davoust, E. and Lass, H.: 1981, Celes. Mech. 24, 63.
Bruno, A.: 1972, "Recherches sur le problême restreint des trois corps, Propietês des solutions pur μ=0", Academie des Sciences, Moscou.
Bruno, A.: 1981, Celes. Mech. 24, 255.
Colombo, G., Franklin, F.A. and Munford, C.M.: 1968, Astron. J. 73, 111.
Contopoulos, G.: 1983a, Celes. Mech. 31, 193.
Contopoulos, G.: 1983b, Physica D.
Contopoulos, G.: 1983c, Astron. Astrophys. 117, 89.

Contopoulos, G. and Michailidis, P.: 1980, Celes. Mech. 22, 403.
Contopoulos, G. and Pinotsis, A.: 1984, Celes. Mech. (to appear).
Davoust, E.: 1983a, Celes. Mech. 31, 241.
Davoust, E.: 1983b, Celes. Mech. 31, 293.
Delibaltas, P.: 1976, Astrophys. Space Science 45, 207.
Delibaltas, P.: 1983, Celes. Mech. 29, 191.
Greene, J. M., MacKay, R.S., Vivaldi, F. and Feigenbaum, M.J.: 1981, Physica 3D, 468.
Griffin, F.L.: 1920, in F.R. Moulton (ed.), Periodic Orbits, Ch.14, Carnegie Inst., Washington.
Guillaume, P.: 1969, Astron. Astrophys. 3, 57-76.
Guillaume, P.: 1975a, Celes. Mech. 11, 213.
Guillaume, P.: 1975b, Celes. Mech. 11, 449.
Hadjidemetriou, J.D.: 1975a, Celes. Mech. 12, 155.
Hadjidemetriou, J.D.: 1975b, Celes. Mech. 12, 255.
Hadjidemetriou, J.D.: 1976a, Astrophys. Space Science 40, 201.
Hadjidemetriou, J.D.: 1976b, in V. Szebehely (ed.) Long Time Predictions in Dynamics, D. Reidel Publ. Co., p. 223.
Hadjidemetriou, J.D.: 1977, Celes. Mech. 16, 61.
Hadjidemetriou, J.D.: 1980, Celes. Mech. 21, 63.
Hadjidemetriou, J.D. and Michalodimitrakis, M.: 1981, "Periodic Planetary-Type Orbits of the General 4-body Problem with an Application to the Satellites of Jupiter", Astron. Astrophys. (to appear).
Hadjidemetriou, J.D.: 1982, Celes. Mech. 27, 305.
Hadjidemetriou, J.D. and Ichtiaroglou, S.: 1983, in "Dynamical Trapping and Evolution in the Solar System", V.V. Markellos and Y. Kosai (eds.), Reidel, P. 141.
Hadjidemetriou, J.D. and Ichtiarogrou, S.: 1984, Astron. Astrophys. 131, 20.
Helleman, R.H.G.: 1980 in "Fundamental Problems in Statistical Mechanics", Vol. 5, E.G.D. Cohen (ed.), North Holland, p. 165.
Helleman, R.H.G. and Bountis, T.: 1978, Lect. Notes in Phys. 93, 353.
Hénon, M.: 1968, Bull. Astron. 3, 377.
Hénon, M.: 1969, Astron. Astrophys. 1, 223.
Hénon, M.: 1974, Celes. Mech. 10, 375.
Hénon, M.: 1976, Celes. Mech. 13, 267.
Hénon, M.: 1977, Celes. Mech. 15, 243.
Henrard, J.: 1980, Celes. Mech. 21, 83.
Hitzl, D.: 1977a, Astron. Astrophys. 54, 47.
Hitzl, D.L.: 1977b, A.I.A.A. 15, 1410.
Hitzl, D. and Hénon, M.: 1977a, Celes. Mech. 15, 421.
Hitzl, D. and Hénon, M.: 1977b, Acta Astronautica, 4, 1019.
Hitzl, D.L. and Levinson, D.A.: 1980, Celes. Mech. 22, 255.
Howell, K.C.: 1984, Celes. Mech. 32, 53.
Howell, K.C. and Breakwell, J.V.: 1984, Celes. Mech. 32, 29.
Ichtiaroglou, S.: 1980a, Astron, Astrophys. 81, 88.
Ichtiaroglou, S.: 1980b, Astrophys. 92, 139.
Ichtiaroglou, S.: 1980c, Astrophys. Space Sci. 67, 111.
Ichtiaroglou, S.: 1981, Astron. Astrophys. 98, 401.
Ichtiaroglou, S. and Michalodimitrakis, M.: 1980, Astron. Astrophys.

81, 30.
Ichtiaroglou, S., Katopodis, K. and Michalodimitrakis, M.: 1980a, Astron. Astrophys. 90, 324.
Ichtiaroglou, S., Katopodis, K. and Michalodimitrakis, M.: 1980b, Astrophys. Space Sci. 73, 445.
Ichtiaroglou, S., Katopodis, K. and Michalodimitrakis, M.: 1978, Astron. Astrophys. 70, 531.
Jorna, S. (ed.): 1978, "Topics in Nonlinear Dynamics, A Tribute to Sir Edward Bullard", AIP.
Kameyer, P.C. 1978, Celes. Mech. 17, 121.
Kameyer, P.C.: 1980, Celes. Mech. 22, 289.
Kameyer, P.C.: 1983, Celes. Mech. 30, 329.
Katopodis, K.: 1979, Celes. Mech. 19, 43.
Katopodis, K. Ichtiaroglou, S. and Michalodimitrakis, M.: Astron. Astrophys. 90, 102.
Kwok, J.H. and Nacozy, P.E.: 1982, Celes. Mech. 27, 27.
Marchal, C.: 1979, in R. Duncombe (ed.), Dynamics of the Solar System, D. Reidel Publ. Co., p. 29.
Markellos, V.V.: 1977a, Monthly Notices Roy. Astron. Soc. 180, 103.
Markellos, V.V.: 1977b, Astron. Astrophys. 61, 195.
Markellos, V.V.: 1979, in Instabilities in Dynamics and Celestial Mechanics, V. Szebehely (ed.), D. Reidel Publ. Co., p. 243.
Markellos, V.V.: 1980, Celest. Mech. 21, 291.
Markellos, V.V.: 1981, Celes. Mech. 25, 3.
Message, J.: 1966, Proc. IAU Symposium No. 25, 19.
Message, J.: 1980, Celes. Mech. 21, 55.
Message, P.J., 1982, Celes. Mech. 28, 107.
Meyer, K.R.: 1981a, J. Dif. Eqs. 39, 2.
Meyer, K.R.: 1981b, Celes. Mech. 23, 69.
Michalodimitrakis, M.: 1978a, Astron. Astrophys. 64, 83.
Michalodimitrakis, M.: 1978b, Astron. Astrophys. 70, 473.
Michalodimitrakis, M.: 1978c, Astrophys. Space Sci. 58, 125.
Michalodimitrakis, M.: 1978d, Astron. Astrophys. 76, 6.
Michalodimitrakis, M.: 1979, Celes. Mech. 19, 263.
Michalodimitrakis, M.: 1979, Astrophys. Space Sci. 65, 459.
Michalodimitrakis, M.: 1980, Astron. Astrophys. 81, 113.
Milani, A.: 1983, (preprint)
Moulton, F.R.: 1910, Ann. of Math. 2, 1.
Moulton, F.R.: 1912, Trans. Amer. Math. Soc. 13, 96.
Palmore, J.: 1976, Ann. of Math. 104, 421.
Palmore, J.: 1982, Celes. Mech. 28, 17.
Perko, L.M.: 1974, SIAM J. Appl. Math. 27, 200.
Perko, L.M.: 1981, Celes. Mech. 24, 155.
Perko, L.M.: 1983, Celes. Mech. 30, 115.
Presler, W.H. and Broucke, R.: 1981a, Comp. Math. Appl. 7, 451.
Presler, W.H. and Broucke, R.: 1981b, Comp. Math. Appl. 7, 473.
Poincaré, H.: 1899, "Les Methods Nouvelles de Mechanique Celeste", Gauthier Villars.
Poincaré, H.: 1902, Bulletin Astronomique, 69, 177.
Richardson, D.: 1980, Celes. Mech. 22, 241.
Robin, I.A.: 1981, Celes. Mech. 23, 97.

Robin, I.A. and Markellos, V.V.: 1980, Celes. Mech. 21, 395.
Schmidt, D.S.: 1972a, in L. Weiss (ed.) Ordinary Differential Equations, NRL-MRC Conference.
Schmidt, D.S.: 1972b, SIAM J. Appl. Math. 22, No.1.
Schubart, J. 1956, Astron. Nachr. 283, 17.
Siegel, C.L. and Moser, J.K.: 1971, "Lectures on Celestial Mechanics", Springer-Verlag.
Sitter, W.de.: 1909, Proc. Acad. Amst. 11, 682.
Smale, S.: 1970, Invent. Math. 11, 45.
Standish, E.M.: 1970, in G.E.O. Giacaglia (ed.), Periodic Orbits, Stability, and Resonances, D. Reidel Publ. Co. Dordrecht, Holland, p. 375.
Standish, E.M.: 1972, Astron. Astrophys. 21, 185.
Stellmacher, I.: 1976, Astron. Astrophys. 51, 117.
Stellmacher, I.: 1977, Astron. Astrophys. 59, 337.
Stellmacher, I.: 1982, Celes. Mech. 28, 351.
Szebehely, V.: 1967, Theory of Orbits, Academic Press, N.Y.
Szebehely, V.: 1970, in G.E.O. Giacaglia (ed.), Periodic Orbits, Stability, and Resonances, D. Reidel Publ. Co., Dordrecht, Holland, p. 382.
Szebehely, V.: 1971, Celes. Mech. 4, 116.
Szebehely, V.: 1972, Celes. Mech. 6, 84.
Szebehely, V.: 1973, in B.T. Tapley and V. Szebehely (ed.), Recent Advances in Dynamical Astronomy, D. Reidel Publ. Co., p.75.
Szebehely, V.: 1974, Celes. Mech. 9, 359.
Szebehely, V. and Feagin, T.: 1973, Celes. Mech. 8, 11.
Szebehely, V. and Zare, K.: 1977, Astron. Astrophys. 58, 145.
Taylor, D.M.: 1981, Astron. Astrophys. 103, 288.
Taylor, D.M.: 1983, Celes. Mech. 29, 51.
Taylor, D.M.: 1983, Celes. Mech. 29, 75.
Wiesel, W.: 1981, Celes. Mech. 23, 231.

THE PROBLEM OF SMALL DIVISORS IN PLANETARY MOTION

Carol A. Williams
Department of Mathematics
University of South Florida
Tampa, Florida 33620

ABSTRACT. Small divisors caused by certain linear combinations of frequencies appear in all analytical planetary theories. With the exception of the deep resonance between Neptune and Pluto, they can be removed at the expense of introducing secular and mixed secular terms, limiting the domain in which the solution is valid. Because of them classical solutions are known not to converge uniformly; Poincaré referred to them as asymptotic. The KAM theory shows that if one is far enough from exact commensurability and has small enough planetary masses, expansions exist which will converge to quasi-periodic orbits. Solutions showing very small divisors are excluded from this region of convergence. The question of whether they are intrinsic to the problem or are just manifestations of the method of solution is not settled. Problems with a single commensurabily that can be isolated from the rest of the Hamiltonian may have solutions with no small divisors. The problem of two or more commensurabilities remains unsolved.

1. APPEARANCE OF SMALL DIVISORS

Small divisors can arise in planetary theory for several reasons. Among these we include the divisors produced by small eccentricity or inclination in the differential equations, the reciprocal of the mutual distances in binary or multiple collisions, and the small divisor produced by the commensurability of two or more natural frequencies of the motion. The last type will appear after formal integration or in the generating functions of canonical transformations. The parameters of the conic sections, used to approximate the solutions initially, often take on values which cause small divisors in the Lagrange equations in the method of the variation of arbitrary constants. These singularities can be removed by the introduction of special variables. An example of this can be found in Brouwer and Clemence (1961). The singularity produced by collision can sometimes be removed with regularization. Szebehely (1967) discusses regularization of the two body problem and the planar restricted three body problem. We will not treat the last two problems outlined above, but will focus on small divisors arising from near commensurability of two or more frequencies

Celestial Mechanics 34 (1984) 395–410. 0008–8714/84.15

of the motion.

The planetary theory treated in this paper assumes purely gravitational attraction between N planets together with the sun, all considered as point masses. In Hamiltonian form there are 3N angle variables, θ, and 3N action variables, Θ. Such a formulation can be developed in Jacobi coordinates, with the rectangular coordinates of each planet referred to the barycenter of the sun and the planets interior to it, the innermost planet referred to the sun.

Let the frequencies of the θ_j be designated by ν_j, and the Hamiltonian by H. Then

$$\nu_j = d\theta_j/dt = \partial H/\partial \Theta_j .$$

With $\underline{\nu}$ defined as a vector of the 3N frequencies and $\underline{k}$, a vector of integers, exact commensurability is defined by

$$\underline{k}\cdot\underline{\nu} = 0, \quad |\underline{k}| \neq 0. \tag{1}$$

Another important case is

$$\nu_j = 0, \tag{2}$$

where ν_j represents a single frequency. Eq.(2) can sometimes be considered a special case of Eq.(1) if new variables are introduced by a suitable linear combination of angular variables. If Eq.(1) or (2) were enforced exactly for all values of the independent variable, t, the Hamiltonian would contain one less degree of freedom for each resonance, (since if $\underline{k}\cdot\dot{\underline{\theta}} = 0$, then $\underline{k}\cdot\underline{\theta}$ = constant.) This last condition allows one of the θ_j to be elminated from the Hamiltonian and its conjugate variable, Θ_j, becomes an ignorable coordinate. The real problem with small divisors is caused by the fact that we are usually interested in solutions not only at, but also in a neighborhood of, exact resonance. Often, conditions (1) or (2) are not exactly satisfied, but only approximately so, for all time. The degree to which one of these conditions is satisfied is characterized by the description deep or shallow, Garfinkel (1966). The demarkation between deep and shallow resonance is given by the square root of a small parameter, ε. One basis for this characterization can be found in Dziobek (1892), Poincaré (1893), Brown and Shook (1933), and in many other standard references. It is a consequence of the fact that if θ is a critical argument, it may sometimes approximately satisfy the differential equation

$$\ddot{\theta} = -\varepsilon \sin\theta, \tag{3}$$

The energy integral of Eq.(3) with total energy C, can be written

$$\dot{\theta}^2 = 2(C+\varepsilon)[1 - (2\varepsilon\sin^2\theta/2)/(C+\varepsilon)] .$$

From this it is clear that the motion can be characterized by $\sqrt{\varepsilon}$.

Small divisors appear at various stages in the development of the solution of the planetary problems. Some examples are given here. The first resonance found analytically among the major planets is called the great inequality between Jupiter and Saturn. The low frequency of this commensurability is $2n_5 - 5n_6$. (The term n_i is the frequency of the mean longitude of the i'th planet from the sun.) The large perturbations which these two planets raise on each other was first explained as due to the resonance by Laplace in 1786. The deepest resonance between orbital frequencies of planets in the solar system is that between Neptune and Pluto, $2n_8 - 3n_9$. This resonance was discovered by Cohen and Hubbard (1965) from a numerical integration of the orbits of the outer planets. Many small divisors appear only at higher orders during the development of the solution, raising the question as to whether or not their source is the method of approximation itself. Two such divisors are mentioned as examples. The small divisor $n_3-2n_4+n_5$ appears in the second order solution for the orbit of Mars derived by Newcomb(1895) and also found by Brumberg et al (1975) and by Bretagnon (1980). In the third order theory, the divisor $3n_5 -8n_6 -2n_7 +7n_8$ is discussed by Duriez (1977) and Simon and Francou (1981). The critical argument associated with this small divisor has a period of 119,000 years; other arguments have been found with longer periods. It seems reasonable to conclude that arbitrarily small divisors could arise at arbitrarily high orders.

Numerical work has indicated that some of the orbits with low order resonances are close to periodic orbits. An example is the work by Nacozy and Diehl (1974) for the Neptune-Pluto system. Periodic orbits with very long periods may be associated with high order resonances. These may be difficult to establish because the orbits may be embedded in stochastic regions of phase space and because of the accumulation of numerical errors in long time intervals of integration. If it were not for these considerations, it seems that there should be no trouble generating orbits numerically which come as close to exact commensurability as the numerical precision will allow. Small divisors come from long period perturbations, and this is how they are recognized numerically. (The longest integrations in the solar system extend for well over one million years, and so far this has been the observation.) If we entertain the possibility, based on these observations from numerical integration, that solutions exist at and near exact commensurability, why has no planetary theory been able to establish such solutions analytically? The concensus of opinion puts the blame on the series expansions used to generate the analysis. The resonances show up as the singularities of the d'Alembert series after integration or canonical transformation. Small divisors appear in other contexts as well, and it is by no means clear that they arise only because of the use of d'Alembert series. Kolmogorov's Theorem and a set of related results by him, Arnol'd, and Moser (referred to as the KAM theory) assure existence of quasi - periodic solutions of the planetary equations and exclude very small divisors from the set of necessary conditions which lead to these solutions. See, for example, Arnol'd (1978). We will discuss these theoretical results further in section 2.

Traditionally some commensurable terms in the analytic solutions have been called inequalities perhaps because the numerator of the coefficients of these terms often could not be established with the precision needed to overcome the numerical smallness of the divisor. The result was that such a term could not predict the perturbation with the required accuracy. The numerical evaluation of this coefficient was made by comparison of theory with observation. Since such a term has a very long period and may appear to be secular over the short time frame of the observations, it may be difficult to separate it from other perturbations. Nevertheless, these terms have been used successfully to determine parameters such as the masses of planets as well as (in artificial satellite theory) parameters in the earth's gravity field.

An example of a small divisor of the type in Eq.(2) might be simply the frequencies of the slowly varying angles $\tilde{\omega}$ and Ω , the longitudes of the perihelion and asending node respectively. To avoid small divisors or secular perturbations, the terms in the disturbing function consisting solely of these arguments are treated separately in a method often called the method of secular perturbations. Brouwer and van Woerkam (1950) treated this problem to second powers in eccentricity and inclination for the first eight planets. Their theory considered only the first approximation to the equations, which resemble Eq.(7). Apparently, as higher order solutions to these equations are considered, the method no longer avoids small divisors.

A problem which has not yet been treated completely analytically and promises to cause some difficulties is the libration of the argument of perihelion of Pluto, ω_9. This libration was first established by Williams and Benson (1971) and confirmed by Kinoshita (1984). A semi-analytical treatment of it appears in Nacozy and Diehl (1974), where they study Pluto's orbit in the context of the restricted problem of three bodies. They show by numerical calculation that one needs to include perturbations by Saturn and Uranus to obtain a libration of this argument, otherwise it will circulate. On the average, $\dot{\omega}_9 = 0$, satisfying the criterion given in Eq.(2). This critical argument may require a special analytical treatment since it appears in the equations together with the deep resonance between Pluto and Neptune, making the whole problem of Pluto one of double resonance.

2. CLASSICAL ANALYTIC SOLUTIONS

An analytic planetary theory is obtained usually by one of two methods, successive approximations or canonical transformations of a Hamiltonian system. If the method of successive approximations is used, no matter what the coordinate system, the integration of the right hand sides of the differential equations requires approximations. The most successful approximation has been to use d'Alembert series, called Poisson series in the recent literature. These are infinite series of sines or cosines of linear combinations of angular variables factored by coefficients which are polynomials of the other variables. The angular variables themselves must be expressed as linear functions of the independent variable and this procedure leads to the presence of small divisors upon integration. Often especially when one needs to obtain an expression for the mean longitude, the small divisor is squared because of the double integration of the mean motion. As the approximation is carried to successively higher orders, small divisors are carried to higher powers. Prior to the avilability of computers to do algebra, few authors went beyond third order and had to be selective in the terms they retained. Since each higher order carries another power of the planetary masses, one would hope that only the very largest terms would be needed and that there would be few of these. To a cetain extent this is true; older theories were successful. Newcomb (1895) went to second order in the theories of the terrestrial planets; Hill (1890), for Jupiter and Saturn, considered third order. Clemence's (1961) theory of Mars carries third order terms. Most of these authors went to at least one higher order when developing the secular perturbations.

Because modern observations carry increased accuracy and older theories decline in precision with increasing time, there is a need for higher precision in planetary theories. The new planetary theories developed at the Bureau des Longitudes carry eight planets simultaneously. The successive approximations of Lagrange's equations are carried to third or higher order in the planetary masses; see, for example, Bretagnon (1982). Brumberg (1975) has gone to second order with successive approximations using a new method with rectangular coordinates.

Another source of truncation error, in addition to that introduced by expansions in powers of the masses, is that introduced by the series expansions of the right hand sides of the equations. It is well known that the irrational ratio of any two real numbers can be approximated arbitrarily closely by the ratio of two relatively prime integers simply by carrying out the continued fraction approximation to a high enough order. The two integers may be very large. This means that the divisor $pn - qn'$ (where the prime refers to the second planet) can be made arbitrarily small even if the frequencies n and n' are not commensurable. The near resonance between Jupiter and Saturn is much better approximated by the combination $60n_5 - 149n_6$. Celestial mechanicians make use of the d'Alembert characteristic to explain why it is unnecessary to retain these very high order terms. Consider the argument of a trigonometric term to be

$$k_1\lambda + k_2\tilde{\omega} + k_3\Omega + k_4\lambda' + k_5\tilde{\omega}' + k_6\Omega' \tag{4}$$

where λ's are mean longitudes. The d'Alembert characteristic states that the lowest power of e, e', sin I, and sin I', appearing in the coefficients, is (respectively) $|k_2|, |k_5|, |k_3|$, and $|k_6|$. Most d'Alembert series are truncated before one gets to powers of the eccentricity and inclination as high as 60 or 149. An important area of research is to determine which of these high order terms should be retained to meet the precision required by a theory. For an interesting discussion of the estimation of the size of a resonance term, see Garfinkel (1982). Chapront and Dvorak (1976) have done some work in estimating the sizes of high order resonances for planetary theory.

The KAM theories give an even stronger indication that very high order small divisors are not important as long as the fundamental frequencies are sufficiently far from exact commensurability. The condition on the frequencies to insure convergence is

$$|\underline{k}\cdot\underline{\nu}| > \gamma|\underline{k}|^{-1} \tag{5}$$

where γ is some positive real number. The other condition for convergence (which may not be satisfied) is the one requiring that the perturbation be small enough; in this context it implies that the masses of the planets be small. This is a very powerful result because it assures us that quasi-periodic solutions exist in spite of the fact that linear combinations of their incommensurable frequencies can be found, with continued fractions, to be as small as we wish. At present, the calculated least upper bound of the perturbation parameter is much smaller than the masses of the planets. However there are numerical indications that the planets are small enough to satisfy the conditions of the theorem, Moser (1973). The planetary frequencies are known, unfortunately, only to a finite number of digits. Since this will always be the case, we may never be able to apply the conditions of the theorem exactly.

For practical applications, to be sure of the numerical convergence far from resonance, fixed frequency expansions are suggested. Some examples of these are given by Moser (1973). In a Hamiltonian formulation, these transformations are different from the usual von Zeipel or Lie series methods used by, for example, Meffroy (1978) and Kamel (1979,1983). Fixed frequency expansions have been done by Merman (1982), Duriez (1979) for the planetary theory, and Ferraz-Mello (1979). Ferraz-Mello studies the Galilean satellites, a system well known for its complicated resonance structures. Even though it is a satellite problem, the equations share some features with the planetary perturbation problem.

The method of von Zeipel used by Brouwer (1959) for the theory of an artificial satellite about an oblate planet, and the Lie series methods of Hori (1966) and Deprit (1969) are examples of canonical transformations which can be applied to the planetary theory in Hamiltonian form as outlined in Section 1. Canonical transformations replace integration with transformation, putting off as long as possible, the integration of the right hand sides of the differential equations with respect to the independent variable. In its place is the computation of generating functions, which require the solution of, hopefully, less complicated partial differential equations. If one has a Hamiltonian $H(\theta, \Theta)$, the partial differential equation for a generating function W starts with a term of the form

$$(\partial H/\partial\Theta)(\partial W/\partial\theta) + \ldots = 0 \tag{6}$$

The θ's represent the angle variables and therefore $\partial H/\partial\Theta$ are the frequencies. If θ is a critical argument, one can see that a small divisor will enter the expression for W.

If it is seen that a term will produce a very small divisor, the trigonometric function can be replaced by a power series in the independent variable before integration. The introduction of such terms at one order will produce mixed secular terms in the next. For this reason, it is desirable to avoid this technique until the highest order possible. The transition from trigonometric to secular terms may also be necessary because of the sheer number of terms producing small divisors. With computer generated perturbations, the number of such terms can be quite large. The choice of when to expand such terms is really a free one and implies that two different theories can have different secular and mixed secular terms while both represent the motion equally well. The solution of this paradox lies with the choice of appropriate values of the constant terms in the series representing the solutions. The introduction of secular terms into a theory is necessitated not only when d'Alembert series are used. In attempting to avoid small divisors in a theory using elliptic functions (discussed in Section 4) secular terms enter the perturbations unavoidably. These theoretical results also lead to the question of the intrinsic nature of small divisors, which so far has no answer.

Duriez (1979) and Laskar (1984) are treating the very important question of whether or not a very precise theory can be generated without secular terms. Such a representation should be possible for quasi-periodic orbits. Difficulties arise in handling large numbers of terms with small divisors. Compared to other planetary theories, these solutions do not yet have the precision of those containing secular terms. However, they should be valid for a longer time interval, and will converge to the orbits of the planets if those orbits are quasi-periodic. To what extent quasi-periodic motions describe the true orbits of the planets depends ultimately on how close the Hamiltonian of Section 1 describes the real solar system and on the true values of the planetary frequencies. As long as these quasi-periodic solutions are not complete, however, it remains true that the maintenance of high precision requires the introduction of secular terms.

3. SOLUTIONS AT SHALLOW RESONANCE

In principle, the problem of shallow resonance can be handled by standard techniques, although the question of convergence of those techniques must be answered by theorems not yet established. In practice, sufficient precision requires keeping high orders in the approximations. The 2:5 commensurability between Jupiter and Saturn is an example of shallow resonance. Hill's (1890) theory of these two planets was developed to third order to represent their motions for 300 years. The Bureau des Longitudes is meeting the challenge of developing high precision analytical ephemerides valid for one thousand years. But in order to produce the theory of Jupiter and Saturn, it is necesary for them to consider terms of at least seventh order in the masses, a much higher order than is needed for other planets. This is the type of difficulty which shallow resonance can produce. For deep resonance, this method of solution seems completely impossible.

The Lindstedt method treats the equation

$$\ddot{x} + n^2 x = \sum_j \varepsilon^j C_j x^j \tag{7}$$

where ε is small, and n and C_j are constants. Equations for critical arguments x in planetary theory can be put in this form. Its solution depends on establishing the correct value of the frequency before integration is introduced. Giacaglia (1972) gives an exact solution to Eq.(7) when x librates, but states that there is no solution when x circulates. As developed by Poincaré (1893), this technique is nearly equivalent to the method of von Zeipel (1916,1918). Because the solution in the von Zeipel-Poincaré method needs mean elements and not initial conditions for its constants of integration, it is a difficult theory to impliment. The Lie series methods of Hori and Deprit remove this complexity. These tools give us a way to calculate the frequencies of the angle variables to any desired order before integration with respect to the independent variable is introduced. Fixed frequency and fixed frequency ratio expansions may be essential here in order that the divisors are not expressed by variable frequencies which can pass through exact resonance at certain stages in the developements.

Sometimes, using an independent variable different from the time may have an advantage. Hill (1890) and Clemence (1961) use Hansen's method. Hansen's method of partial anomalies has been suggested for the orbit of Pluto. Poincaré (1893) stated that Gylden's method would be more accurate than Lindstedt's close to resonance. The method of quadratic convergence developed by Howland (1979) and Howland and Richardson (1984) may be useful since the perturbations decrease with the square of the small parameters as higher orders are calculated. This should aid considerably by providing faster numerical convergence so that fewer higher order terms will be required to maintain high precision.

4. DEEP RESONANCE

4.1 One resonant term

If there is one critical argument in a Hamiltonian it may be possible to isolate it. First, a new variable must be introduced which is given by an appropriate combination of angles required to create the critical argument. This may necessitate the introduction of an additional degree of freedom, a technique used successfully by Brouwer (1963) in finding an integral of the motion for asteroids resonant with Jupiter. This was discussed for a general problem by Jefferys (1976). The isolation of the critical term may be accomplished in one of two ways. One method removes all of the other variables in the system using either an integration of the short period terms in the Lagrange equations as was done by Brown (1912) for the libration of asteroids, or a von Zeipel method to eliminate them as was done by Hori (1960) for artificial satellites, producing the first solution connecting the regimes of libration and circulation. This technique is described for a more general problem by Garfinkel (1976) in a theory of libration. Lie series techniques can be used instead of the von Zeipel method. (At the present time it is not clear that they are producing equivalent results). If the solution to the ideal resonance problem (Garfinkel, Jupp, and Williams, 1971), or any other single resonance solution, can be applied to the critical argument, then this part of the Hamiltonian may also be eliminated first, as is done by Garfinkel (1977) in his theory of Trojan asteroids.

The Bohlin (1889) technique has been used successfully to solve problems with one critical argument. This method uses expansions in powers of the square root of a small parameter to avoid small divisors. In a Hamiltonian context, start with a Hamiltonian of the form

$$H(\theta,\Theta) = H_0(\Theta) + \varepsilon H_1(\theta, \Theta) \tag{8}$$

with ε a small parameter and θ the critical argument. Eq.(6) describes the first terms in the equation for the generating function W. The Bohlin technique enables one to set up the transformations in such a way that the equation for the determination of W begins with the term

$$(\partial^2 H_0/\partial\Theta^2)/(\partial W/\partial\theta)^2 + \dots \tag{9}$$

$\partial W/\partial\theta$ and subsequently W itself are found by taking a square root; hence, factored by $\sqrt{\varepsilon}$. It is possible that $\partial^2 H/\partial\Theta^2$ is also small and the method suggests a cube root, and so on. Unfortunately, the n'th root of ε will approach unity, quickly jeopardizing the convergence of the method if roots much greater than the second are required. These problems will not be well suited to the Bohlin technique.

The Bohlin-von Zeipel method was applied to a problem of one small divisor by Garfinkel (1966), Jupp(1969,1970), and Garfinkel, Jupp and Williams (1971). The problem is called the ideal resonance problem (IRP), and its Hamiltonian, F, has the form

$$F(x,y) = B(y) + \varepsilon A(y) \sin^2 x . \tag{10}$$

$\partial B/\partial y = 0$ is assumed to have one and only one root in the vicinity of the solution. A further assumption is that A and B'' are sufficiently far from zero. This condition is called the normality condition. The topology of the solution is that of a simple pendulum. The solution is global in the sense that by a continuous variation of a single parameter, one can trace the solution from exact resonance through libration, the separatrix, and into circulation. In the limit the resonance solution can be matched to the classical solution. Jupp (1982) discusses the developement of the Bohlin-Lie series method for the ideal resonance problem. A difficulty in these solutions is the loss of an order of magnitude, a problem which seems to be handled more easily by the von Zeipel method. But the use of Lie series may eliminate the need for a regularizing function, required in the other method. Hence, one can interpret the remark that the equivalence of the two methods is not immediately obvious. The application of this method to the Hamiltonian of Eq.(10) (Jupp 1972,1974) is not global in the sense given above. Deprit and Richardson (1982) duplicate some of these results by means of simplifying canonical transformations. Deprit (1981a) suggests simplifying the complexity of a Hamiltonian as much as possible by closed-form transformations before expansions are introduced. Concepts such as Deprit's may be particularly useful for the problem of more than one resonance.

Some of the problems containing only one critical argument do not have the same features as Eq.(10). One way to generalize is to replace $\sin^2 x$ by a more complicated function of x. This problem was considered by Giacaglia (1970)and Garfinkel (1976). If the normality condition is not satisfied, Garfinkel (1962,1966) calls the problem abnormal. His paper of 1972 describes a way of removing the singularity, $AB'' = 0$, from the solution of the IRP. If there is more than one root of $\partial B/\partial y = 0$, the problem may have a topology different from that of the simple pendulum. Such problems have been addressed by Henrard and Lemaitre (1983) and Ferraz-Mello (1984). A useful generalization has been to consider a single resonance within the context of two or more degrees of freedom. Such a problem is treated by Ferraz-Mello (1978) using an extension of Delauney's method to treat many degrees of freedom. Moore(1983,1984) applies a Bohlin-von Zeipel algorithm to a problem of two degrees of freedom in which the single critical argument cannot be isolated. These generalizations may be the logical steps needed to understand the mathematical link between the problems of single and double resonance.

4.2 Two or more critical arguments

Numerical studies of double resonance problems are giving new insights into the phase space structure of these dynamical systems. These studies generally take the form of coupled perturbed harmonic oscillators. See, for example, Hitzl (1975), Sanders (1977) and Contopoulos (1978). It is not clear whether such results will be directly applicable to the planetary theory, but they certainly will shed some light on the problem of small divisors. The classical paper of Hénon and Heiles (1964) develops an important method for obtaining theoretical insights from numerical computations of surfaces of section. Existence of surfaces displaying closed invariant curves suggests orbital stability, isolating integrals, periodic orbits, and perhaps integrability of the dynamical system. Regions where the invariant curves break down are associated with instability, stochasticity, chaotic motion and destruction of isolating integrals.

The fact that stable periodic orbits exist for planetary type problems is an encouraging factor when considering the complex question of the existence of solutions with exact multiple resonances. Hadjidemetriou and Michalodimitrakis (1978) find families of stable periodic orbits in two dimensions for planetary type systems containing the sun and two or more planets. The orbits are not stable for some resonances. Hadjimetriou (1981) finds stable periodic orbits for Jupiter and Saturn with a resonance of 2:5 and with fairly representative eccentricities. Markellos (1981) has found three dimensional periodic orbits for the general three body problem and Nacozy and Diehl (1974) found a periodic orbit of the third kind in the circular restricted three body problem which is close to the orbit of Pluto. While all of these results are encouraging, they do not demonstrate that the point mass, ten body planetary problem has stable periodic solutions close to the observed orbits of the planets.

Linear stability is the principle indicator used to analyze periodic orbits. Oseledets (1968) introduces Liapunov characteristic numbers (LCN) as a means of studying stability vs. stochasticity. Froeschle and Scholl (1981) examine the stability of the Kirkwood gaps this way. Jefferys and Zhao-Hua Yi (1983) apply the theory of LCN's to the restricted three body problem. If the maximum LCN = 0, the orbits are stable in the sense that they have closed invariant curves around them. The maximum LCN of an orbit is equal to the value of the Kolmogorov entropy, an indicator of stochasticity and the complexity of the algorithms needed to generate that orbit. The numerical work of Gonczi (1984) addresses this question.

Stable periodic orbits in problems with few degrees of freedom are important because they are surrounded by closed invariant curves. It seems reasonable that they are also thus separated from stochastic regions in more complicated problems. Periodic orbits may also represent exact commensurability of the fundamental frequencies and that therefore quasi-periodic solutions which have closed invariant curves exist very near to them if they are stable. Even if such periodic solutions to the planetary equations do exist, solutions very near to them that are developed by present techniques may not converge by the results of the KAM theory.

As mentioned in Section 1, it has often been suggested that the small divisors appearing at higher orders in the developments come from the need to use d'Alembert series and the concomitant need for linear combinations of angles that are, in turn, written as linear functions of the independent variables. Several authors have tried to avoid their use. Some success is expected since numerical integrations of the same equations show no singularities due to small divisors. Semi-analytic theories have been developed by Goodrich and Carpenter (1966) and by Chapront (1984). Goodrich and Carpenter use Chebyshev polynomials in multiples of the angle

$$g^* = (n/q)\ t = (n'/p)\ t \tag{11}$$

For a resonant orbit, $pn - qn' = 0$, they have no small divisors and convergence is rapid. Chapront (1984) uses what he calls Legendre expansions containing terms of the form $t^j \cos nt$ and $t^j \sin nt$. He develops a semi-analytic theory for Pluto which is also rapidly convergent, but valid for only a short interval of time.

Williams, van Flandern and Wright (1983) develop a first order solution to the Lagrange equations in two dimensions. For circular orbits, the expressions for da/dt and $d\varepsilon^{I}/dt$ depend only on the synodic angle and the first order perturbations can be integrated in closed form. These results are equivalent to those of Richardson (1982). The remaining two equations for $h = e \cos \tilde{\omega}$ and $k = e \sin \tilde{\omega}$ involve a second frequency. To obtain the first order perturbations, one must integrate terms of the form

$$\int \frac{\exp(\sqrt{-1}\ w\phi)}{\sqrt{1-\kappa^2 \sin^2 \phi}} \tag{12}$$

where $\kappa^2 < 1$ is a constant parameter depending on the semi-major axes of the planets and w is like a winding number. Eq.(12) is integrable in closed form if w is an integer. As it turns out, w will be an integer if and only if the orbital frequencies of the two planets have the relation

$$n/n' = N/(N+1), \quad N \text{ even}$$
$$n/n' = N/(N+2), \quad N \text{ odd} \tag{13}$$

In other cases, we let $w = N + \delta$, $|\delta| \leq 1/2$, and introduce the Fourier series

$$\cos \delta\phi = \frac{\sin \delta\pi}{\delta\pi} \sum_{J=0}^{\infty} \frac{(-1)^J 2\delta^2}{\delta^2 - J^2} \cos J\phi \tag{14}$$

The terms in Eq.(14) are free from small divisors, but the integration of Eq.(12) with Eq.(14) substituted for the trigonometric terms of the numerator introduces a secular term.

If progress is to be made in analysis, one perhaps should start with the general three body problem (sun and two planets). Even then, the equations will have to be simplified. One way to do that is to remove complexities with simplifying canonical transformations. An example is the elimination of the parallax by Deprit (1981b). Another method which I am considering is to simplify the expression for the reciprocal of the distance between planets (Δ). Begin by writing

$$\Delta^2 = [m_O r/(m_O+m) - r']^2 \cos^2(S/2) + $$
$$+ [m_O r/(m_O+m) + r']^2 \sin^2(S/2) \tag{15}$$

where r and r' are the distances of the two planets from the sun, S is the synodic angle, m_O the mass of the sun, and m the mass of the disturbing planet. Define

$$u_O = m_O r/(m_O+m) - r' \quad \text{and} \quad v_O = m_O r/(m_O+m) + r' \tag{16}$$

Then a sequence of canonical transformations which, for the coordinates, is given by

$$u_i = (u_{i-1} + v_{i-1})/2 \quad \text{and} \quad v_i = \sqrt{u_{i-1} v_{i-1}} \tag{17}$$

will transform Δ into

$$\Delta^2 = u_i^2 \cos^2\theta_i + v_i^2 \sin^2\theta_i \tag{18}$$

Since, in this process, $u_i \to v_i$ as $i \to \infty$, we have $\Delta \to u_i$(or v_i). In practice, i does not have to be too large for convergence.

Another simplifying approach is to find integrals of the motion. The search for isolating integrals can usually be undertaken either numerically or analytically, making this a powerful tool. The early papers by Marchal and Saari (1975), Szebehely (1977) and Zare (1977) began a new set of investigations on c^2h stability, a useful method to determine if there are bounded regions in phase space for certain types of motion (Hill curves). As Szebehely (1984) reminds us, this type of stability is independent of resonance and therefore may be useful to study it. Their work has already shed light on the structure of the phase space for the three body problem and some of its extensions. Systems of coupled perturbed harmonic oscillators have also been examined for the existence of isolating integrals. Froeschle and Schneidecker (1974) find, for a particular Hamiltonian, that the system has either N-1 or 0 isolating integrals for N degrees of freedom. Gustavson (1966), also for a particular Hamiltonian with N degrees of freedom, suggests only N-r+1 formal integrals can be constructed if there are r resonance relations. Contopoulos (1975,1978) obtains similar results and suggests that the isolating integrals break down not at resonance but on the boundaries between resonances. The destruction of the integrals, and hence of the invariant curves, occurs due to the interaction of resonances. It is at these boundaries that we also find homoclinic and heteroclinic orbits.

It may be possible to study multiple resonances by multiple application of single resonance solutions. If so, in the light of Contopoulos' remarks, these solutions would have to be local and could not represent orbits in the stochastic regions except for very localized volumes of phase space. If the separatrices of the resonance regions overlap, such an approach would likely be altogether impossible. The calculation of resonance widths can help to determine if this is the case. Wisdom (1980) has done this for the circular restricted three body problem using the technique of isolating the critical argument.

At the present time, no global solution has been found for any double or multiple resonance problem and it remains one of the outstanding unsolved problems of Celestial Mechanics.

5. CONCLUSIONS

Many single resonance problems have been solved analytically by expressions which do not contain small divisors. Although these solutions have not been proven to be convergent, nevertheless they represent the data reasonably well. Solutions probably exist for all single resonance problems for which the original problem can be reduced to one degree of freedom. This has also not been rigorously demonstrated; many versions of it remain unsolved. Single resonances which cannot be isolated from other variables in the problem are also being studied. The problem of a double or multiple resonance still exists. If such resonances represent periodic orbits surrounded arbitrarily closely by stochastic regions, it may be impossible to obtain global solutions with any set of finite algorithms.

As for the planetary orbits, we can now generate solutions analytically, for all planets but Pluto, which contain secular and mixed secular terms, but can compete somewhat with numerical integration. Purely quasi-periodic solutions have not been so fully developed, but they have been started. Pluto introduces deep resonance into the system, and with it the need for new analytical tools. Coupled with the fact that Pluto's orbit displays double resonance, this becomes a very challenging problem.

ACKNOWLEDGMENTS

I wish to thank Silvio Ferraz-Mello for pointing out some of the work done with fixed frequency expansions. I also thank Andre Deprit, Boris Garfinkel, Tom van Flandern, and Ken Seidelmann for helpful discussions of some aspects of planetary theory. I am especially grateful to Mrs. Brenda Corbin, Librarian of the U.S.Naval Observatory for her invaluable help with obtaining some of the reference material.

The Office of Sponsored Research of the University of South Florida supported part of the research and travel, and the National Science Foundation, through the office of the American Astronomical Society, provided travel support. The Center for Mathematical Services of the University of South Florida gave a great deal of assistance in the preparation of the manuscript. I am extremely indebted to these institutions.

REFERENCES

Arnol'd, V.I.: 1978, Mathematical Methods of Classical Mechanics, Springer-Verlag, New York, Appendix 8.

Bohlin, K.P.: 1889, 'Uber eine neue Annaherungmethode in der Storungtheorie', Ak. Handl. Bihang. (Afd. 1 Stockholm).

Bretagnon, P.: 1980, Astron. Astrophys. 84,329.

Bretagnon, P.: 1982, Astron. Astrophys. 114,278.

Brouwer, D.: 1959, Astron. J. 64,378.

Brouwer, D.: 1963, Astron. J. 68,152.

Brouwer, D. and Clemence, G.: 1961, Methods of Celestial Mechanics, Academic Press, New York, pp.287f.

Brouwer, D. and van Woerkom, A.J.J.: 1950, Astron. Pap. Am. Eph. 13,2.

Brown, E.W.: 1912, Mon. Not. Roy. Astr. Soc. 72,609.

Brown, E.W. and Shook, C.A.: 1933, Planetary Theory, Dover Edition, 1964,Chap.8.

Brumberg, V.A., Evdokimova, L.S. and Skripnichenko, V.I.: 1975, Sov. Astron. 19,255.

Chapront, J.: 1984, these proceedings.

Chapront, J. and Dvorak, R.: 1976, Mitt. Astron. Ges. 40,214.

Clemence, G.M.: 1961, Astron. Pap. Am. Eph. 16,2.

Cohen, C.J. and Hubbard, E.C.: 1965, Astron. J. 70,10.

Contopoulos, G.: 1975, Dynamics of Stellar Systems, I.A.U. Symp. #69, D.Reidel, 209.

Contopoulos, G.: 1978, Celes. Mech. 17,167.

Deprit, A.: 1969, Celes. Mech. 1,12.

Deprit, A.: 1981a, Bull. Amer. Astr. Soc. 13,2,570.

Deprit, A.: 1981b, Celes. Mech. 17,167.

Deprit, A. and Richardson, D.: 1981, Private Communication.
Duriez, L.: 1977, Astron. Astrophys. 54,93.
Duriez, L.: 1979, 'Approche d'une Theorie Generale Planetaire en Variables Elliptiques Heliocentriques', These, Univ. Sci. Tech., Lille, France.
Ferraz-Mello, S.: 1978, Comptes Rendues Ac. Sci., Paris, Series A, 286, 969.
Ferraz-Mello, S.: 1979, Dynamics of the Galilean Satellites, Univ. de Sao Paulo, Instituto Astronomico e Geofisico, Sao Paulo, Brasil.
Ferraz-Mello, S.: 1984, these proceedings.
Froeschle, C. and Scholl, H.: 1981, Astron. Astrophys. 93,62.
Froeschle, C. and Schneidecker, J.P.: 1974, Stability of the Solar System and of Small Stellar Systems, I.A.U. Symp. #62, D. Reidel, 297.
Garfinkel, B.: 1962, Proc. First Int'l. Symp. the on Use of Artif. Sat. for Geodesy (Washington,D.C.).
Garfinkel, B.: 1966, Astron. J. 71,657.
Garfinkel, B.: 1972, Celes. Mech. 6,151.
Garfinkel, B.: 1976, Celes. Mech. 13,229.
Garfinkel, B.: 1977, Astron. J. 82,368.
Garfinkel, B.: 1982, Celes. Mech. 28,275.
Garfinkel, B., Jupp, A. and Williams, C.A.: 1971, Astron. J. 76,157.
Giacaglia, G.E.O.: 1970, Periodic Orbits, Stability, and Resonances, D.Reidel, 515.
Giacaglia, G.E.O.: 1972, Perturbation Methods in Non-Linear Systems, Springer-Verlag, New York, Chap.2.
Gonczi, R.: 1984, these proceedings.
Goodrich, E.F. and Carpenter, L.: 1966, 'Computation of General Planetary Perturbations for Resonance Cases', Goddard Space Flight Center, Rept.# X-643-66-133.
Gustavson, F.G.: 1966, Astron. J. 71,670.
Hadjidemetriou, J.D.: 1981, Celes. Mech. 23,277.
Hadjidemetriou, J.D. and Michalodimitrakis, M.: 1978, Dynamics of Planets and Satellites and Theories of their Motion, I.A.U. Colloq. # 41, D.Reidel, 263.
Hénon, M. and Heiles, C.: 1964, Astron. J. 69,73.
Hill, G.W.: 1890, Astron. Pap. Am. Eph. 4.
Hitzl, D.L.: 1975, Astron. Astrophys. 40,147.
Hori, G.: 1960, Astron. J. 65,291.
Hori, G.: 1966, Publ. Astron. Soc. Japan 18,287.
Howland, R.A.: 1979, Celes. Mech. 19,95.
Howland, R.A. and Richardson, D.L.: 1984, Celes. Mech. 32,99.
Jefferys, W.H.: 1976, Astron. J. 81,132.
Jefferys, W.H. and Zhao-Hua Yi: 1983, Celes. Mech. 30,85.
Jupp, A.: 1969, Astron. J. 74,35.
Jupp, A.: 1970, Mon. Not. Roy. Astr. Soc. 148,197.
Jupp, A.: 1972, Celes. Mech. 5,8.
Jupp, A.: 1974, Celes. Mech. 8,523.
Jupp, A.: 1982, Celes. Mech. 26,413.
Kamel, O.M.: 1979, Astrophys. Sp. Sci. 64,227.
Kamel, O.M.: 1983, Moon Plts. 28,221.

Kinoshita, H.: 1984, these proceedings.
Laskar, J.: 1984, these proceedings.
Marchal, C. and Saari, D.G.: 1975, Celes. Mech. 12,115.
Markellos, V.V.: 1981, Celes. Mech. 25,319.
Meffroy, J.: 1978, Moon Plts. 19,3.
Merman, G.A.: 1982, Celes. Mech. 27,225.
Moore, P.: 1983, Celes. Mech. 30,31.
Moser, J.: 1973, Stability of the Solar System and of Small Stellar Systems, I.A.U. Symp. # 62, D.Reidel,1.
Nacozy, P. and Diehl, R.E.: 1974, Celes. Mech. 8,445.
Newcomb, S.: 1895, Astron. Pap. Am. Eph. 5,2.
Oseledets, V.I.: 1968, Tr. Mosk. Mat. Obs. 19,179.
Poincaré, H.: 1893, Methodes Nouvelles de la Mecanique Celeste, Volumes 1,2,and 3, Dover Edition, 1957.
Richardson, D.L.: 1982, Celes. Mech. 26,187.
Sanders, J.: 1977, Celes. Mech. 16,421.
Simon, J.L. and Francou, G.: 1981, Astron. Astrophys. 114,125.
Szebehely, V.: 1967, Theory of Orbits, Academic Press, New York, Chap.3.
Szebehely, V.: 1977, Celes. Mech. 15,107.
Szebehely, V.: 1984, these proceedings.
von Zeipel, H.: 1916, Arkiv. Mat. Astron. Fysik. 11,1,7.
von Zeipel, H.: 1918, Arkiv. Mat. Astron. Fysik. 12,9.
von Zeipel, H.: 1918, Arkiv. Mat. Astron. Fysik. 13,3.
Williams, C.A., van Flandern, T. and Wright, E.: 1983, Bull. Am. Astron. Soc. 15,869.
Williams, J.G. and Benson, G.S.: 1971, Astron. J. 76,167.
Wisdom, J.: 1980, Astron. J. 85,1122.
Zare, K.: 1977, Celes. Mech. 16,35.

THE IDEAL RESONANCE PROBLEM
A COMPARISON OF TWO FORMAL SOLUTIONS I

A. H. Jupp
D.A.M.T.P., University of Liverpool, P.O.Box 147,
Liverpool, L69 3BX, England

and

A.Y. Abdulla
Dept. of General Sciences, Military Technical College,
P.O. Box 478, Baghdad, Iraq

ABSTRACT. Garfinkel's solution of the Ideal Resonance problem derived from a Bohlin-von Zeipel procedure, and Jupp's solution, using Poincaré's action and angle variables and an application of Lie series expansions, are compared. Two specific Hamiltonians are chosen for the comparison and both solutions are compared with the numerical solutions obtained from direct integrations of the equations of motion. It is found that in deep resonance the second-mentioned solution is generally more accurate, while in the classical limit the first solution gives excellent agreement with the numerical integrations.

This article represents a summary of a much more extensive programme of research, the complete results of which will be published in a future article.

1. INTRODUCTION

The Ideal Resonance Problem is characterised by the Hamiltonian system of equations

$$-F = B(y) + 2\mu^2 A(y)\sin^2 x,$$
$$\dot{x} = F_y, \quad \dot{y} = -F_x. \tag{1}$$

The resonance manifests itself through the function B(y), which is such that its first derivative $B_y (\equiv B^{(1)})$ vanishes for some value of y, say y_0. Further it is assumed that μ is small, and that in the neighbourhood of y_0, A $B^{(2)}$ is O(B).

System (1) represents a perturbed simple pendulum, with the phase plane partitioned by the separatrix into regions of libration and circulation. Since the system comprises just one degree of freedom a solution may readily be obtained from quadrature. Nevertheless, it is desirable to have an analytic solution which can then be taken as the

Celestial Mechanics **34** (1984) 411–423. 0008–8714/84.15

basis for deriving solutions to more complex resonance problems. An analogy may be made with the two-body problem, which is used as a first approximation to many more sophisticated problems, such as those encountered in artificial satellite theory and planetary theory. Recently Moore (1983) has taken advantage of the Ideal Resonance Problem formulation to construct a formal solution of a resonance problem with two degrees of freedom. The 1:1 resonance associated with the motion of the Trojan asteroids has been the subject of an in-depth study by Garfinkel. His analysis is based upon an extension of the original Ideal Resonance theory, and his most recent publication (1983) sheds new light on this fascinating problem. Further applications are mentioned in Garfinkel's review paper (1982).

In an important series of articles, Garfinkel develops an asymptotic solution of (1), making use of the well-known techniques of Bohlin and von Zeipel. For the purposes of this article the most pertinent papers are Garfinkel (1966), Garfinkel et al (1971) and Garfinkel and Williams (1974). Hereafter these will be referred to as Papers I, II and III respectively. The last provides a definitive second-order solution in terms of standard elliptic functions and integrals.

An alternative and independent method of solution is presented by Jupp (1969, 1970, 1972); hereafter Papers I', II' and III'. This second approach employs an adaptation of some earlier work of Poincaré (1893). With the aid of a series of preliminary transformations and action and angle variables the resonant feature of the problem is essentially removed from the Hamiltonian; the remaining problem can then be handled using either the standard von Zeipel perturbation procedure or one based on Lie series expansions. The formal solution, presented in terms of elliptic functions and integrals, is in the form of asymptotic expansions in powers of μ.

Each of the two solutions just outlined has certain merits and some shortcomings. It would seem appropriate at this juncture to enumerate these essential differences.

(1) In Garfinkel's method, there is an unavoidable loss of an order of magnitude (in μ) on differentiation with respect to the momentum. This in turn leads to the requirement that, in order to construct the solution to order n, the new Hamiltonian F' and the generating function S must be determined to orders $n+2$ and $n+1$ respectively. In contrast, this phenomenon is not a feature of Jupp's solution. Indeed, the aforementioned preliminary transformations are such that the nth-order solution provides the momentum y accurate to order $n+1$ in μ; thus there is a gain of one order of magnitude in μ for y.

(2) The generating function of the canonical transformation is non-periodic in Garfinkel's formulation but periodic in that of Jupp.

(3) Garfinkel's solution is global and free from singularities; it is therefore appropriate for the libration region, the separatrix and all the circulation region. The latter incorporates 'deep' and 'shallow' circulation and the classical limit. On the other hand Jupp's original solution is generally singular at the separatrix and is inappropriate for shallow resonance and the classical limit. The first of these deficiencies has very recently been removed and the revised solution will be published in the near future.

(4) It would seem that an attempt to solve the general initial-value problem using Garfinkel's solution leads to an impasse, whereas employing the alternative solution presents no difficulty. These features of the two approaches will be described in a future article.

In this short article the two aforementioned theoretical solutions are compared with the solutions obtained from direct numerical integrations of the differential equations of motion. To this end two different Hamiltonians are chosen. In §2 the first-order theoretical solutions are given for the libration and circulation regions. The principal features of the subsequent comparison of each solution with the direct numerical solution are outlined in §3, where there is also given a brief report on the numerical procedures adopted.

In a future fuller article the results will be extended to second order and there, for the first time, the second-order circulation solution (Jupp) will be presented. Also, the general initial-value problem, referred to earlier, will be analysed in the lengthened version.

2. FIRST-ORDER SOLUTIONS

(i) GARFINKEL'S SOLUTION

With reference to Paper II constants ω_o, α, k, a_i, b_i and c_1 are defined by

$$\omega_o = |4AB^{(2)}|^{\frac{1}{2}}\mu, \quad = -B^{(1)}/\omega_o, \; k = \min(|\alpha|, |\alpha^{-1}|), \qquad (2)$$
$$a_i = A^{(i)}/A, \; b_i = B^{(i)}/B^{(2)}, \; c_1 = -\tfrac{1}{2}(a_1 + b_3).$$

Here the superscript (i) indicates differentiation of order i with respect to the momentum y. These coefficients are evaluated at $y = y'$; y' being a constant dependent upon the initial conditions.

The first-order libration solution is

$$x = \sin^{-1}(k\,\mathrm{sn}u), \quad y = y' + b_1(\mathrm{cn}u - 1) \qquad (3)$$

$$\text{with} \quad u = w + (b_1 b_3/3k)\sin^{-1}(k\,\mathrm{sn}w), \qquad (4)$$

$$\text{and} \quad w = \omega_o t(1 + b_1 c_1)\,\mathrm{sgn}\alpha. \qquad (5)$$

The modulus of the Jacobi elliptic functions is $k = |\alpha| < 1$, and t is the symbol for time.

The first-order circulation solution is

$$x = \sin^{-1}(\mathrm{sn}u) = \mathrm{am}u, \; y = y' + b_1(\mathrm{dn}u - 1), \qquad (6)$$

$$\text{with} \quad u = w + b_1[b_3\mathrm{am}^*w/3 + (c_o\phi_2/k'^2)(Z(w) - k^2\,\mathrm{sn}w\,\mathrm{cn}w/\mathrm{dn}w)] \qquad (7)$$

$$\text{and} \quad w = \alpha\omega_o t[1 + b_1(c_1 + b_3\pi/6K + c_o\phi_2 E/k'^2 K)].$$

In (6) - (8), am* is the periodic part of the amplitude function am,

Z is the Jacobi zeta function and K and E are the complete elliptic integrals of the first and second kind respectively. The complementary modulus k' is defined by $k'^2 \doteq 1 - k^2 = 1 - \alpha^{-2}$, and the constant c_o is evaluated from

$$c_o = \frac{1}{2}a_1 + \frac{1}{6}b_3. \tag{9}$$

The function ϕ_2 is the second-order part of the regularising function ϕ and is obtained, as shown in Garfinkel and Williams (1974), from

$$\phi_2 = \exp[-(k/k')^6]. \tag{10}$$

The role of the regularising function has been fully described in Paper II.

(ii) JUPP'S SOLUTION

The first-order libration solution may readily be extracted from Equations (2), (3) and (53)-(55) of Paper III'. In order to compare the two solutions more easily, certain new notations will henceforth be adopted. First, x and y will be interchanged, so that $A_o^{(1)}$ now indicates the value of the ith derivative of A evaluated at y_o, where y_o is the simple zero of $B^{(1)} = 0$. In place of the definitions (3) they are now, for $i = 0, 1 \ldots,$

$$\tilde{a}_i = i!A_i = A_o^{(i)}/A_o, \quad \tilde{b}_{i+2} = (i+2)!B_{i+2} = B_o^{(i+2)}/B_o^{(2)}. \tag{11}$$

Thus, the tilde indicates evaluation at $y=y_o$, which leads to the following relations between the coefficients in the two theories:

$$\tilde{a}_i = [a_i]_{y'\to y_o}, \quad \tilde{b}_i = [b_i]_{y'\to y_o}, \quad \tilde{c}_i = [c_i]_{y'\to y_o} \tag{12}$$

The first-order libration solution now becomes

$$x = \sin^{-1}(k\,\mathrm{sn}w) + \frac{1}{3}\mu p_o \tilde{b}_3 k\,\mathrm{cn}w \sin^{-1}(k\,\mathrm{sn}w), \tag{13}$$

$$y = y_o + \mu p_o k\,\mathrm{cn}w - \frac{1}{6}\mu^2 p_o^2 k[k(\tilde{b}_3\mathrm{cn}^2 w + \tilde{a}_1\mathrm{sn}^2 w) + 2\tilde{b}_3\mathrm{sn}w\,\mathrm{dn}w \sin^{-1}(k\,\mathrm{sn}w)], \tag{14}$$

$$w = w_{oo} - \tilde{\omega}_o t. \tag{15}$$

Here, as before, $p_o^2 = 4A_o/B_o^{(2)}$ and so, in view of (2),

$$\mu p_o B_o^{(2)} = \tilde{\omega}_o. \tag{16}$$

The constant modulus k of the elliptic functions and the constant w_{oo} are to be determined from the initial conditions.

The first-order circulation solution is given in its corrected form in Equations (9)-(11) of the first author's 1974 paper. There the modulus of the elliptic functions is 1/k. Writing, instead, $\kappa = 1/k$ and $\kappa'^2 = 1-\kappa^2$, the solution, in terms of the revised notation, is

$$x=\pm \mathrm{am}w + \frac{\mu p_o}{\kappa}\left[\frac{\tilde{b}}{3}\,\mathrm{dn}w\,\mathrm{am}^*w + \frac{\pi}{2K\kappa'^2}\tilde{C}_o(\mathrm{dn}w Z(w) - \kappa^2\,\mathrm{sn}w\,\mathrm{cn}w)\right], \tag{17}$$

$$y=y_o \pm \frac{\mu p_o}{\kappa}\mathrm{dn}w - \mu^2 p_o^2\left[\frac{\tilde{b}_3}{6}\left(\frac{1}{\kappa^2} + 2\mathrm{sn}w\,\mathrm{cn}w\,\mathrm{am}^*w\right) + \tilde{c}_3\mathrm{sn}^2 w + \right.$$
$$\left. - \frac{\pi}{2K\kappa^2\kappa'^2}\tilde{C}_o\{\mathrm{sn}w\,\mathrm{cn}w\, Z(w) - (\frac{\kappa'^2}{\kappa^2} + \mathrm{sn}^2 w)\mathrm{dn}w\}\right], \tag{18}$$

$$w=w_{oo} - \frac{\tilde{\omega}_o t}{\kappa}\left[1 \pm \frac{\mu p_o \pi}{2K\kappa}\left(\frac{\tilde{b}_3}{3} - \tilde{C}_o\frac{E}{K\kappa^2\kappa'^2}\right)\right]. \tag{19}$$

The double sign is necessary to take account of both regions of circulation in the (x,y) phase plane. The coefficients $\tilde{C}_o$ and $\tilde{c}_3$ are defined by

$$\tilde{C}_o = (\tilde{b}_3 - 3\beta_1\kappa^2)/6 = (\tilde{b}_3 + 3\tilde{c}_3\kappa^2)/6, \tag{20}$$

$$\tilde{c}_3 = \tilde{a}_1/2 - \tilde{b}_3/6. \tag{21}$$

It is readily apparent that the solutions of Garfinkel and Jupp have a certain similarity; yet they possess distinctly different properties. In Garfinkel's solution the small parameter of the asymptotic representations is b_1, which is a direct measure of the closeness of the resonance to exact commensurability; it is a function of y' which, in turn, depends upon the initial conditions. In circulation, as $k \to 0$, the appropriate contribution from Φ_2 takes over when b_1 is no longer small enough to ensure the asymptotic nature of the solution. The small parameter in Jupp's asymptotic solutions is μ, an absolute constant of the given problem. Since the original system (1) is one of second order, any general solution should possess two arbitrary constants of integration. In Garfinkel's theory one is y' and the other, x'(0), is chosen to be zero. The two arbitrary constants in Jupp's theory are, on the other hand, k(or κ) and w_{oo}. The determination of these arbitrary constants will be discussed in a later publication. The demarcation between libration and circulation corresponds to $|\alpha| = k = 1$ in Garfinkel's theory. In view of the definition of Φ_2 it is clear that $\Phi_2/k'^2 \to 0$ as $k \to 1$. Consequently, both the libration and the circulation solution lead to the solution for the separatrix;

$$x = \sin^{-1}(\tanh u),\quad y = y' + b_1(\mathrm{sech}\,u - 1),$$
$$u = w + (b_1 b_3/3)\,\sin^{-1}(\tanh u), \tag{22}$$
$$w = \omega_o t(1 + b_1 c_1)\,\mathrm{sgn}\alpha.$$

On the contrary, whilst Jupp's first-order libration solution is nonsingular at $k=1$, the circulation solution is singular, since $K\kappa' \to 0$ as $\kappa \to 1$. In fact, it can readily be shown that only the expression for w, as given by (19), is singular; the formulae for x and y are not. Further, as the classical limit is approached $k \to 0$ (Garfinkel) and $\kappa \to 0$ (Jupp); in the first case the circulation solution is non-singular, while in the second it is not.

In practical terms, formula (10) for Φ_2 is required only in the range $0.4 \leqslant k \leqslant 0.8$. For, as can be seen from Fig.1, Φ_2 is very close

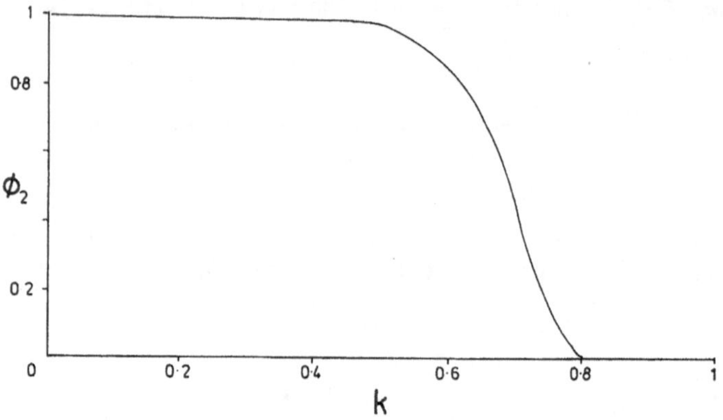

Fig. 1: The behaviour of ϕ_2 (Garfinkel's solution)

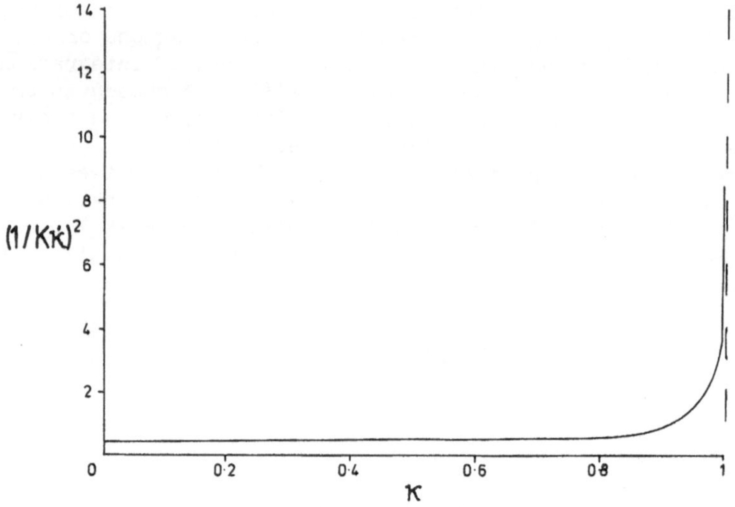

Fig. 2: The behaviour of $(K\kappa')^{-2}$ (Jupp's solution)

to unity for $k < 0.4$ and very close to zero for $k > 0.8$. In other words, the role of ϕ_2 is not practically significant in deep resonance.

Fig.2 illustrates the behaviour of $(K\kappa')^{-2}$, for the range $0 \leqslant \kappa \leqslant 1$. It is clear that although singular at $\kappa = 1$ (on the separatrix), the value is as small as 12.5 for $\kappa = 0.995$. In consequence, it would seem that the circulation solution described by Equations (17)-(19) may be used in practice throughout the deep resonance circulation regions except for the range $0.995 < \kappa < 1$. A very similar situation prevails at second order, where the libration solution admits a singularity of the same type. Thus, it is just a very small region about the separatrix which must be excluded from the theory.

3. PRINCIPAL RESULTS OF THE COMPARISON

The two analytical solutions given in §2 were compared with numerical solutions for the Hamiltonians

$$\begin{aligned} -F_1 &= y + 1/2y^2 + 2\mu^2(y+1)\sin^2 x, \\ -F_2 &= y^4/4 - y^3/3 + y^2/2 - y + 2 + 2\mu^2(y^2 + 3y - 2)\sin^2 x. \end{aligned} \tag{23}$$

In both cases $B'(1) = 0$, and so for Jupp's theory, $y_0 = 1$. The small parameter μ^2 was given the values 10^{-2}, 10^{-3} and 10^{-4}.

The elliptic functions and integrals sn, cn, dn, am, E, K and Z appearing in the formal solutions were calculated from algorithms given by Bulirsch (1965), based on the standard Landen and Gauss transformations. The computed values were checked with those given by Milne-Thomson (1970), and in every case very good agreement was achieved.

The numerical solutions were obtained using, first, a standard 4th-order Runge-Kutta method and, second, sub-routine Do2 AHF of the Computer Laboratory at Liverpool University. The two solutions were in excellent agreement. In addition, as a further check, the value of the Hamiltonian was computed for $x(t)$ and $y(t)$ from (23) and compared with the initial value.

In view of the difficulty referred to in §1 concerning the initial-value problem, in every case, the starting value for x was chosen as zero; i.e. $x(0) = 0$. (This choice leads to the relation $x'(0) = 0$; which is an implicit feature of Garfinkel's solution.)

Figs. 3-10 illustrate the departures of the formal solutions from the numerical ones for a variety of initial conditions (in y). The notation XG1 refers to the value of x computed from Garfinkel's first-order theory. Similarly YJ1 is the value of y calculated from Jupp's first-order solution. The symbol N denotes the value obtained from direct numerical integration of the equations of motion. It is worth noting that, once the various programmes have been set up, the computer time required to calculate x and y from the formulae given in §2 is a small fraction of that needed for the numerical integration.

Taking Garfinkel's solution first, one would expect XN − XG1 and YN − YG1 to be $O(b_1^2)$. With the choice $x(0) = 0$, it is readily seen from (3)-(8) that $y' = y(0) + O(b_1^2)$, and so, accurate to first order,

$$b_1 = B^{(1)}(y(0))/B^{(2)}(y(0)) \tag{24}$$

The value of b_1, computed from this formula, is given for each of the cases illustrated. On the other hand, in view of (13)-(19) one would expect XN - XJI = $O(\mu^2)$ and YN - YJI = $O(\mu^3)$; the gain of an order of magnitude being taken into consideration.

One feature of the figures requires immediate comment; namely that the plotted differences increase with time t. This is not unexpected as the formulae for w, which are linear in $\omega_0 t$(or $\tilde{\omega}_0 t$), are calculated only up to terms in b_1 (or μ). Consequently, for large t, the neglected terms can make a significant contribution to w, and thence to x and y. In view of the periodic nature of the systems this is of no practical significance except possibly for motions very close to the separatrix where the motion of the 'particle' along its trajectory in the phase plane is, on average, very slow. In general, the solution over a single period suffices to determine the solution for any t. In this regard the first-order formula for the period (for libration)

$$T = 4K/\omega_0, \tag{25}$$

evaluated at $y' = y(0)$, must be used. Clearly, as the separatrix is approached $k \to 1$ and K and therefore T tend to infinity. The libration periods are given for each case depicted. For circulation, of course, x comprises the sum of a secular part and a periodic part.

Figs. 3-10 represent just a very small selection from a whole range of numerical comparisons, yet they are quite representative of the general findings. Accordingly, the following general observations may be made about Garfinkel's (G) and Jupp's (J) solutions. In the light of what has just been written the comments refer to only the first few periodic fluctuations, not necessarily the whole range of t.

(1) The over-all behaviour is the same for F_1 as for F_2, and for each value of μ.
(2) Close to the centre of libration G and J are in very good agreement with the numerical solution (N) and the errors are well within the estimates for J and close to them for G. [Fig.5].
(3) Half-way between the centre and the separatrix J remains very close to N, the difference keeping well inside the error estimates. On the other hand N - G is greater than $O(b_1^2)$, and considerably so for x. [Figs. 3 and 6].
(4) Very close to the separatrix, but within libration neither J nor G is satisfactory, with G worse than J. [Fig.7].
(5) In circulation, but very close to the separatrix the situation is the same as (4).
(6) Within deep resonant circulation, but away from the separatrix J is very good but G very poor. [Fig.8].
(7) In shallow resonance J and G are unsatisfactory, especially for x. [Figs.4 and 9].
(8) At the classical limit, far from the resonance, G provides excellent agreement with N while J is very poor indeed. [Fig.10].

All the observations, but one, concerning Jupp's solution fit the theoretical predictions; that is, the solution is good for deep reson-

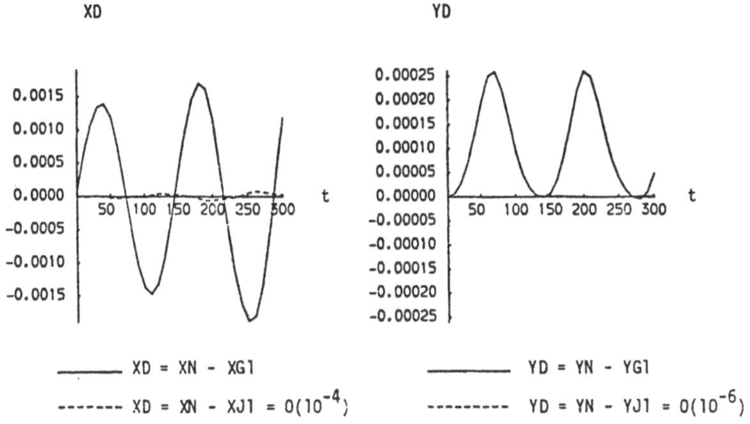

Fig. 3: Hamiltonian F_1, $\mu^2 = 10^{-4}$, y(0) = 1.007(L),

$b_1^2 = 5 \times 10^{-5}$, k = 0.43, T = 162

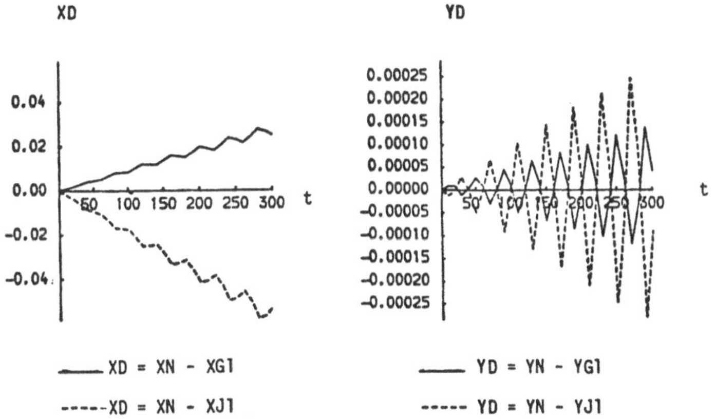

Fig. 4: Hamiltonian F_1, $\mu^2 = 10^{-4}$, y(0) = 1.030(C),

$b_1^2 = 8 \times 10^{-4}$, k = 0.55

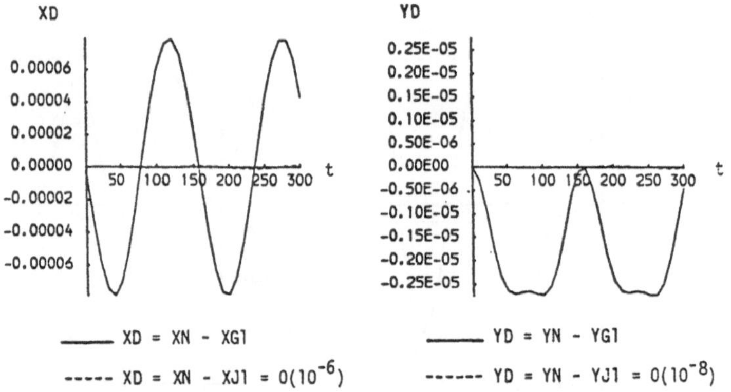

Fig. 5: Hamiltonian F_2, $\mu^2 = 10^{-4}$, y(0) = 1.001(L), $b_1^2 = 10^{-6}$, k = 0.05, T = 156

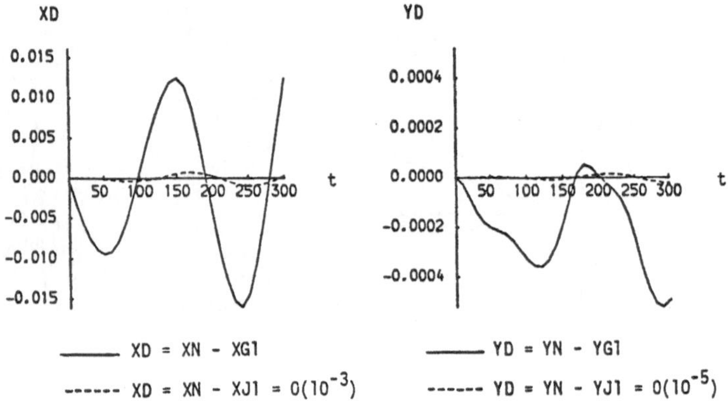

Fig. 6: Hamiltonian F_2, $\mu^2 = 10^{-4}$, y(0) = 1.010(L), $b_1^2 = 10^{-4}$, k = 0.50, T = 165

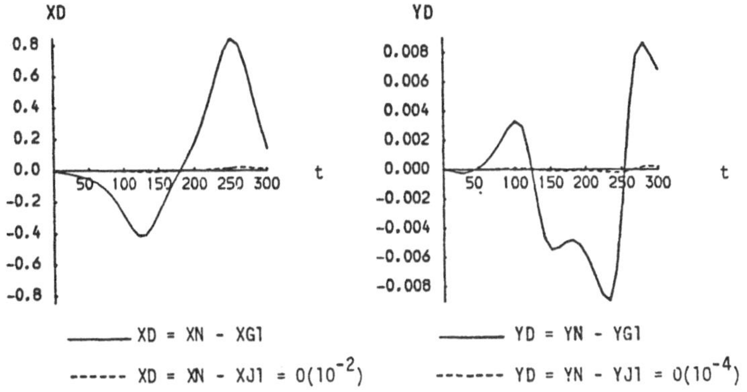

Fig. 7: Hamiltonian F_2, $\mu^2 = 10^{-4}$, $y(0) = 1.019(L)$,

$b_1^2 = 3.5 \times 10^{-4}$, $k = 0.95$, $T = 247$

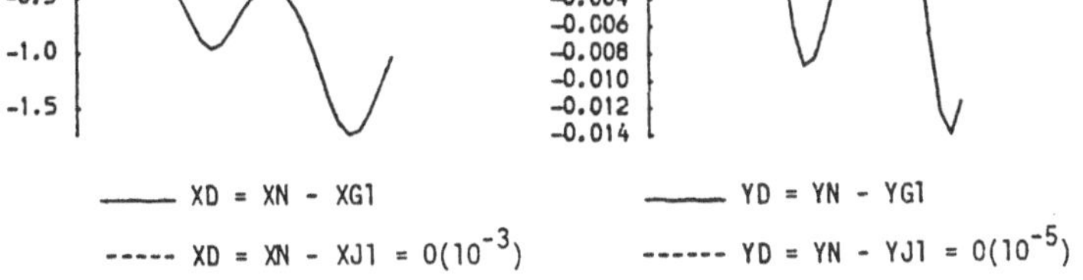

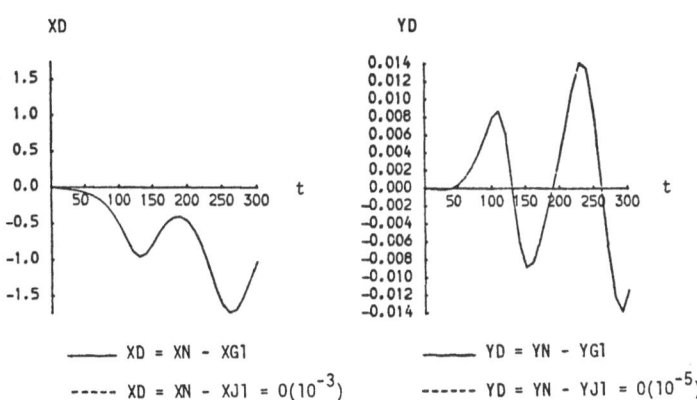

Fig. 8: Hamiltonian F_2, $\mu^2 = 10^{-4}$, $y(0) = 1.021(C)$,

$b_1^2 = 4.2 \times 10^{-4}$, $k = 0.95$

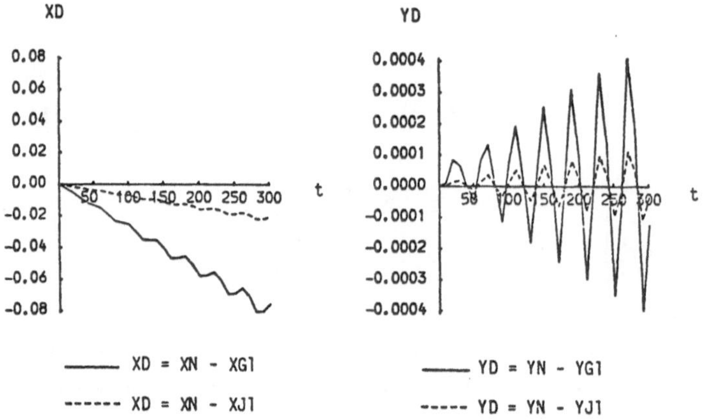

Fig. 9: Hamiltonian F_2, $\mu^2 = 10^{-4}$, y(0) = 1.040(C),

$b_1^2 = 1.5 \times 10^{-3}$, k = 0.52

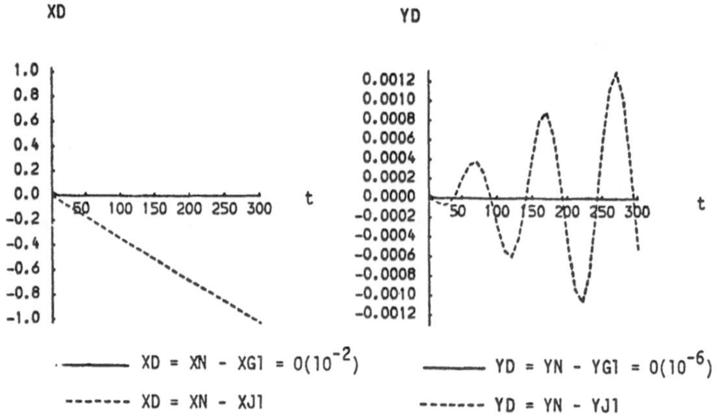

Fig. 10: Hamiltonian F_2, $\mu^2 = 10^{-4}$, y(0) = 1.150(C)

$b_1^2 = 1.7 \times 10^{-2}$, k = 0.16

ance (libration and circulation), but invalid for the vicinity of the separatrix and away from deep resonance. The one failure of the theoretical predictions is the observation that the solution appears to break down on approaching the separatrix rather sooner than expected.

On the contrary, some of the results arising from Garfinkel's solution are not readily explained. Very close to the libration centre ($k \approx 0$) and at the classical limit ($k \approx 0$) the solution is excellent, but elsewhere it is not at all satisfactory. During the current investigation a number of minor errors, mostly typographical, were traced, before which the results were even worse. It is possible that there are other errors still to be traced which would provide a remedy. Alternatively, it may be that the explanation is more profound; the loss of accuracy may be an intrinsic feature of the global solution. Perhaps the price to be paid for the elegance of the global solution is a correspondingly less accurate solution locally. Of course, the accuracy of the global solution may be improved by using a higher-order theory, the algorithm of which is described in II.

In summary, it would seem that a given accuracy can be achieved either by combining local solutions (i.e. J for libration and deep circulation) of a particular order or by using a higher-order global solution (G). While the latter technique is more demanding algebraically than the former it is arguably aesthetically more attractive.

REFERENCES

Bulirsch, R.: 1965, Numerische Mathematik 7, 78.

Garfinkel, B.: 1966, Astron. J. 71, 657. (I).

Garfinkel, B.: 1982, Celest. Mech. 28, 275.

Garfinkel, B.: 1983, Celest. Mech. 30, 373.

Garfinkel, B., Jupp, A.H. and Williams, C.: 1971, Astron. J. 76, 157.(II).

Garfinkel, B. and Williams, C.: 1974, Celest. Mech. 9, 105. (III).

Jupp, A.H.: 1969, Astron. J. 74, 35. (I').

Jupp, A.H.: 1979, Monthly Notices Roy. Astron. Soc. 148, 197 (II').

Jupp, A.H.: 1972, Celest. Mech. 5, 8. (III').

Jupp, A.H.: 1974, Celest. Mech. 8, 523.

Milne-Thomson, L.M.: 1970, Jacobean Elliptic Function Tables, Dover.

Moore, P.: 1983, Celest. Mech. 30, 31.

Poincaré, H.: Les méthodes Nouvelles de la Méchanique Céleste, 2, 206, Gauthier-Villars, Paris.

TROJAN ORBITS IN SECULAR RESONANCES

R. Bien and J. Schubart
Astronomisches Rechen-Institut
Mönchhofstrasse 12-14
D-6900 Heidelberg 1
Federal Republic of Germany

ABSTRACT. A near equality between the nodal rates of suitably defined Trojan orbits and Jupiter represents an important type of a secular resonance. This case is realized by the model Sun-Jupiter-Saturn-Trojan, referred to the invariable plane. A second theoretical example is based on the elliptic three-body problem Sun-Jupiter-Trojan, where the vanishing nodal rate of a special Trojan orbit and the vanishing rate of Jupiter's longitude of perihelion define a secular resonance.

We investigate the perturbations in the asteroidal inclinations and the nodes and consider the possibility of a libration.

1. INTRODUCTION

A secular resonance can occur, when a basic period that appears in the motion of a mass-less body, is nearly equal to a "secular" period in the motion of the major planets. For instance, a typical secular period is given by the revolution of Jupiter's and Saturn's node along the invariable plane. This period is about 50 000 yr (see Cohen et al.,1973). Therefore, a minor body having almost the same nodal period would be close to a secular resonance. An interesting application of this case is the orbit of the Lost City meteorite, discussed by Williams (1975). The secular resonance causes an unusually large perturbation of the orbital inclination of the meteorite with a period of 10^6 yr.

In the present paper we investigate some effects of secular resonances on Trojan orbits. Besides the possibility of a secular resonance, every Trojan orbit is in a 1/1 resonance with Jupiter, i.e. Jupiter and the Trojans have approximately the same mean motion around the Sun.

We study two types of secular resonance. The first one is suggested by the case of the Lost City meteorite: A close equality of the nodal rates of Jupiter, Saturn and the Trojan. A few years ago, Yoder (1979) has pointed to the possible importance of this type for the long-term evolution of Trojan orbits. The second type of secular resonance is suggested by a special example given by Bien and Schubart (1983): The nodal rate of a Trojan orbit equals nearly the mean value of the rate of

Celestial Mechanics **34** (1984) 425–434. 0008–8714/84.15

Jupiter's longitude of perihelion.

The aim of our investigation is to find an indication of the long-period effects that are caused by a secular resonance and a preliminary conclusion concerning the orbital stability.

2. DEFINITIONS

We introduce the following heliocentric orbital elements:

a = semi-major axis	
e = eccentricity	
ℓ = mean longitude	referred to a
$\tilde{\omega}$ = longitude of perihelion	reference plane
Ω = longitude of the node	and a zero-point
ω = argument of perihelion	specified below
i = inclination	

The subscripts 0, J and S refer to starting values, Jupiter and Saturn, respectively. Furthermore, we define $\mu = \ell - \ell_J$. Throughout the present paper we consider only preceding Trojans. a is given in units of $(a_J)_0$; the masses correspond to the IAU (1976) System of Constants. For the integrations we used the n-body program by Schubart and Stumpff (1966); the integrations start at a moment comparatively close to the present.

3. RESONANCE OF THE NODAL RATES

We consider the four-body model Sun-Jupiter-Saturn-Trojan. The reference plane is the invariable plane defined by the major bodies. The longitudes are reckoned from the ascending node of the invariable plane on Jupiter's initial orbital plane.

In Fig.1 the relationship between $\dot{\Omega}$ and i_0 for a set of orbits with $a_0 = 1$, $e_0 = 0.11$, $\tilde{\omega}_0 - (\tilde{\omega}_J)_0 \approx 60°$, and $\mu_0 \approx 35°, 40°, 45°$ is given. Note that $\mu_0 \approx$ minimum of μ. $\dot{\Omega}$ results from an integration over 1 500 yr and therefore gives only a first approximation to a mean value. For $\mu_0 \approx 35°$, the corresponding $\dot{\Omega}$, i_0 curve crosses the level $\dot{\Omega} = \dot{\Omega}_J$ near $i_0 = 18°$. In other words, for $\mu_0 \approx 35°, i_0 \approx 18°$ we are near a resonance of the nodal rates. In addition, one notices that for $\mu_0 \approx 40°, i_0 \approx 24°$ and $\mu_0 \approx 45°, i_0 \approx 13°$ a 1/2 ratio of the rates can occur.

The case $\mu_0 \approx 35°, i_0 = 18°$ has been integrated for the values $\Omega_0 \approx 0°$, $+45°$, $+90°$, $-45°$, $-90°$, $-135°$, $\pm 180°$ over some 10^4 yr. In Fig.2 a rough sketch demonstrates the trend of the evolution of i for these seven examples. One finds a zero-effect in the slope of i for $\Omega_0 \approx 0°, \pm 180°$, and maximum slopes of opposite sign for $\Omega_0 \approx +90°, -90°$. For comparison, in Fig.3 we show the real variations of i over 60 000 yr for the case $\Omega_0 \approx -90°$.

We selected three cases of special interest, namely $\Omega_0 \approx 0°, +90°, -90°$, and extended the integrations up to $\pm 283\,000$ yr. The result of i versus time appears in Fig.4a-c. This time span of more than 0.5×10^6 yr is sufficiently large to show changes in the trend of the slopes. The

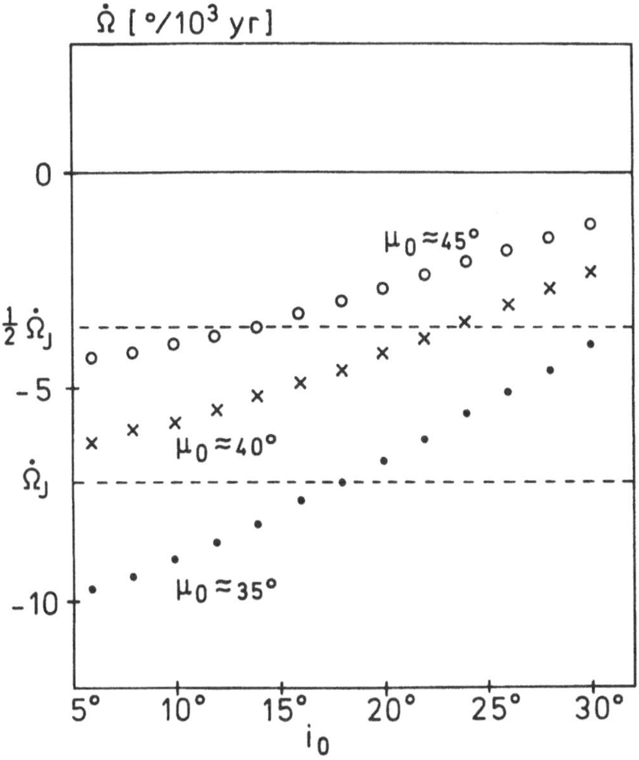

Figure 1. $\dot{\Omega}$ depending on i_0 for three different values of $\mu_0 \approx$ minimum of μ. The two dotted horizontal lines represent $\dot{\Omega}_J/2$ and $\dot{\Omega}_J$.

variation of i is several degrees. However, a better insight into the evolution of these orbits is given in Fig.5. Here we have used a polar diagram, with i as the polar radius and the slowly varying argument $\Omega - \Omega_J$ as the polar angle. In two of the examples ($\Omega_0 \approx \pm 90°$) it appears that $\Omega - \Omega_J$ does not evolve monotonously, but a libration of $\Omega - \Omega_J$ around 180° is indicated, in analogy to simple cases of resonances between orbital periods (cf. Schubart, 1966). We cannot exclude the possibility that these librations are temporary features of the evolution. However, if these features are permanent librations we estimate their period to be in the order of 10^6 yr. The interval covered by the integration does not allow to draw conclusions about a libration in case of the third example ($\Omega_0 \approx 0°$). At the moment, it seems to us that the other orbital elements are much less affected by the secular resonance.

Finally, we mention that we have no indication of an instability of the orbits, since a libration of $\Omega - \Omega_J$ would represent a quasi-periodic behaviour. Even a change between libration and non-libration does not imply very strong changes in the mean values of characteristic

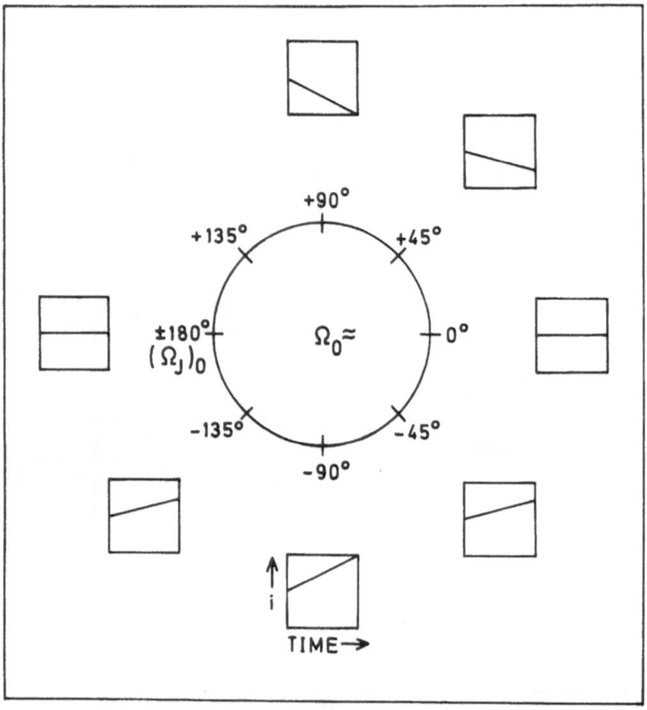

Figure 2. Trend in the evolution of i for various values of Ω_0 in case of resonant nodal rates, $\mu_0 \approx 35°$.

orbital elements.

4. NODAL RATE OF TROJAN AND MEAN RATE OF JUPITER'S LONGITUDE OF PERIHELION

In a former paper (Bien and Schubart, 1983) we have mentioned an example with $i_0 = 40°$ and with a slow mean variation of $\Omega - \tilde{\omega}_J$, i.e. a case where $\dot{\Omega} \approx$ mean of $\dot{\tilde{\omega}}_J$. In Fig.6 we demonstrate our former result that the mean of i with respect to the visible periodicities decreases monotonously in this case. The calculations are based on the model Sun-Jupiter-Saturn-Trojan, where both Jupiter and Saturn move in the reference plane.

This example suggests to investigate the possibility of librations of the mean of $\Omega - \tilde{\omega}_J$. Since this argument is important in the elliptic restricted three-body problem as well ($\tilde{\omega}_J = \text{constant}$), we study this possibility in the model Sun-Jupiter-Trojan. The reference plane is Jupiter's orbital plane, the zero for reckoning the longitudes is given by putting $\tilde{\omega}_J = 0°$. Since now $\dot{\tilde{\omega}}_J = 0$, we have to select starting values with $\dot{\Omega} \approx 0$. For $\mu_0 = 47°$ and several choices of Ω_0 the level $\dot{\Omega} = 0$ is reached at inclinations near 27° or 28°, if $a_0 = 1$, $e_0 = 0.11$, $\tilde{\omega}_0 = 60°$.

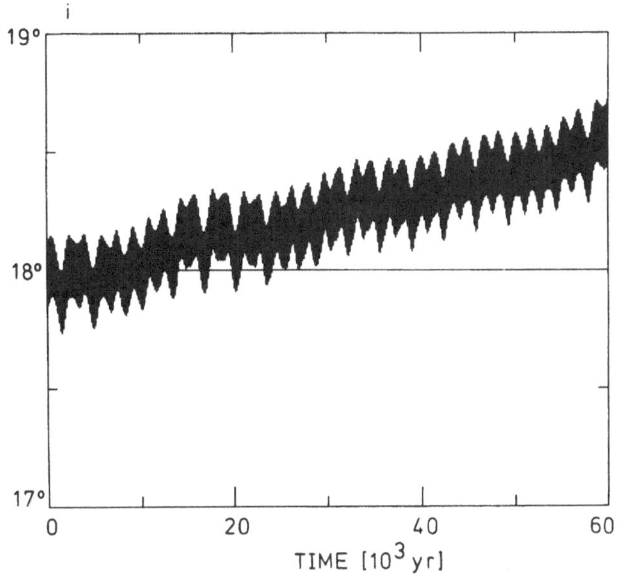

Figure 3. The evolution of i over 60 000 yr for $\mu_0 \approx 35°$, $\Omega_0 \approx -90°$.

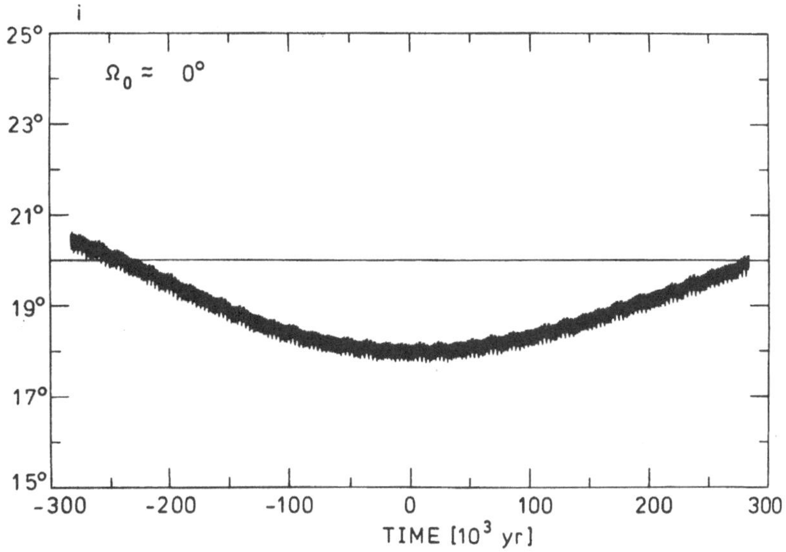

Figure 4a. $\Omega_0 \approx 0°$: The evolution of i over the whole integration interval (±283 000 yr).

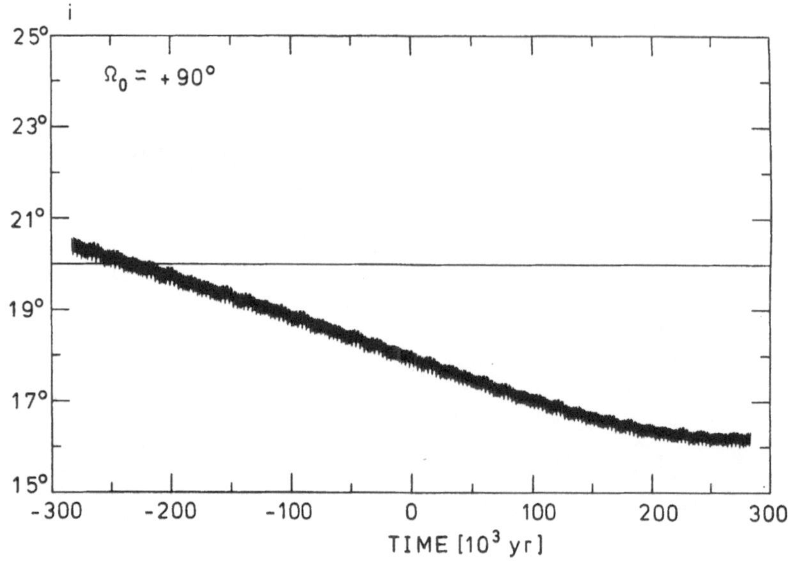

Figure 4b. Ω_0 ≈ +90° : The evolution of i over the whole integration inverval (±283 000 yr).

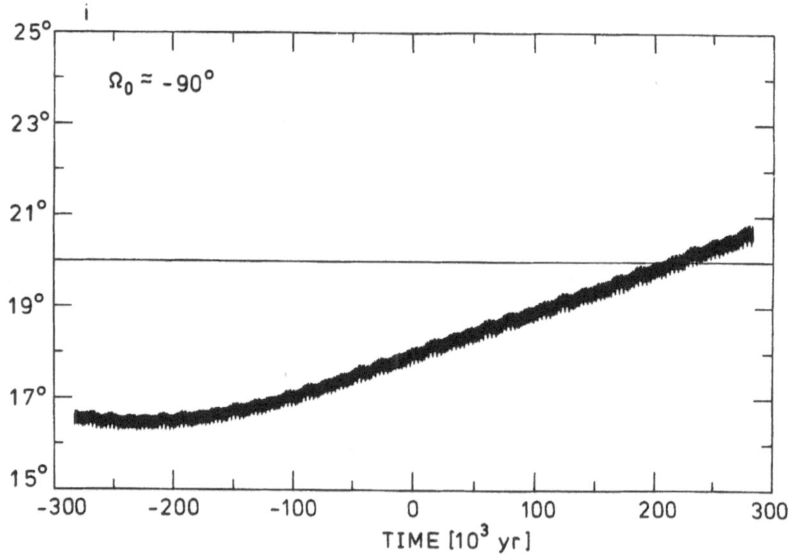

Figure 4c. Ω_0 ≈ -90° : The evolution of i over the whole integration interval (±283 000 yr).

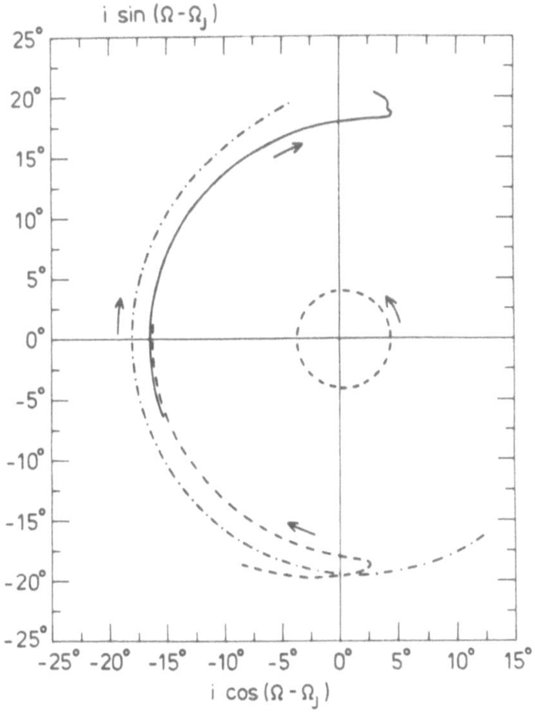

Figure 5. $i \sin(\Omega - \Omega_J)$ versus $i \cos(\Omega - \Omega_J)$ for $\Omega_0 \approx 0°$ (-·-·-·-), $\Omega_0 \approx +90°$ (- - - - -), and $\Omega_0 \approx -90°$ (———) over ±283 000 yr. The dotted circle around the origin represents a non-resonant case with $\mu_0 \approx 53°$, $i_0 \approx 4°$. Small arrows indicate the direction of motion.

Jupiter starts at $\ell_J = 0°$ with a constant $e_J = 0.048$.

For the choices $\Omega_0 = 25°$, 70°, 115°, 160° with appropriate values of i_0 we have integrated the orbits at least over 75 500 yr. Short additional integrations indicate that the starting values $\Omega_0 + 180°$, $\omega_0 + 180°$ (instead of Ω_0, ω_0) lead to identical results in the other orbital elements according to a symmetry of the problem. A rough sketch of the trend of the evolution of i is given in Fig.7, using the property of symmetry. In other words, we can restrict our discussion to the range $\Omega_0 = 0°$ to 180°. We note a zero-effect for $\Omega_0 = 25°$, 115°, and a positive and negative slope for $\Omega_0 = 70°$ and 160°, respectively.

We extended the integrations of the cases $\Omega_0 = 70°$, 160° (where $i_0 = 27{:}^{\circ}27$ and $27{:}^{\circ}95$, respectively) over a longer time interval. As a result, the mean of $\Omega - \tilde{\omega}_J$ indicates a libration: The argument reaches a minimum in case of $\Omega_0 = 70°$, and a maximum in case of $\Omega_0 = 160°$. In Fig.8 the mean variations of i and $\Omega - \tilde{\omega}_J$ are plotted against each other for $\Omega_0 = 70°$ using Labrouste's method (see the paper by Schubart and Bien

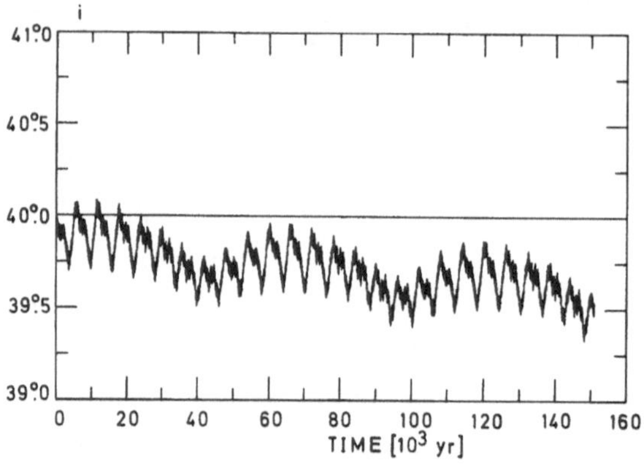

Figure 6. The evolution of i over 151 000 yr for $i_0 = 40°$, $\mu_0 = 53°$. $\Omega - \tilde{\omega}_J$ has a slow mean variation. Jupiter and Saturn move in the reference plane.

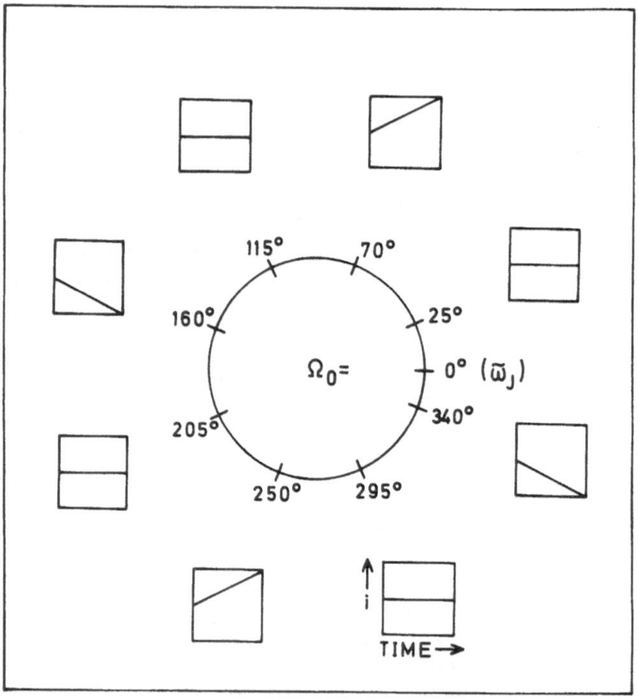

Figure 7. Resonance of $\dot{\Omega}$ and $\dot{\tilde{\omega}}_J$ in the elliptic three-body problem: Trend in the evolution of i for various values of Ω_0, if $\Omega - \tilde{\omega}_J = \Omega$ has a slow mean motion; $\mu_0 = 47°$.

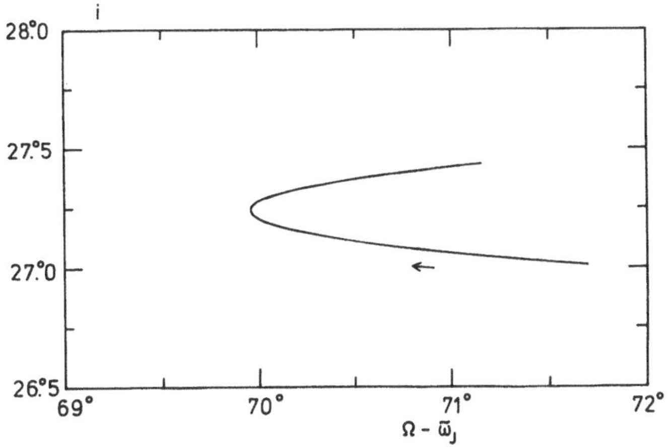

Figure 8. i versus $\Omega - \tilde{\omega}_J$ over ±140 000 yr. Periods smaller than 6 000 yr are eliminated (see text). $\Omega_0 = 70°$, $\mu_0 = 47°$; Jupiter is the only perturbing planet and moves in the reference plane.

given at this conference). Tentatively, we assume that the argument $\Omega - \tilde{\omega}_J$ can librate around values in the vicinity of 115°. This behaviour may depend on the starting values, in particular on μ_0. The periods of libration are expected to be one or several 10^6 yr. The variation of i appears to be smaller than in the case of resonant nodal rates. This may be due to the fact that μ_0 is different. Again it appears that the other orbital elements are less affected by the secular resonance.

5. CONCLUSION

We have investigated two types of secular resonance in Trojan orbits, i.e. secular resonances in a resonance. In analogy to known cases of orbit - orbit resonance we conclude that the arguments $\Omega - \Omega_J$ and $\Omega - \tilde{\omega}_J$ can librate with very large periods (some 10^6 yr), although our computations cover only parts of the supposed periods. According to this analogy, the orbital evolution of our examples of secular resonances can be quasi-periodic. At least, we did not find indications that the type of motion is completely changed. In particular, one may assume that the example with $i_0 = 40°$ mentioned above (see Fig.6) is also under the influence of a secular resonance and that the mean variations of i follow a very long period that causes either a revolution or a libration of the mean of $\Omega - \tilde{\omega}_J$.

It will be interesting to derive numerically mean values of $\dot{\Omega}$ of the real Trojan asteroids and to study their distribution with respect to the secular resonances.

REFERENCES

Bien,R. and Schubart,J.: 1983, *Dynamical Trapping and Evolution in the Solar System,* Ed.: V.V. Markellos and Y. Kozai, IAU Coll. No.74, pp.153-161.
Cohen,C.J., Hubbard,E.C., and Oesterwinter,C.: 1973, *Astron. Papers American Ephemeris Washington,* Vol.22, part 1.
Schubart,J.: 1966, *The Theory of Orbits in the Solar System and in Stellar Systems,* Ed.: G. Contopoulos, IAU Symp. No.25, pp.187-193.
Schubart,J. and Stumpff,P.: 1966, *Veröffentl.Astron. Rechen-Inst. Heidelberg* No.18.
Williams,J.G.: 1975, *J. Geophys.Res.* **80**, pp.2914-2916.
Yoder,C.F.: 1979, *Icarus* **40**, pp.341-344.

CRITICAL INCLINATION OF TROJAN ASTEROIDS

Bálint Érdi
Department of Astronomy
Eötvös University
Budapest, Hungary

ABSTRACT

The author's earlier solution for Trojan asteroids is developed further. It is shown that depending on the amplitude of libration around the Lagrangian point L_4, there is a critical inclination which determines the sign of the variation of the ascending node. If the orbital inclination of a Trojan is smaller than the critical one, then the ascending node decreases and otherwise it increases. The variation of the eccentricity and of the longitude of the perihelion has also a dependence on the critical inclination.

1. INTRODUCTION

An interesting result has been found recently by Bien and Schubart (1983) on Trojan asteroids. They have studied Jupiter's long-periodic effects in the motion of Trojans by numerical integration and have obtained that the mean value of $\Omega^{\cdot}$ (the rotational rate of the ascending node) is either negative or positive depending on their orbital inclination i. Actually, $\Omega^{\cdot}$ increases with i and there is a critical value of i at which the sign of $\Omega^{\cdot}$ changes from negative to positive. $\Omega^{\cdot}$ also depends on the librational amplitude of the asteroids around the Lagrangian points L_4 or L_5: when it increases $\Omega^{\cdot}$ decreases and the decrease of $\Omega^{\cdot}$ is smaller for large i.

This work has been motivated by the above results. The purpose of this paper is to study the critical inclination of Trojan asteroids analytically, using the author's earlier solution for the Trojan problem (Érdi, 1978, 1981). First it is suitable to give a short account of that solution.

2. A THEORY OF TROJANS

I have studied the motion of Trojan asteroids under the assumptions that a Trojan is perturbed only by Jupiter, and Jupiter's orbit

Celestial Mechanics 34 (1984) 435–441. 0008–8714/84.15

around the Sun is an ellipse. Then the equations of motion are

$$\frac{d^2r}{dv^2} - r\left(\frac{d\alpha}{dv}\right)^2 - 2r\frac{d\alpha}{dv} = \frac{1}{1+e_J\cos v}\left[r - \frac{1-\mu}{R_1^3}r + \mu\left(\frac{\cos\alpha - r}{R_2^3} - \cos\alpha\right)\right],$$

$$\frac{d}{dv}\left(r^2\frac{d\alpha}{dv} + r^2\right) = \frac{\mu r\sin\alpha}{1+e_J\cos v}\left(1 - \frac{1}{R_2^3}\right), \qquad (1)$$

$$\frac{d^2z}{dv^2} + z = \frac{z}{1+e_J\cos v}\left(1 - \frac{1-\mu}{R_1^3} - \frac{\mu}{R_2^3}\right),$$

where r, α, z are the cylindrical coordinates of the asteroid (r and α are polar coordinates in the orbital plane of Jupiter, α counted from the direction of Jupiter, and z is perpendicular to it), v is the true anomaly of Jupiter, e_J is the eccentricity of Jupiter's orbit, μ is the mass of Jupiter divided by the total mass of the Sun-Jupiter system, and

$$R_1 = \sqrt{r^2+z^2}, \quad R_2 = \sqrt{1+r^2-2r\cos\alpha+z^2}.$$

Both r and z are dimensionless, the instantaneous Sun-Jupiter distance serving as unit distance.

I have looked for a solution of Equations (1) in the form of a three-variable asymptotic expansion

$$r = 1 + \sum_{n=1}^{N} \varepsilon^n r_n(v, u, \tau) + O(\varepsilon^{N+1}),$$

$$\alpha = \alpha_0(u, \tau) + \sum_{n=1}^{N} \varepsilon^n \alpha_n(v, u, \tau) + O(\varepsilon^{N+1}), \qquad (2)$$

$$z = \varepsilon^{\frac{1}{2}}\left[\sum_{n=0}^{N} \varepsilon^n z_n(v, u, \tau) + O(\varepsilon^{N+1})\right],$$

where

$$\varepsilon = \sqrt{\mu}, \quad u = \varepsilon(v - v_0), \quad \tau = \varepsilon^2(v - v_0),$$

and v_0 is the epoch. This solution is a generalization of Kevorkian's two-variable solution for the planar problem (Kevorkian, 1970).

The method of the solution consists of substituting (2) into Equations (1) and then solving the system of partial differential equations, obtained from equating the co-efficients of the same powers of ε on both sides of Equations (1), for the unknown functions r_n, α_n, z_n. An important feature of the solution is that the arbitrary functions appearing in the solution of the partial differential equations are specialized so that the solution for r, α, z should not contain secular terms. Applying this method I have determined α_0, z_0, r_1, α_1, z_1, r_2, α_2 and the main perturbations of Trojan asteroids.

3. FURTHER DEVELOPMENT OF THE THEORY

To study the perturbations in the ascending node Ω more thoroughly it is necessary to extend the solution one order of ε further and to determine z_2, r_3, α_3, and to investigate the solution for z_3. The equations for z_2, r_3 and α_3 have been given in an earlier paper (Érdi, 1978). The equation for z_3 is

$$
\begin{aligned}
\frac{\partial^2 z_3}{\partial v^2} + z_3 = & -2\frac{\partial^2 z_2}{\partial v \partial u} - 2\frac{\partial^2 z_1}{\partial u^2} - 2\frac{\partial^2 z_1}{\partial v \partial \tau} - 2\frac{\partial^2 z_0}{\partial u \partial \tau} + \\
& + \Big\{ 3r_3 + 3r_1 r_2 + \tfrac{3}{2} z_1^2 + 3z_0 z_2 - \tfrac{15}{4}(2r_1 + z_0^2)(r_1^2 + 2r_2 + 2z_0 z_1) + \\
& + \tfrac{35}{16}(2r_1 + z_0^2)^3 - 3r_1 - \tfrac{3}{2} z_0^2 + e_1^2\left(\tfrac{1}{2} + \tfrac{1}{2}\cos 2v\right)\left(3r_1 + \tfrac{3}{2} z_0^2\right) - \\
& - e_1 \cos v \left[3r_2 + \tfrac{3}{2} r_1^2 + 3z_0 z_1 - \tfrac{15}{8}(2r_1 + z_0^2)^2 + 1\right] \Big\} z_0 + \\
& + \left[3r_2 + \tfrac{3}{2} r_1^2 + 3z_0 z_1 - \tfrac{15}{8}(2r_1 + z_0^2)^2 + 1 - e_1 \cos v \left(3r_1 + \tfrac{3}{2} z_0^2\right)\right] z_1 + \\
& + \left(3r_1 + \tfrac{3}{2} z_0^2\right) z_2 + 2^{-3/2}(1 - \cos\alpha_0)^{-3/2}(e_1 z_0 \cos v - z_1) + \\
& + 2^{-5/2}(1 - \cos\alpha_0)^{-5/2}\left(r_1 + \tfrac{1}{2} z_0^2 - r_1 \cos\alpha_0 + \alpha_1 \sin\alpha_0\right) 3z_o ,
\end{aligned}
\qquad (3)
$$

where e_1 is defined by $e_J = \varepsilon e_1$.

To calculate the right-hand side of Equation (3) one should know first z_2 and r_3 (the other functions are already known). To calculate r_3 one should know α_3 too, and I have determined all of them. These are given by rather lengthy expressions consisting of different periodic terms and constants. Using this solution for z_2 and r_3 and calculating the right-hand side of Equation (3) it can be shown that the conditions for z_3 not to have secular terms are

$$\frac{\partial}{\partial u}\left[\lambda_2\cos(\nu_2-\nu_0)\right]+Q_1(u,\ \tau)+\frac{\partial}{\partial\tau}(\lambda_{10}\cos\nu_{10})=0,$$

$$\frac{\partial}{\partial u}\left[\lambda_2\sin(\nu_2-\nu_0)\right]+Q_2(u,\ \tau)+\frac{\partial}{\partial\tau}(\lambda_{10}\sin\nu_{10})+Q_3(\tau)+\lambda_0A_4=0. \quad (4)$$

Here λ_2 and ν_2 are unknown functions of u and τ, ν_0 is a known function of u and τ, $Q_1(u,\ \tau)$ and $Q_2(u,\ \tau)$ are known functions of u and partly unknown functions of τ which however can be integrated according to u without secular terms, λ_{10} and ν_{10} are unknown functions of τ, $Q_3(\tau)$ is a function of τ and λ_0 and A_4 are constants. The unknown functions λ_{10}, ν_{10} and λ_2, ν_2 appear in the solutions for z_1 and z_2 which are

$$z_1=\lambda_1\cos(v+\nu_1)-\frac{1}{2}\lambda_0\rho_1\cos(2v+\nu_0+\psi_1)+\frac{3}{2}\lambda_0\rho_1\cos(\nu_0-\psi_1),$$

where

$$\lambda_1\sin\nu_1=\lambda_{10}\sin(\nu_{10}+\nu_0)-\frac{1}{2}\lambda_0\frac{\partial\alpha_0}{\partial u}\sin\nu_0+\lambda_0q_1\cos\nu_0-\frac{1}{2}\lambda_0b_3\cos\nu_0,$$

$$\lambda_1\cos\nu_1=\lambda_{10}\cos(\nu_{10}+\nu_0)-\frac{1}{2}\lambda_0\frac{\partial\alpha_0}{\partial u}\cos\nu_0-\lambda_0q_1\sin\nu_0+\frac{1}{2}\lambda_0b_3\sin\nu_0,$$

and

$$z_2=\lambda_2\cos(v+\nu_2)-e_1\frac{\partial\alpha_0}{\partial u}\lambda_0\cos(2v+\nu_0)-\frac{1}{3}\frac{\partial\alpha_0}{\partial u}\lambda_0\rho_1\cos(2v+\psi_1+\nu_0)-$$

$$-\frac{1}{2}\rho_2\lambda_0\cos(2v+\nu_0+\psi_2)-\frac{1}{8}\lambda_0^3\rho_1\cos(2v+\psi_1+\nu_0)-$$

$$-\frac{5}{16}\lambda_0^3\rho_1\cos(2v+3\nu_0-\psi_1)+\frac{3}{8}\lambda_0\rho_1^2\cos(3v+2\psi_1+\nu_0)-$$

$$-\frac{1}{2}\rho_1\lambda_1\cos(2v+\psi_1+\nu_1)+\frac{3}{2}\rho_2\lambda_0\cos(\psi_2-\nu_0)+\frac{\partial\alpha_0}{\partial u}\lambda_0\rho_1\cos(\psi_1-\nu_0)+$$

$$+\frac{21}{16}\lambda_0^3\rho_1\cos(\psi_1-\nu_0)+e_1\frac{\partial\alpha_0}{\partial u}\lambda_0\cos\nu_0+\frac{3}{2}\rho_1\lambda_1\cos(\psi_1-\nu_1).$$

The paper (Érdi, 1981) contains explicit expressions for the functions α_0, ν_0, ρ_1, ψ_1, ρ_2, ψ_2, q_1, b_3 of the above solutions.

Actually, Equations (4) can be used to determine λ_{10}, ν_{10} and λ_2, ν_2 as the function of τ or u. The constant expression λ_0A_4 however produces a problem. In separating Equations (4) according to u and τ, it gives a secular term either in the solution for λ_{10} or in λ_2 and accordingly in z_1 or z_2.

This difficulty can be solved by introducing a new variable w by the relation

$$w = \varepsilon^3(v - v_0),$$

and looking for the solution of Equations (1) in the form of a four-variable expansion. Introducing w all the previous results remain unchanged and only in Equation (3) will appear an additional term $-2\partial^2 z_0/\partial v \partial w$ in the right-hand side which produces a term $\partial\lambda_0/\partial w$ in the first and a term $\lambda_0 \partial\nu_0/\partial w$ in the second equation of (4). (It is supposed now that λ_0 which has been a constant in the earlier solution may depend on w.) Thus Equations (4) can be separated into the following systems

$$\frac{\partial}{\partial u}\left[\lambda_2 \cos(\nu_2-\nu_0)\right] = -Q_1(u, \tau), \quad \frac{\partial}{\partial u}\left[\lambda_2 \sin(\nu_2-\nu_0)\right] = -Q_2(u, \tau),$$

and

$$\frac{\partial}{\partial \tau}(\lambda_{10}\cos\nu_{10}) = 0, \quad \frac{\partial}{\partial \tau}(\lambda_{10}\sin\nu_{10}) = -Q_3(\tau),$$

and

$$\frac{\partial \lambda_0}{\partial w} = 0, \quad \frac{\partial \nu_0}{\partial w} = -A_4 \tag{5}$$

These equations can be solved by integration. For our later purposes the most interesting is the solution of the last system. It follows that λ_0 is constant (it does not depend on any of the variables v, u, τ, w). Using the earlier result (Érdi, 1981)

$$\nu_0 = \alpha_0 - \frac{1}{2}A_3\tau + \nu_{01},$$

where α_0 is known function of u and τ, A_3 is constant and ν_{01} is also constant but this latter may now depend on w, it follows from the second equation of (5) that

$$\nu_{01} = -A_4 w.$$

Thus

$$\nu_0 = \alpha_0 - \frac{1}{2}A_3\tau - A_4 w + \nu_{02}, \tag{6}$$

where ν_{02} is a constant of integration.

4. CRITICAL INCLINATION

The importance of Equation (6) is that the main perturbation of Ω is given by $\alpha_0 - \nu_0$ (Érdi, 1981). Using the relations

$$A_3 = -\frac{3}{2}\ell^2 + \frac{3}{2^5}\ell^4 + O(\ell^6),$$

$$A_4 = \lambda_0^2\left(\frac{3}{2^4} + \frac{35}{2^7}\ell^2 + O(\ell^4)\right),$$

$$\sin^2 i = \varepsilon\lambda_0^2 + O(\varepsilon^2),$$

it can be shown that the main perturbation in Ω is

$$\Omega = \left[-\frac{3}{2^2}\ell^2 + \frac{3}{2^6}\ell^4 + \sin^2 i\left(\frac{3}{2^4} + \frac{35}{2^7}\ell^2\right)\right]\tau. \tag{7}$$

Here ℓ is a constant parameter, characteristic for the amplitude of libration around L_4 or L_5.

It follows from Equation (7) that Ω decreases if $i<i_c$ and increases when $i>i_c$ where i_c is the critical inclination in radians

$$i_c = 2\ell\left(1 - \frac{3}{2^5}\ell^2 + O(\ell^4)\right). \tag{8}$$

For most of the known Trojan asteroids $0<\ell<0.5$. The value $\ell=0$ corresponds to the point L_4. When $\ell=0.5$ the libration around L_4 takes place between the limits $37^\circ\!.2<\alpha_0<95^\circ\!.6$ (α_0 is counted from Jupiter). It can be seen from Equation (8) that i_c increases with ℓ. Figure 1 shows the mean value of $\dot{\Omega}$ obtained from Equation (7) for different values of i and ℓ. The curves show the same features which have been described by Bien and Schubart (1983).

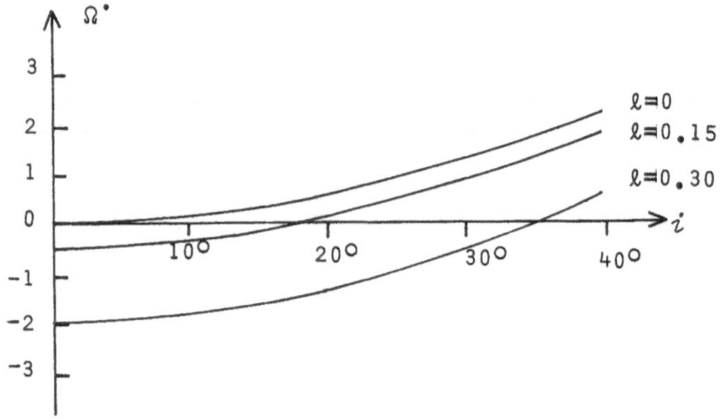

Figure 1. $\dot{\Omega}$ in degree/10^3 yr is plotted against i for $\ell=0$, 0.15 and 0.3.

The existence of the critical inclination affects also the behaviour of the "ψ_1, ψ_2 curve". Here $\psi_1=e\cos\tilde{\omega}$, $\psi_2=e\sin\tilde{\omega}$ and the curve shows geometrically the long-periodic variation of the eccentricity e and of the longitude of the perihelion $\tilde{\omega}$ of Trojan asteroids. It has been shown by numerical integration (Bien and Schubart, 1983) and also analytically (Érdi and Váradi, 1983) that the "ψ_1, ψ_2 curve" can be well approximated by a rotating ellipse. The ellipse rotates with the same rate as $\dot{\Omega}$. Thus the ellipse rotates backwards if $i<i_c$ and forwards if $i>i_c$ and e and $\tilde{\omega}$ vary accordingly.

REFERENCES

Bien, R. and Schubart, J.: 1983, in V. V. Markellos and Y. Kozai (eds.), 'Dynamical Trapping and Evolution in the Solar System', D. Reidel Publ. Co., Dordrecht, Holland, pp. 153-161.
Érdi, B.: 1978, Celes. Mech. 18, 141.
Érdi, B.: 1981, Celes. Mech. 24, 377.
Érdi, B. and Váradi F.: 1983, in C.-I. Lagerkvist and H. Rickman (eds.), 'Asteroids, Comets and Meteors', Uppsala Universitet, Uppsala, pp. 155-159.
Kevorkian, J.: 1970, in G. E. O. Giacaglia (ed.), 'Periodic Orbits, Stability and Resonances', D. Reidel Publ. Co., Dordrecht, Holland, pp. 286-303.

AN APPLICATION OF LABROUSTE'S METHOD TO QUASI-PERIODIC ASTEROIDAL MOTION

J. Schubart and R. Bien
Astronomisches Rechen-Institut
Mönchhofstrasse 12-14
D-6900 Heidelberg 1
Federal Republic of Germany

ABSTRACT. In an earlier paper the authors have applied Labrouste's method to single orbital elements in order to isolate periodicities. However, in practice the investigation of two-dimensional vectors, where the components are combinations of orbital elements, can be useful. In the present paper we apply Labrouste's method to vectorial components of this type and represent the results by two-dimensional graphs. Examples refer to the Trojan case of asteroidal motion.

1. INTRODUCTION

In a former paper (Bien and Schubart, 1983a; hereafter called paper A) we have described a method by Labrouste (e.g. Labrouste and Labrouste, 1936) which we applied to two examples of asteroidal motion. The method allows to isolate single periods in a quasi-periodic process. For instance, let $x(t)$ be an orbital element as a function of time t. By Labrouste's method it is possible to separate in $x(t)$ effects of short period, long period, and effects that are constant or linear in t. Moreover, $x(t)$ can be a function of one or several orbital elements.

As a generalization we consider two functions of this type, say $x(t)$, $y(t)$. Practical examples are

(a) $x(t) = \Omega$, $y(t) = i$
(b) $x(t) = i \cos(\Omega - \Omega_J)$, $y(t) = i \sin(\Omega - \Omega_J)$
(c) $x(t) = e \cos(\tilde{\omega} - \tilde{\omega}_J)$, $y(t) = e \sin(\tilde{\omega} - \tilde{\omega}_J)$

where i, e, Ω, $\tilde{\omega}$ denote inclination, eccentricity, ascending node, longitude of perihelion, respectively, of an asteroid. The subscript J refers to an orbital element of Jupiter. Mathematically $\{x(t), y(t)\}$ is a time-dependent two-dimensional vector which can be represented by the motion of a point in a plane.

In the present paper we apply Labrouste's method to the two components of vectors of this type with the aim to describe the principal periodicities in x, y diagrams. The treatment of these vectors includes more than a simple generalization of the one-dimensional technique:

Celestial Mechanics **34** (1984) 443–452. 0008–8714/84.15

If the variations are quasi-periodic, the motion of the point representing the vector can be decomposed into a sequence of epicycles. Incidentally, the following examples represent new contributions to the question of the long-period behaviour of Trojan asteroids; see Bien and Schubart, 1983b (hereafter called paper B).

2. QUALITATIVE DESCRIPTION OF THE METHOD

Let the vectorial components x(t), y(t) be quasi-periodic functions of t, i.e. both x(t), y(t) can be represented by a finite sum of sine terms with arguments that depend linearly on t. The same frequencies will appear in both components in general.

It is sufficient to consider Labrouste's method for one component, say x(t), which is supposed to be given at equidistant instants of t. For numerical reasons the whole time interval should be as large as possible. In order to isolate a band of frequencies, a filter profile is defined, i.e. a function of frequency in a certain finite interval. This function takes the value 1 inside the band and drops to zero outside. The filter is described by a Fourier approximation with a sufficiently large number of terms, say 300 or 1000. The Fourier coefficients are used as weights in the definition of a moving average of the discrete values of x(t). In other words, each "inner" value of x(t) is replaced by a weighted mean of neighbouring values. The new sequence of discrete values describes a new quasi-periodic function "x(t)" which differs from x(t) only in the amplitudes of the sine terms: The amplitudes practically disappear for frequencies outside the band and its immediate neighbourhood. For frequencies of the band the amplitudes are only slightly changed, depending on the degree of Fourier approximation. It is an important property of the method that frequencies and phases are unchanged. The mathematical formalism as used by us is published in paper A. For the reader's convenience the basic formulae are repeated in an Appendix.

Suppose that y(t) has "passed" the same filter. Then "x(t)", "y(t)" can be interpreted as a new vector in the x,y plane.

3. APPLICATIONS

3.1. Example 1

The most simple case for an application of the method of the present paper is given, when both x(t) and y(t) are orbital elements, and "y(t)" is plotted versus "x(t)". An example is type (a) of the Introduction, which has played a role in our recent studies of the dynamics of the Trojan asteroids (see Bien and Schubart, Fig.8, this conference), where Ω is replaced by the difference between Ω and the constant $\tilde{\omega}_J$.

3.2. Example 2

Our second example refers to type (c) of the Introduction: $x(t) = \psi_1 = e \cos(\tilde{\omega} - \tilde{\omega}_J)$, $y(t) = \psi_2 = e \sin(\tilde{\omega} - \tilde{\omega}_J)$. We have numerically integrated the elliptic restricted three-body problem Sun-Jupiter-Trojan (for the integration method see Schubart and Stumpff, 1966). The integration covers 75 500 yr. The Trojan starts with $i = 30°$, $e = 0.05$, $\tilde{\omega} - \tilde{\omega}_J = 60°$. The other starting values represent standard values taken from paper B.

In Fig.1 the original ψ_1, ψ_2 relationship is plotted over approximately one revolution around the point $\psi_1 = 0.025$, $\psi_2 = 0.040$, which lasts about 4800 yr. Every three years a point is plotted and connected with the preceding point by a straight line. There is a strong influence by effects of short period (e.g. periods of about 12 yr), but the "mean figure" is a kind of ellipse. This result has suggested to us to eliminate periods smaller than 1000 yr. By applying our method on the whole interval of integration we get Fig.2. Obviously, the point moves counterclockwise along an ellipse which shows a trend to rotate in the same sense (compare the ψ_1, ψ_2 curve of (1208) Troilus in paper B). We mention that the determination of a moving average cannot give information at the beginning and the end of the integration interval. As a consequence, the curve in Fig.2 covers only 30 000 yr.

We assume that the center of the ellipse of Fig.2 is in motion due to the action of a very long period. Érdi and Varadi (1983) have found such an effect in their analytical work. Actually, if we eliminate not only periods smaller than 1 000 yr, but also periods larger than 14 000 yr (together with constant effects), Fig.3 results: An ellipse rotating around the origin remains. Since we know that the rotation period of the ellipse is equal to the period of Ω (cf. paper B), we represent the

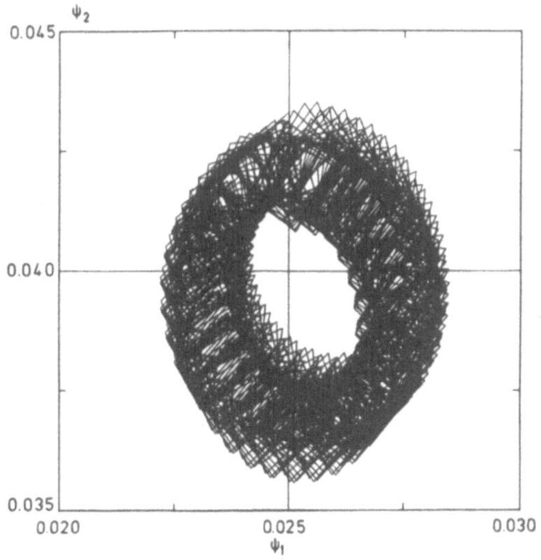

Figure 1. Original ψ_1, ψ_2 relationship of example 2 over about 5 000 yr. Every three years a point is plotted and connected with the preceding point.

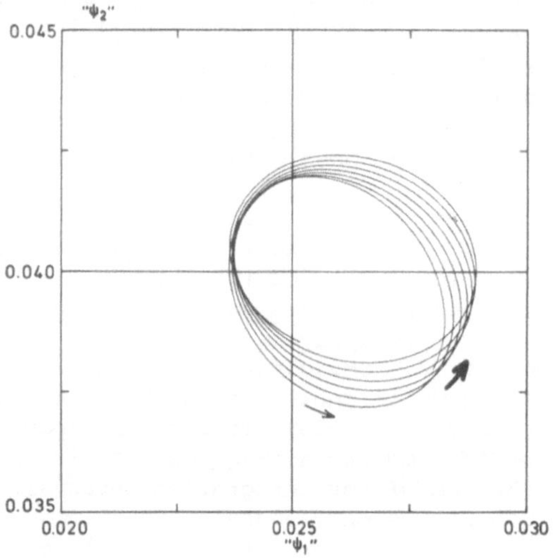

Figure 2. Example 2: The curve corresponds to 30 000 yr; periods smaller than 1 000 yr are eliminated. Thin arrow: motion along the ellipse; thick arrow: rotation of the ellipse. For the meaning of "ψ_1", "ψ_2" see the text.

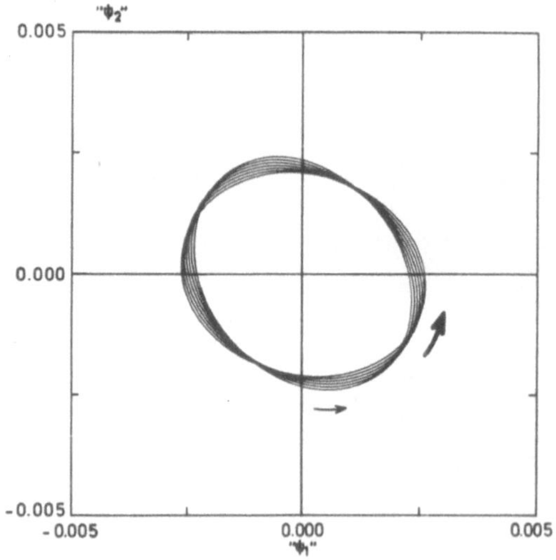

Figure 3. Example 2: Same as Fig.2, but the position and motion of the origin is eliminated.

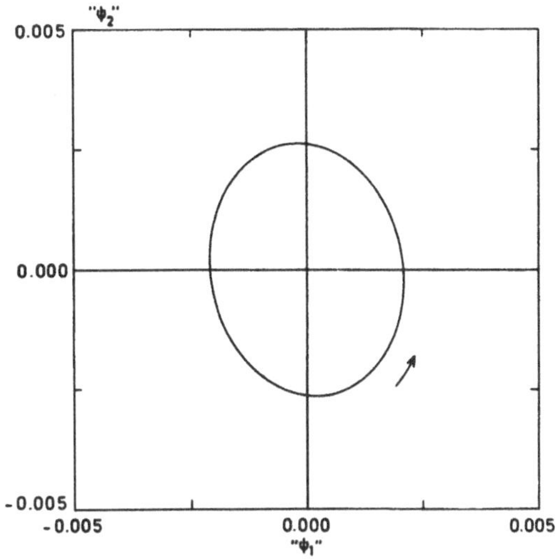

Figure 4. Example 2: The rotation of the ellipse is compensated by introducing a rotating coordinate system.

vector ("ψ_1", "ψ_2") in a coordinate system rotating with the angular velocity of $\dot{\Omega}$ before we apply the same filter. As a success of this procedure, Fig.4 shows a non-rotating ellipse. It is remarkable that eccentricity and size of the ellipse do not change over 30 000 yr. A secular change of the shape of the ellipse would be visible after applying the filter.

We suppose that the ellipse is transformed to suitable normal coordinates and can be described as follows

$$"\psi_1" = C \cos \alpha t^*, \quad "\psi_2" = D \sin \alpha t^* \quad (C > D > 0)$$

where C, D are constants, α is a frequency, and t^* is the time counted from an instant with "ψ_2" = 0. Then

$$"\psi_1" = \frac{C+D}{2} \cos \alpha t^* + \frac{C-D}{2} \cos \alpha t^*$$

$$"\psi_2" = \frac{C+D}{2} \sin \alpha t^* - \frac{C-D}{2} \sin \alpha t^*$$

In other words, the vector ("ψ_1", "ψ_2") is given by the sum of two vectors with constant length which rotate in opposite directions (i.e. with angular velocities $+\alpha$ and $-\alpha$).

In a non-rotating coordinate system, these two vectors rotate with nearly the same angular velocities, since $\dot{\Omega}$ is comparatively small. Again we introduce a rotating coordinate system with a new angular velocity near $+\alpha$, before we apply another new filter. Then one vector rotates with velocity near -2α, whereas the other one rotates very slowly.

Note that originally constant and long-period effects now appear with frequencies near the absolute value of $-\alpha$. We define a filter in order to eliminate all frequencies outside a narrow band which is centered around 2α. Then the vector which rotates with velocity -2α is separated from the other effects mentioned above and appears as a circle in the "ψ_1", "ψ_2" plane; see Fig.5, small circle (radius $(C-D)/2$). Conversely, a coordinate system rotating with nearly $-\alpha$ and the application of the same filter gives the other vector; large circle in Fig.5 (radius $(C+D)/2$).

In a final step we investigate in a fixed coordinate system the motion of the center of the ψ_1, ψ_2 curve. A filter is used where all periods smaller than about 10 000 yr are more or less eliminated; the result is shown in Fig.6. If we assume that the curve is an arc of a circle, we can estimate the angular velocity of the motion around its center and find a value close to $2\,\dot{\Omega}$; compare this with Érdi and Varadi's (1983) result.

3.3. Example 3

Our third example refers to the vector (ψ_1, ψ_2) of another Trojan orbit that we have integrated earlier (paper B) together with Jupiter and Saturn, both moving in the reference plane. The Trojan starts with $i = 40°$, $e = 0.11$, and the other starting values are equal to those of example 2. Example 3 is close to a secular resonance, compare the paper

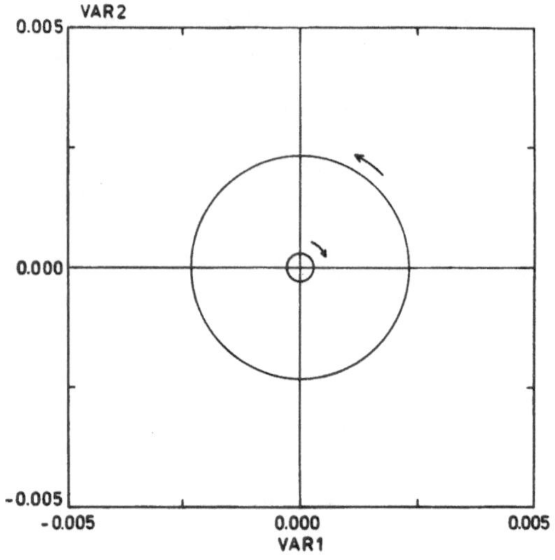

Figure 5. Example 2: The figure is actually the composition of two graphs which refer to special rotating coordinates VAR1, VAR2, see text. However, it is demonstrated that two vectors of constant length rotate in opposite directions. In the original non-rotating coordinates the sum of these vectors generates the ellipse of Fig.3.

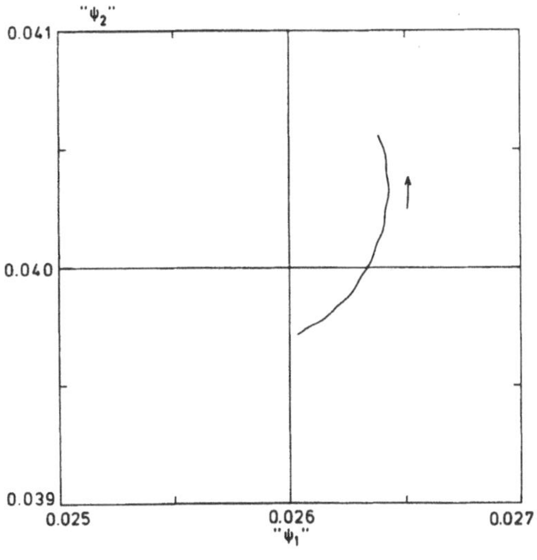

Figure 6. Example 2: Motion of the origin of the ψ_1, ψ_2 curve over about 30 000 yr. Note that the filter does not entirely eliminate one of the shorter periods.

by Bien and Schubart given at this conference.

Fig.7 is analogous to Fig.3 of example 2 and corresponds to an interval of about 105 000 yr. The striking difference is, that the point rotates in one direction, but the ellipse does not. The possibility of an alternation of foreward and backward rotation of such an ellipse under the action of Saturn has already been mentioned in paper B. Apparently, the major axis of the ellipse does not change. Fig.7 demonstrates that a frequency can pass our filter, although it is modulated by a very long period. Incidentally, we point to the possibility to separate neighbouring frequencies which have passed a first filter, by the application of additional filters (see paper A).

Fig.8 shows the motion of the center of the ψ_1, ψ_2 curve over 80 000 yr; all periods smaller than about 7 000 yr are eliminated. In principle, Fig.8 is analogous to Fig.6, but a quite different and more important effect dominates. One revolution along the curve corresponds to about 54 000 yr, a period which is known from the variations of Jupiter's eccentricity. This relation is already indicated in paper B. However, it is a new result that the center moves on a narrow ellipse and not more or less on a straight line through the origin with an inclination of about 60° to the "ψ_1" axis.

4. CONCLUSION AND OUTLOOK

We have demonstrated, that the application of Labrouste's method enables us to separate various periodic effects in the motion of minor bodies.

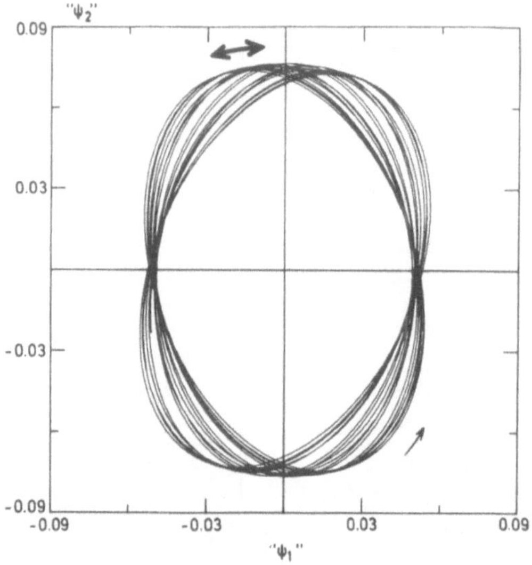

Figure 7. Example 3: The figure is analogous to Fig.3. However, the thick symbol indicates that the ellipse swings foreward and backward several times during about 105 000 yr. One complete revolution of a point along the ellipse lasts 6 000 yr.

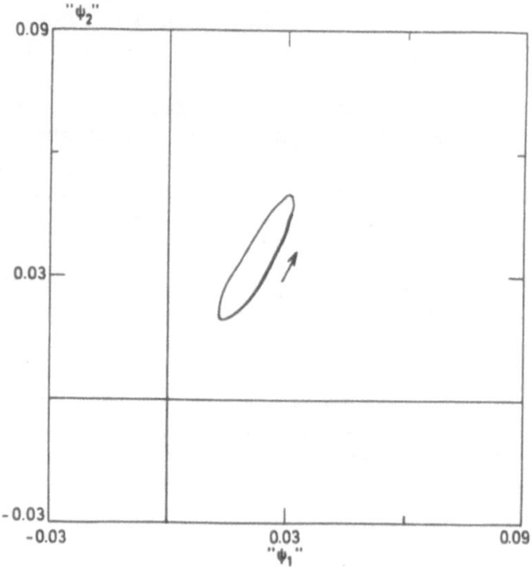

Figure 8. Example 3: Motion of the origin of the ψ_1, ψ_2 curve. Due to the action of Saturn, the variations are much stronger than in the analogous Fig.6. Time interval: 80 000 yr.

If an orbit allows such a separation one may expect that the orbit will show a stable behaviour over a time interval which is longer than the actual integration interval.

In the present paper we have restricted ourselves to qualitative results only. However, we plan to use Labrouste's method for deriving quantitative results as well.

APPENDIX

We consider the quasi-periodic function

$$x(t) = A_0 + \sum_{i=1}^{M} A_i \sin(\alpha_i t + \beta_i)$$

and want to determine A_0, A_i, α_i, and β_i. Furthermore, we introduce in the interval $0 \leq \alpha \leq \pi$ a function $\kappa(\alpha)$ of frequency α (a so-called filter) by the Fourier approximation

$$\kappa(\alpha) = \sum_{\nu=0}^{n} K_\nu \cos \nu\alpha$$

using a suitable unit of time. Then

$$"x(t)" = S(t,n) = \sum_{\nu=0}^{n} K_\nu \cdot s(t,\nu)$$

with

$$s(t,\nu) = \frac{1}{2} [x(t+\nu) + x(t-\nu)]$$

can be written as

$$"x(t)" = A_0 \cdot \sum_{\nu=0}^{n} K_\nu + \sum_{i=1}^{M} \left(\sum_{\nu=0}^{n} K_\nu \cos \nu\alpha_i \right) \cdot A_i \cdot \sin(\alpha_i t + \beta_i)$$

$$= \kappa(0) A_0 + \sum_{i=1}^{M} \kappa(\alpha_i) A_i \sin(\alpha_i t + \beta_i)$$

Comparing "x(t)" and x(t) we note that the amplitudes A_0, A_i are modified, but the frequencies α_i and phases β_i are unchanged. For more details, compare paper A. Note: The properties of $\kappa(\alpha)$ reappear according to symmetry and periodicity for $\alpha > \pi$.

REFERENCES

Bien,R. and Schubart,J.: 1983a, *Asteroids, Comets, Meteors*, Ed.: C.-I. Lagerkvist and H. Rickman, Uppsala University, pp.161-166 [= paper A].

Bien,R. and Schubart,J.: 1983b, *Dynamical Trapping and Evolution in the Solar System*, Ed.: V.V. Markellos and Y. Kozai, IAU Coll.No.74,

pp.153-161 [= paper B].
Érdi,B. and Varadi,F.: 1983, *Asteroids, Comets, Meteors,* Ed.: C.-I. Lagerkvist and H. Rickman, Uppsala University, pp.155-159.
Labrouste,H. and Mrs. Labrouste : 1936, *Terrestrial Magnetism and Atmospheric Electricity,* 41, pp.15-28 and 105-126.
Schubart,J. and Stumpff,P.: 1966, *Veröffentl.Astron. Rechen-Inst. Heidelberg* No.18.

A NOTE ON RESONANCE IN REGULAR VARIABLES AND AVERAGING

S. Ferraz-Mello
Departamento de Astronomia
Universidade de São Paulo
01051, São Paulo, Brazil

and

W. Sessin
Departamento de Astronomia
Instituto Tecnológico de Aeronáutica
12200, São José dos Campos, Brazil

ABSTRACT. A point relative to the application of the method of Hori to resonant systems is considered: For systems having one degree of freedom the topology of the phase plane of the auxiliary Hori's system is unaltered in the process of construction of a formal solution. The transformed Hamiltonian may not lead to singular points other than those included in the auxiliary Hori's Hamiltonian.

1. INTRODUCTION

Several recent papers (Henrard and Lemaitre, 1983; Sessin and Ferraz-Mello, 1984; Lemaitre, 1984) have treated the problem of a perturbed regular Hamiltonian in the case where the flow of the perturbed motions shows a bifurcation due to a null frequency (i.e. a resonance) in the unperturbed Hamiltonian. The results obtained by Sessin and Ferraz-Mello (1984) in the study of a particular case of planetary resonance suggested us to make a thorough study of the problem of the resonance in regular (i.e. non singular) variables (Ferraz-Mello, 1984). The aim of this note is to present a general result that extrapolates the particular situations considered in earlier papers, which implies the existence of a limitation of the method of Hori and, perhaps, of averaging methods, in general.

Generally we consider the Hamiltonian

$$F = F_0(I) + \sum_{q\geq 1} \varepsilon^q F_q(I,\theta) \tag{1}$$

where $I\varepsilon R^{+n}$, $\theta\varepsilon T^n$ are the action-angle variables associated to F_0 and ε is a small positive parameter. We assume that the undisturbed Hamilto-

Celestial Mechanics **34** (1984) 453–457. 0008–8714/84.15

nian is regular in a domain about the origin. For the present case it is sufficient to consider one degree of freedom (i.e. n=1) and to restrain F_o to its leading terms

$$F_o = \omega^o I + \frac{1}{2} I^2 . \tag{2}$$

The disturbing part of the Hamiltonian is composed of the terms

$$\varepsilon^q F_q = \varepsilon^q \sum_{q' \geq 1} V_q^{q'}(k,h) \tag{3}$$

where $V_q^{q'}$ is the part of F_q homogeneous of degree q' with respect to the regular variables

$$k = \sqrt{2I} \cos\theta, \quad h = \sqrt{2I} \sin\theta . \tag{4}$$

The one-degree-of-freedom cases where the F_q are reduced to F_1 and to only a term $-(\sqrt{2I})^q \cos q$ (q=1,2 or 3) have thoroughly been studied by Jefferys (1966), Henrard and Lemaitre (1983), Lemaitre (1984) and Ferraz-Mello (1984).

2. ON FORMAL SOLUTIONS WITH HORI'S METHOD

The construction of formal solutions of the dynamical system defined by the Hamiltonian (1) where regular variables are used instead of action-angle variables may be made by means of Hori's perturbation method with unspecified canonical variables (Hori, 1966). The generic equation of Hori's theory is

$$\{X^*, B_q\} = X_q^* - \psi_q(k^*, h^*) \tag{5}$$

where

$$\{X^*, B_q\} = \sum \left\{\frac{\partial X^*}{\partial k_i^*}\frac{\partial B_q}{\partial h_i^*} - \frac{\partial X^*}{\partial h_i^*}\frac{\partial B_q}{\partial k_i^*}\right\} ; \tag{6}$$

B_q and X_q^* are the homogeneous parts of degree q with respect to k*,h* and the small parameter, of the Lie transform and of the new Hamiltonian, respectively; ψ_q is a known function when the equations for the preceding orders were solved; k_i^* and h_i^* are the variables resulting from the transformation and X* is the part of lower degree of homogeneity in F (Ferraz-Mello, 1984). In the construction, special attention is paid to the fact that the derivatives with respect to k_i^*, h_i^* alter the order of the functions, since k_i^*, h_i^* are small quantities in the domain under study (neighbourhood of the origin). When ε is taken as a small parameter, the identification of the terms in Hori's equation is made by collecting terms of the same degree of homogeneity with respect to k, h,ε. In this case

$$X^* = X_2^* = \frac{1}{2} \sum \omega_i^o (k_i^2 + h_i^2) + \varepsilon V_1^1 (k,h) \tag{7}$$

which is separable in n parts

$$\frac{1}{2}\,\omega_i^o\,(k_i^{\,2} + h_i^{\,2}) - \varepsilon\tau_i k_i \qquad (8)$$

where plane rotations were made to eliminate the linear terms in h_i. The solution of the corresponding Hori auxiliary equations are off-centered harmonic motions

$$\begin{aligned} k_i^* &= C_i \cos(\omega_i^o t + C_i') + \varepsilon\tau_i/\omega_i^o \\ h_i^* &= C_i \sin(\omega_i^o t + C_i') \ . \end{aligned} \qquad (9)$$

The orbits are circles around S_2 (see figure 1).

In the case of a central resonance ($\omega^o = 0$) an orbit starting close to the origin may reach distances up to $O(\varepsilon^{1/3})$ from the origin. Therefore $\varepsilon^{1/3}$ is taken as the small parameter and the terms are collected following their degree of homogeneity with respect to $k,h,\varepsilon^{1/3}$. The results are

$$X^* = X_4^* - \frac{1}{8}\,\Sigma\Sigma(k_i^2+h_i^2)\,(k_j^2+h_j^2) + \varepsilon\, V_1^1(k,h)\ . \qquad (10)$$

This case is separable only if $n_{ij}^o=0$ for $i\neq j$ and the solutions are ovals a little more asymmetric than those shown in fig. 1(a) and the solutions are written in terms of elliptic functions.

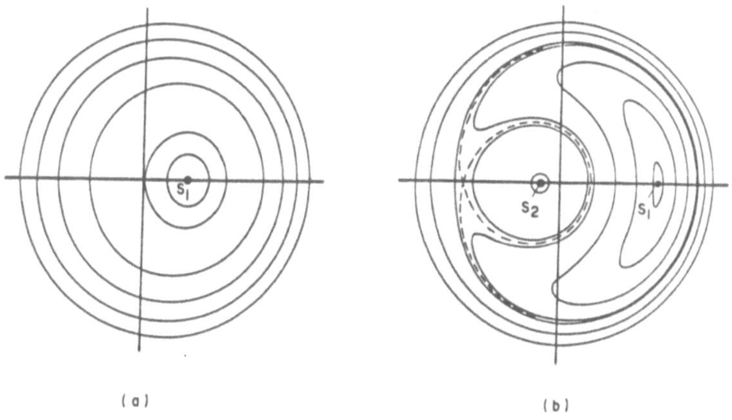

Figure 1 - Orbits defined by the Hamiltonian $F = \omega^o I + I^2/2 - \varepsilon\tau k$ in the plane (k,h). (a) Orbits in the case $\omega^o > -1.5(\varepsilon\tau)^{2/3}$; (b) Orbits in the case $\omega^o < -1.5\,(\varepsilon\tau)^{2/3}$.

3. A TOPOLOGICAL CONSTRAINT

Let us consider again the motion in the neighbourhood of the origin in the case of a non-central resonance (ω^o finite) and where (1) has only one degree of freedom and is restricted to

$$F = \omega^o I + \frac{1}{2} I^2 - \varepsilon\tau\, k \, . \tag{11}$$

The only part of F that will be considered in the auxiliary equations is $\omega^o I - \varepsilon\tau k$ while $I^2/2$ will act only at higher orders. In this case, equations (6) are easily solved up to a given order and we may construct the transformed Hamiltonian as

$$F^* = \Sigma\, \varepsilon\, X^*_q \, . \tag{12}$$

Following the rules of the method of Hori, the X^*_q are to be determined by the averaging

$$X^*_q = \frac{\omega^o}{2\pi} \int_0^{2\pi/\omega^o} \psi_q(k^*,h^*)dt \; ; \tag{13}$$

therefore X^*_q may depend on k^*,h^* only through

$$C^2 = k^{*2} + h^{*2} - 2\,\frac{\varepsilon\tau}{\omega^o}\, k^* + \frac{\varepsilon^2\tau^2}{\omega^{o2}} \, . \tag{14}$$

The transformed Hamiltonian equations are

$$\frac{dk^*}{dt} = -\frac{\partial F^*}{\partial h^*} = \frac{-1}{C}\,\frac{\partial F^*}{\partial C}\, h^*$$
$$\frac{dh^*}{dt} = \frac{\partial F^*}{\partial k^*} = \frac{1}{C}\,\frac{\partial F^*}{\partial C}\,(k^* - \frac{\varepsilon\tau}{\omega^o}) \tag{15}$$

whose solutions are off-centered harmonic motions with the same eccentricity as that of the auxiliary equations. The only difference lies in the fact that the frequencies are now given by

$$\omega^o = \frac{1}{C}\,\frac{\partial F^*}{\partial C} \tag{16}$$

a constant, since C is also a first integral of the transformed equations. The actual calculation gives

$$\omega^* = \omega^o + \frac{1}{2}\,(C^2 + 2\,\varepsilon^2\tau^2/\omega^{o2}) + o(\varepsilon^4) \, . \tag{17}$$

On the other hand, the series ΣB_q may be constructed indefinitely without heavy algebraic troubles (the fact that here n=1 rules out the possibility of having small divisors).

There results that the Lie transform generated by B, leads necessarily to a Hamiltonian system that has the same phase space topology as that of the auxiliary system.

We may note that if ω^0 is negative and small, for values of C large enough we may have a change in the sign of ω^*. As a conclusion, we find that the results are only valid for small values of C. If we force and push the solutions to finite values of C, the resulting flow does not display the libration zone. The whole libration zone is substituted by a zero-frequency solution. In other words, a surgery is made in the actual phase space in order to make it homeomorphic to the phase space of the auxiliary system.

4. CONCLUSION

The study of the simplest cases of first-order resonance allows one to see that the topology of the phase space of the Hori auxiliary system affects the topology of the phase space of the averaged equations, and, in the particular case of one degree of freedom, is kept unaltered in the process of construction of a formal solution. Therefore it may be necessary to engineer the Hamiltonian of the auxiliary system in order to introduce in it the features we want to study quantitatively through formal series expansions.

REFERENCES

Ferraz-Mello, S.: 1984, *Celes. Mech.* (submitted). Papers I and II.
Jefferys, W.: 1966, *Astron. J.* **71**, 306.
Henrard, J. and Lemaitre, A.: 1983, *Celes. Mech.* **30**, 197.
Lemaitre, A.: 1984, *Celes. Mech.* **32**, 109.
Sessin, W. and Ferraz-Mello, S.: 1984, *Celes. Mech.* **32**, 307.

ON THE BROWN CONJECTURE

BORIS GARFINKEL
Department of Astronomy, Yale University, New Haven, CT, U.S.A.

ABSTRACT. In 1911 E. W. Brown conjectured that the family of long-periodic orbits in the Trojan case of the restricted problem of three bodies terminates in an asymptotic orbit that passes through the Langrangian point L_3 for $t = \pm \infty$. The paper refutes this conjecture analytically, thereby confirming the previously published numerical refutation by Henrard (1983).

1. INTRODUCTION

The background for the reading of the paper is furnished by the author's previously published theoretical solution, referred to here as Parts I (1977), II (1978), III (1980), and IV (1983) of the Theory of the Trojan Asteroids, as well as Part V, now in press.

In IV, p. 373 our family $\mathcal{L}(\alpha)$ is described by the equations of the form,

$$G = G(\lambda, \alpha), \quad \lambda = \lambda(t,\alpha), \quad z = z(t,\alpha). \qquad (1)$$

Here G is the angular momentum, canonically conjugate to the mean synodic longitude λ, while the complex z synthesizes the eccentric canonical variables ζ and η of Poincaré, as

$$z = \zeta + i\eta, \quad i = \sqrt{-1}$$

The family parameter α has been normalized so that the tadpole branch of the family lies in the range

$$\sigma \leq \alpha \leq 1.$$

The right hand members of (1) are series in powers of $\sqrt{m}$, where m is the small mass-parameter.

In the refutation of the conjecture, a crucial role is played by the third equation of (1), which can be written explicitly as

$$z = m(1 + f(\lambda) + g\ (t)) + O(m^{3/2}), \qquad (2)$$

with

Celestial Mechanics 34 (1984) 459–463. 0008–8714/84.15

$$f \equiv \left(\frac{1}{4s} - 2s^2\right) \quad (1 - 2ic/s), \qquad (3)$$

$$s \equiv \sin\frac{\lambda}{2} \qquad c \equiv \cos\frac{\lambda}{2},$$

and g(t) denoting the epicyclic term, satisfying the linear differential equation,

$$\frac{dg}{dt} - ig + f\,(\lambda) = 0.$$

Its solution is of the form

$$g(t) = [g(o) + J(t)]e^{it} \equiv R(t)e^{i(t+\delta)}, \qquad (4)$$

where

$$J(t) \equiv \int_o^t e^{-it}\, df$$

$$R(t) \equiv |g(t)|, \qquad \delta \equiv \arg(g(t)) \qquad (5)$$

Eq. (2) can be rewritten as

$$z = \bar{z}\,(\lambda(t)) + mg(t) + \ldots, \qquad (6)$$

where the bar marks the so-called "mean" orbit, deprived of the rapidly oscillating epicyclic term,

$$\bar{z} \equiv m(1 + f) + \ldots \qquad (7)$$

2. THE LIMITING CASE $\alpha = 1$

For an asymptotic orbit of infinite period, corresponding to $\alpha = 1$, the second equation of (1) takes the form (See V, §2),

$$\sin\frac{\lambda}{2} = 1 - \frac{1}{2}(3-\sqrt{2})\ \mathrm{sech}^2 u, \qquad (8)$$

$$u = pt + u_*(t),$$

where the characteristic exponent p and the asymptotic part u_* of u satisfy the relations

$$p = (21m/8)^{1/2}, \qquad u_* = \frac{1}{2}\mu \tanh u + \ldots \qquad (9)$$

$$\mu = 0.0372$$

The terminal state of the motion corresponds to $t = \pm\infty$, with

$$u = \pm\infty,\ \lambda = \pi,\quad f = -7/4,$$

in view of (8) and (3), while (7) becomes

$$\bar{z}\ (\infty,1) = -\frac{3}{4}m + O(m^{3/2})$$

As shown in I, p. 375 , the latter point is the projection of L_3 into the z-plane, and the terminus of the orbit, in view of (6), therefore deviates from L_3 by the epicyclic term mg(±∞). In the light of (4), this terminus is the circle,

$$mg(\pm\ \infty) = mR(\pm\ \infty)e^{it} \qquad (\delta = o) \qquad (10)$$

centered on L_3.

That

$$R(+\ \infty) = R(-\ \infty) \equiv R \qquad (11)$$

follows from the conservation of the Jacobi constant C, related to R by

$$C = C_3 - m^2R^2(\pm\ \infty), \qquad (12)$$

where C_3 is the value of C on the zero-velocity curve through L_3 (See V, §12). From (11), (10), and (4) we then deduce

$$|g(\pm\ \infty)| = R = |g(o)+J(\infty)| = |g(o)+J(-\ \infty)| \qquad (13)$$

Furthermore, a direct calculation of J(±∞) in (5) with the aid of (3), (8), and (9) yields (See V)

$$J(\pm\ \infty) = (1 \pm \mu)\bar{J} \qquad (14)$$

$$\bar{J} \equiv (-28 + 481\ i)m + O(m^2),$$

where the constant μ of (9) reflects the <u>asymmetry</u> of J(t) with respect to t = 0. A substitution from (14) converts (13) into

$$R = |g(o) + (1 + \mu)\bar{J}| = |g(o) + (1 - \mu)\bar{J}| \qquad (15)$$

By the algebra of complex numbers, we now deduce

$$g(o) = (\lambda i - 1)\bar{J},$$

$$R = (\lambda^2 + \mu^2)^{1/2}\ |\bar{J}|, \qquad (16)$$

$$\min R = \mu|\bar{J}| \equiv R_* = 18m + \ldots > o,$$

where λ is an arbitrary real number.

From, (10) it then follows that

$$|g(\pm\ \infty)| = R \geq R_* > o \qquad (17)$$

i.e. the epicyclic term cannot vanish for $t = \pm\infty$, so that the orbit does not pass through L_3. Instead, in view of the nature of sech u, the orbit approaches asymptotically a "limit-cycle" mRe^{it}. The latter can be identified as a member of the family of short-periodic orbits around L_3 (Szebehely, 246).

3. THE FAMILY OF HOMOCLINIC ORBITS

By varying λ in (16) we generate an entire family of orbits parametrized by R, for $\alpha = 1$. In virtue of the equality (11), this is a family of homoclinic orbits, doubly asymptotic to short-periodic orbits around L_3. This result is in agreement with the numerical integrations carried out by Henrard (1983).

The homoclinic family can be viewed as a subfamily $\mathcal{L}(1,R)$ of our bi-parametric long-periodic family $\mathcal{L}(\alpha,R)$. That only two parameters suffice to describe the latter family is due to the omission of the trivial parameter t_o, or δ in (10), and to the constraint imposed on the initial value g(o) by the first equation of (16). This constraint eliminates one of the four parameters in the general-solution of fourth-order differential system. Of course, such a reduction in the number of parameters could be expected a priori, inasmuch our long-periodic family is a particular, rather than the general solution of the problem of the motion. Indeed, we had already chosen one parameter by setting $e' = o$ for the unperturbed eccentricity, in order to remove from our solution the short-periodic terms (cf I, 373).

Inasmuch as the parameter R of the homoclinic family is bounded away from zero in virtue of (17), it follows that the interval

$$0 \leq R < R_*$$

is inaccessible to homoclinic orbits and that the Brown conjecture regarding L_3 is, therefore, false! From (12), (11), and (16) we deduce the inequality

$$\max C = C_3 - m^2R_*^2 < C_3. \tag{18}$$

This result confirms the Henrard conjecture (1983, p. 122) regarding the Jacobi constant of the homoclinic orbits.

4. THE HISTORY OF THE BROWN CONJECTURE

E. W. Brown conjectured (1911) that the family of long-periodic orbits around L_4 terminates in a asymptotic orbit passing through L_3 for $t = \pm\infty$. This conjecture is refuted by equations (17) and (18); the non-vanishing of the epicyclic term, merely suggested by the author in I, p. 375, has now been rigorously established. This analytical refutation of the Brown conjecture regarding L_3 confirms the previously published numerical refutation by Henrard (1938).

The conjecture has had a long and a turbulent history. The Age of Acceptance, 1911-1970, was marked by a wide belief in the opinion of the world-authority on the Lunar Theory, and even "proofs" by analysis (Thüring 1932, Stumpff 1963), as well as by numerical integration (Rabe 1961).

The Age of Controversy, 1970-1984, was inaugurated by Deprit and Henrard (1970). The first "refutation" of the conjecture was Deprit's discovery that the long-periodic family around L_4 terminates by bifurcation at a critical value of α, long before α reaches the value $\alpha = 1$. However, this "refutation" was rebutted by the author (IV - 1983), who introduced his disjointed family $\mathcal{L}(\alpha)$ extending to the vicinity of $\alpha = 1$, and beyond.

The second "refutation" by the author in I - (1977, p. 375) remained unconvincing because the crucial inequality $g(\infty) \neq 0$ was not proved!

The third refutation was delivered by Henrard (1983), who tested the conjecture numerically in its equivalent form, asserted by Brown on p. 446; "...unstable solutions in the vicinity of L_3 are portions of an asymptotic orbit around L_4". What Henrard did was to continue these solutions analytically a short distance from L_3, to furnish the starting values for a very precise numerical integration. He discovered that the incoming and the outgoing solutions do not meet in phase-space, no matter how far extended! With this numerical refutation, now confirmed by the author's analysis, the Age of Controversy has come to an end.

To give Brown full credit, his second and third conjectures must be mentioned. The second conjecture , asserting the existence of the horseshoe-shaped orbits, has been amply confirmed by numerical integrations (Rabe 1961, Taylor 1981), by analysis (I-1977), and by direct observations of co-orbiting satellites of Saturn (1981). His third conjecture, regarding the termination of the horseshoes at L_2, may be true after all, and will be the subject of the forthcoming Part VI.

REFERENCES

Brown, E. W. 1911, Monthly Notices, Roy. Astron. Soc. 71,438.
Garfinkel, B. 1977, Astron J. 82,368 (Part I)
Garfinkel, B. 1978, Cel. Mech. 18,259 (Part II)
Garfinkel, B. 1980, Cel. Mech. 22,267 (Part III)
Garfinkel, B. 1983, Cel. Mech. 30,373 (Part IV)
Garfinkel, B. Cel. Mech. (in press) (Part V)
Deprit, A. and Henrard, J. 1970 Periodic Orbits, Stability, and Resonances, ed. G. E. O. Giacaglia, p. 1 (Reidel Co.)
Henrard, J. 1983, Cel. Mech. 31, 115.
Rabe, E. 1961, Astron. J. 66,500.
Stumpff, K. 1963, NASA Tech. Note D 1416.
Szebehely, V. 1967, Theory of Orbits, (Academic Press).
Taylor, D. 1981, Astron. & Astroph. 103,288.
Thüring, B. 1931, Astron. Nachr. 238,357.

LIST OF PARTICIPANTS

Reinhold Bien, Astronomisches Recheninstitut, Mönchhofstrasse 12-14, D-6900 Heidelberg, F.R.G.

Nicole Borderies, Observatoires du Pic-du-Midi et de Toulouse, 14, Avenue Belin, 31400 Toulouse, France

Pierre Bretagnon, Bureau des Longitudes, 77, Avenue Denfert-Rochereau, F 75014, Paris, France.

Willibald Brugger, Institut fur Astronomie, Universität Graz, Universitätsplatz 5, A 8010 Graz, Austria.

Jean Chapront, Bureau des Longitudes, 77, Avenue Denfert-Rochereau, F 75014, Paris, France.

Magda Delva, Institut für Astronomie, Universität Graz, Universitätsplatz 5, A 8010 Graz, Austria.

Raynor L. Duncombe, Department of Aerospace Engineering, University of Texas, Austin, TX 78712, U.S.A.

Rudolf Dvorak, Institüt für Astronomie, Universität Wien, Türkenschanzstrasse 17, A 1180 Wien, Austria.

Heinrich K. Eichhorn, Department of Astronomy, University of Florida, Gainesville, FL 32605, U.S.A.

Bálint Erdi, Department of Astronomy, Eötvös University, Budapest, Hungary.

Konradin Ferrari d'Occieppo, Innstrasse 17/5, A 6020 Innsbruck, Austria.

Sylvio Ferraz-Mello, Universidade de Sao Paulo, Instituto Astronomico e Geofisico, Avenida Miguel Stefano, 4200, Caixa Postal 30.627, 01000 Sao Paulo, Brasil.

Walter Fricke, Astronomisches Rechen-Institut, Mönchhofstrasse 12-14, D 6900 Heidelberg, F.R.G.

Christiane Froeschle, Observatoire de Nice, Boite Postale 139, F 06003, Nice Cedex, France.

Claude Froeschle, Observatoire de Nice, Boite Postale 139, F 06003, Nice Cedex, France.

Boris Garfinkel, Department of Astronomy, Yale University, Box 6666, New Haven, CT 06511, U.S.A.

N. Gonzci, Observatoire de Nice, Boite Postale 139, F 06003, Nice, Cedex, France.

John Hadjidemetriou, Department of Theoretical Mechanics, University of Thessaloniki, Thessaloniki, Greece.

Arnold Hanslmeier, Institut für Astronomie, Universität Graz, Universitätsplatz 5, A 8010, Graz, Austria.

Hermann F. Haupt, Institut für Astronomie, Universität Graz, Universitätsplatz 5, A 8010 Graz, Austria.

Jacques Henrard, Facultes Universitaires Notre Dame de la Paix, Départment de Mathematique, Rempart de la Vierge, 8, B 5000, Namur, Belgium.

Takeshi Inoue, Kyoto Sangyo University, Kyoto 603, Japan.

Alan Jupp, Department of Applied Mathematics and Theoretical Physics, The University of Liverpool, P.O. Box 147, Liverpool L69 3BX, United Kingdom.

Jean Kovalevsky, Centre d'Etudes et de Recherches Géodynamiques et Astronomiques, Avenue Copernic, F 06130 Grasse, France.

Hiroshi Kinoshita, Tokyo Astronomical Observatory, Osawa Mitaka, 181 Tokyo, Japan.

Jacques Laskar, Bureau des Longitudes, 77, Avenue Denfert-Rochereau, F 75014, Paris, France.

Anne Lemaitre, Facultes Universitaires Notre Dame de la Paix, Department de Mathematique, Rempart de la Vierge, 8, B 5000, Namur, Belgium.

Herbert Lichtenegger, Institut fur Astronomie, Universitat Graz, Universitatsplatz 5, A 8010 Graz, Austria.

Christian Marchal, Office National d'Etudes et de Recherches Aerospatiales, 29, Avenue de la Division Leclerc, F 92320 Chatillon-sous-Bagneux, France.

P. James Message, Department of Applied Mathematics and Theoretical Physics, The University of Liverpool, P.O. Box 147, Liverpool, L69 3BX, United Kingdom.

Francois Mignard, Centre d'Etudes et de Recherches Géodynamiques et Astronomiques, Avenue Copernic, F 06130, Grasse, France.

Andrea Milani, Dipartimento di Matematica, Universita di Pisa, Piazza dei Cavalieri 2, I 56100 Pisa, Italy.

Michele Moons, Facultes Universitaires Notre Dame de la Paix, Department de Mathematique, Rempart de la Vierge 8, B 5000, Namur, Belgium.

Bruno Morando, Bureau des Longitudes, 77, Avenue Denfert-Rochereau, F 75014, Paris, France.

Gerald Rabl, Institut fur Astronomie, Universität Graz, Universitätsplatz 5, A 8010 Graz, Austria.

Margret Rotter, Institut fur Astronomie, Universität Graz, Universitätsplatz 5, A 8010 Graz, Austria.

Freidrich Schmeidler, Universitäts-Sternwarte Munchen, D 8000, Munchen, F.R.G.

Joachim Schubart, Astronomisches Rechen-Institut, Monchhofstrasse 12-14, D 6900, Heidelberg, F.R.G.

Kenneth Seidelmann, United States Naval Observatory, Washington, D.C., 20390, U.S.A.

Klaus Strassmeier, Institut fur Astronomie, Universitat Graz, Universitätsplatz 5, A 8010, Graz, Austria.

Victor Szebehely, Department of Aerospace Engineering, University of Texas, Austin, TX 78712, U.S.A.

William Thuillot, Bureau des Longitudes, 77, Avenue Denfert-Rochereau, F 75014, Paris, France.

Werner Tscharnuter, Institut fur Astronomie, Universitat Wien, Turkenshanzstrasse 17, A 1180, Wien, Austria.

Carol A. Williams, Department of Mathematics, University of Florida, Tampa, FL 33620, U.S.A.

Gunther Wuchterl, Institut fur Astronomie, Universitat Wien, Turkenshanzstrasse 17, A 1180 Wien, Austria.

ANNOUNCEMENT

The Stability of Planetary Systems

Editors: R.L. Duncombe, R. Dvorak, P.J. Message

Please note that a hardbound edition of these special issues of Celestial Mechanics, Vol. 34 Nos 1-4 (September-December 1984), is available from the publishers.

ISBN: 90-277-1961-6. Prices: Dfl. 240,- / $96.00 / £60.95.

TABLE OF CONTENTS

(Volume 34)

ARTICLES

TABLE OF CONTENTS